THE HISTORY OF CHARTERED SURVEYORS IN SINGAPORE

The first hundred years: 1868–1968

THE HISTORY OF CHARTERED SURVEYORS IN SINGAPORE

The first hundred years: 1868–1968

Bill Jones

World Scientific

NEW JERSEY • LONDON • SINGAPORE • BEIJING • SHANGHAI • HONG KONG • TAIPEI • CHENNAI • TOKYO

Published by

World Scientific Publishing Co. Pte. Ltd.
5 Toh Tuck Link, Singapore 596224
USA office: 27 Warren Street, Suite 401-402, Hackensack, NJ 07601
UK office: 57 Shelton Street, Covent Garden, London WC2H 9HE

Library of Congress Cataloging-in-Publication Data
Names: Jones, Bill (William James), author.
Title: The history of Chartered Surveyors in Singapore : the first hundred years: 1868–1968 / Bill Jones.
Description: New Jersey : World Scientific, [2023] | Includes bibliographical refererences.
Identifiers: LCCN 2022055932 | ISBN 9789811268441 (hardcover) |
ISBN 9789811268465 (ebook other) | ISBN 9789811268458 (ebook)
Subjects: LCSH: Royal Institution of Chartered Surveyors--History. |
Surveyors--Singapore--History--19th century. | Surveyors--Singapore--History--20th century. |
Surveying--Singapore--History--19th century. | Surveying--Singapore--History--20th century.
Classification: LCC TA527.S55 .J66 2023 | DDC 526.9095957--dc23/eng/20230105
LC record available at https://lccn.loc.gov/2022055932

British Library Cataloguing-in-Publication Data
A catalogue record for this book is available from the British Library.

Supported by

National Heritage Board

The views expressed here are solely those of the author in his private capacity and do not in any way represent the views of the National Heritage Board and/or any government agencies.

For any available supplementary material, please visit
https://www.worldscientific.com/worldscibooks/10.1142/13202#t=suppl

Desk Editors: Gregory Lee/Amanda Yun
Design and layout: Chin Choon Ng

Preface

Since its independence, Singapore has attracted an increasing number of tourists from all over the world, with many going on tours of the city, during which they will enjoy the sight of its diverse range of magnificent buildings. Singapore citizens, residents and foreigners working here experience the splendour and magnificence of the Singapore-built environment every day. When recognising the beauty of the Raffles Hotel, the size and complexity of the Marina Bay Sands complex or the efficiency of Singapore's built infrastructure, most people associate their creation with the skills of architects and engineers. However, during the period of time covered by this book, the role of the chartered surveyor was integral to the creation of a successful and sustainable urban environment. In Singapore, Chartered Surveyors played an important role in the development and transformation of Singapore. The story of these Chartered Surveyors is largely unrecognised, so the aim of this book is to redress the balance, given that they paved the way for future generations of Singaporeans to become Chartered Surveyors.

From the second half of the 19th Century through to the 1920s, Chartered Surveyors in Singapore were mostly British "Expatriates" or military personnel with a small number of Eurasians. However, after a period in the 1940s and 1950s when local talent struggled to break through the Expatriate stranglehold on the profession, the mix from 1958 onwards was much more diverse, and there have been well-known (and many less well-known) Singaporean Chartered Surveyors who played a key role in the development of today's Singapore. This book cannot be the complete story, so this is not an attempt to list every single Chartered Surveyor who worked in Singapore in the period up to the late 1960s, as that would be an impossible task. The omission or lack of a story in relation to any one individual is not intentional. However, it is hoped that the descendants of anyone missing will be able to contribute to future editions of the book, stories of the role they played in the development of Singapore. The history of Singapore, its administrative affairs and governance structures during these years have been the subject of many excellent books, online material and publications, so this book will not repeat what has been said in print already; instead, it will focus on individual stories of the role that Chartered Surveyors and the Royal Institution of Chartered Surveyors (RICS) played, along with their achievements.

From the days of the Straits Settlements to independence, the work of Chartered Surveyors in Penang, Kuala Lumpur, Malacca, Johor and Singapore was entwined. However, this book does not capture the story of Chartered Surveyors in Malaysia — this will hopefully be captured in book form in the future. I hope that in reading this book, you will get a stronger appreciation of the role Chartered Surveyors play and their contributions made to Singapore. I would be delighted to hear from anyone reading this book who has any corrections or additional information (including photographs) that you can share with me for future editions of this book. My email address is bjrics4780@yahoo.com.sg.

As a final word, the research conducted to produce this book is solely the work of the author. Furthermore, the views expressed in this book are solely those of the author in his private capacity and do not in any way represent the views of the RICS, the National Heritage Board of Singapore or any Government agencies.

Bill Jones
March 2022

Foreword

by Clement Lau, FRICS

Since its earliest known settlement, the area today known as Singapore has consistently been influential in global affairs. Its location at the confluence of several major land and naval communication routes makes it a natural hub for trade and interaction between peoples, and for over a millennium, this strategic positioning has made Singapore a major player in the region and beyond.

In this context, Singapore — literally meaning "lion city" — has embodied an outward-looking, dynamic spirit that today sees it as a leading contemporary economy and a vibrant society. With commercial and cultural connections to every corner of the globe, it is hardly surprising that Singapore enjoys economic success. However, one of its more notable features is that it combines this economic prowess with being one of the most livable cities in the world, and professionals working in the built environment have played a crucial role in this story.

Focusing on its modern history, luminary figures like George Drumgold Coleman and John Turnbull Thomson were pivotal to ensuring sound city planning and implementation in Singapore, while Chartered Surveyors like Alfred Lermit, Harvey Chevallier and Henry L. Coghlan made significant and innovative contributions to building up this "garden city" we enjoy today.

Looking ahead, like the rest of the world, Singapore is facing up to the challenge presented by climate change and its net zero by 2050 target. It will be necessary for all of us to be integrated, positive and supportive of collaboration — however, I am confident that Singapore will continue to push the boundaries by hosting leading thinking and practice in the built environment. We should draw encouragement from the history so well described in this publication and face the future confident that we can deliver more healthy, resilient and sustainable spaces for communities everywhere, in Singapore and around the world, for years to come.

Clement Lau, FRICS
President
Royal Institution of Chartered Surveyors

Foreword

by Seah Choo Meng

This is indisputably the only publication that chronicles the historical development and evolution of the Surveying profession in Asia.

Bill Jones has painstakingly assembled and narrated the brief life stories of the various personalities from the pre-war period to present times. It is a remarkable record of the lives of Surveyors from early days, and in this manner, the history of the development of Surveyors of the three disciplines — Land Surveyors, Valuation Surveyors and Quantity Surveyors — is portrayed in a proper and recorded chronology of past events.

My late father, Bill Seah Mong Hee, the first Chinese Chartered Quantity Surveyor, had started his career in 1938 in an innocuous office premise situated along Battery Road — the Bank of China now occupies the site of that office.

The work of my late father, and the development of the three core surveying disciplines in Singapore over many years, laid the foundations for the profession to make its contribution to the growth and success of the nation.

The accounts related by Bill Jones are, to my knowledge, a reflection of the historical events and life stories, and I have no doubt he has taken much time and effort to verify the correctness of the facts and events.

My sentiment is this will indeed be the first history book of Surveyors and I would like to sincerely congratulate Bill Jones for his tremendous and laudable effort in the compilation and editorial work for the book.

Seah Choo Meng
Senior Advisor, GCEO's Office
Surbana Jurong Group

About the Author

William James ("Bill") Jones is a Singapore Permanent Resident.

He qualified as a Chartered Valuation Surveyor in 1986 whilst working in private practice in the North-West of England.

He joined Royal Dutch Shell Plc in January 2002 and through Shell, he came to work, live and settle permanently in Singapore.

In March 2008, he first became directly involved with the Royal Institution of Chartered Surveyors (RICS) Singapore as the inaugural RICS Singapore Working Group Chair (2008–09), as well as being the RICS Singapore representative on the RICS Asia-Pacific Board, later known as the RICS Asia World Regional Board.

He was appointed as the full-time Singapore-based RICS Managing Director for the South-East Asia World Region in October 2011, and then later, he became the RICS Director of Standards for the Asia-Pacific World Region.

He became an RICS Fellow in 2012 and continues to work in the property industry in Singapore.

He is married with two grown-up children and lives in Pasir Ris, Singapore.

He continues his involvement with RICS as an APC Assessor, Licensed Assessor Trainer and APC Auditor for the South-East Asia region and over the years, he has been a mentor and counsellor to many RICS APC candidates in that region. He is also an elected member of the RICS Governing Council with effect from 2020 to 2023.

Reader's Guide

The book entitled *Building a Name: The History of the Royal Institution of Chartered Surveyors* by Jennifer North tells us that during its history, the R.I.C.S. has undergone several name changes, usually to correspond with an increase in status.

The names by which the Institution has been known are set out below and will correspond with what is referenced inside the Book:

15 June 1868: The Institution of Surveyors

26 August 1881: The Surveyors' Institution (following the granting of the Royal Charter)

27 October 1930: The Chartered Surveyors' Institution (by permission of the Privy Council)

3 July 1947: The Royal Institution of Chartered Surveyors (by personal grant of the King)

Also, during the period covered by this book, the use of "R.I.C.S." for the Royal Institution of Chartered Surveyors was prevalent, though it was gradually replaced with "RICS" from the late 1980s onwards. The author has chosen to use the former throughout, apart from situations where the author refers to an event — or an individual — from the 1990s onwards or where the author makes a reference to the present day.

The book also refers throughout to Chartered Surveyors but specifically makes reference to Chartered Land Surveyors, Chartered Quantity Surveyors and Chartered Valuation Surveyors. The reason is this: throughout its existence, members of the RICS have had specialist qualifications to become chartered in their field of expertise. These qualifications have for some years now been known as Chartered Alternative Designations (CADs). These references are made in some Chapters to CADs in order to make the distinction between the various specialisms and the individuals who practised in the various fields of specialism in Singapore over the years covered in the book.

List of Abbreviations

There are many abbreviations in this book and whilst some will be familiar to most readers, some will not. The author has included a full list to ensure that all readers of this book can cross-check any abbreviations that they are not familiar with:

ACT: Australian Capital Territory

AIQS: Australian Institute of Quantity Surveyors

AMDGW and **AMWD**: Directorate-General of Works (and the shorter version: "Works Department"), Air Ministry — a UK Government Ministry with responsibility for managing the affairs of the Royal Air Force.

ANZUK: The name of a tripartite force formed by Australia, New Zealand and the United Kingdom (UK) to defend the Asia Pacific region after the UK withdrew its forces from the region in the 1970s.

APC: The Assessment of Professional Competence, the process established by the RICS for ensuring that candidates meet the high standards of professionalism required to become a Chartered Surveyor.

APL: American President Lines, a Freight and Shipping Company (also known as APL Logistics).

ARIBA: Associate of the Royal Institute of British Architects

ARICS: Associate Member of the Royal Institution of Chartered Surveyors (a designation that is called MRICS as of 2022).

BBC: British Broadcasting Corporation

BOAC (and BA): British Overseas Airways Corporation, which merged with British European Airways (BEA) in 1971 to become BA (British Airways).

CBE: Commander of the Order of the British Empire, a British award of chivalry.

CCA: Centre for Contemporary Arts. Part of NTU (see below).

CDL: City Developments Limited

CE: Common Era, now used instead of AD (Anno Domini) as a year notation for the Gregorian calendar.

CEM (now UCEM): College of Estate Management (now called the University College of Estate Management)

CIAEA: Chartered Institute of Auctioneers and Estate Agents

CID: Criminal Investigation Department — a branch of the Police force in the UK.

CIOB: Chartered Institute of Building

CMG: Companion, the Order of St. Michael and St. George is a British order of chivalry. See also KCMG.

CPF: The Central Provident Fund of Singapore

CPG: CPG Corporation, which is the corporatised entity of the former Public Works Department of Singapore.

CRE: Chief Royal Engineer

DSO: The Distinguished Service Order is a military decoration of the UK (and previously in parts of the Commonwealth) awarded for distinguished service by members of the armed forces during wartime.

EDB: The Singapore Economic Development Board

EEA&C: The Eastern Extension, Australasia and China Telegraph Company

FARELF: Far East Land Forces, set up with its Headquarters in Singapore in 1947 to develop an infantry division made up of Gurkhas, as well as advise and oversee the re-establishment of volunteer forces and the Malay Regiment in consultation with the local colonial governments.

FEPOW: Far East Prisoners of War

FMS: Federated Malay States

FRICS: A Fellow of the Royal Institution of Chartered Surveyors (a designation that remains as at 2022).

FSI (and PASI): Fellow (and Professional Associate) of the Surveyors' Institution

GLC: Greater London Council

GP: General Practitioner (a medical doctor)

GPO: General Post Office

GST: Goods and Services Tax, Singapore

HDB: Housing & Development Board, Singapore

HKIS: The Hong Kong Institute of Surveyors

HKVDC: Hong Kong Volunteer Defence Corps

HMS and H.M.S.: His (or Her) Majesty's Ship, which is the Naval ship prefix used by some monarchies.

HQ: Headquarters

IRAS: The Inland Revenue Authority of Singapore

ISM: See RISM

JTC: Jurong Town Corporation

KCMG: Knight Commander, the Order of St. Michael and St. George, a British order (award) of chivalry that was originally awarded to those holding commands or high position as far back as the Napoleonic Wars but later extended to holders of similar office or position across the entire British Empire.

MAS: Monetary Authority of Singapore

MBE: Member of the Most Excellent Order of the British Empire, a British order (award) of chivalry.

MC: The Military Cross, which is granted in recognition of "an act or acts of exemplary gallantry during active operations against the enemy on land" to all members of the British Armed Forces of any rank.

MCC: Middlesex County Council

MCDS: Ministry of Community Development and Sports — originally called the Ministry for Community Development.

MICE: Member of the Institution of Civil Engineers

MINDEF: Ministry of Defence

M. Inst. CE: Member of the Institution of Civil Engineers.

MND: Ministry of National Development, Singapore

MRT: Singapore's Mass Rapid Transit system

MV: Motor Vessel (ship prefix)

NSW: New South Wales, a State in Australia

NTU: The Nanyang Technological University, a Singapore research University

NUS: The National University of Singapore

NZIV: New Zealand Institute of Valuers (now part of the Property Institute of New Zealand)

OBE: The Most Excellent Order of the British Empire, a British order (award) of chivalry.

OCBC: The Oversea-Chinese Banking Corporation, Limited, is a multinational banking and financial services corporation headquartered in Singapore.

PAM: Pertubuhan Akitek Malaysia, the Malaysian Institute of Architects.

PASI: See FSI

POW: Prisoner of war — a person held captive by an aggressor during or after a war or conflict.

PSA: The Port of Singapore Authority, formed in 1964. It was corporatised in 1997 and became known as PSA Corporation (and regulated by the Maritime and Port Authority of Singapore (MPA)).

PWD: The Public Works Department of Singapore.

QS: Quantity Surveyor, one of the professional disciplines (groups) within the RICS.

RAF: The Royal Air Force, which is the United Kingdom's air and airspace force.

REVA(S): The Real Estate Valuers Association (Singapore), formed in 1963.

RIAI: The Royal Institute of the Architects of Ireland

RIBA (and FRIBA): Royal Institute of British Architects (FRIBA is the title used by a Fellow of that Institute).

RISM (and ISM): The Royal Institution of Surveyors Malaysia.

Roman Numerals in the book are: IV = 4, VIII = 8, XV = 15, IX = 9, XII = 12, XIII = 13, XVI = 16 and XXVIII = 28

RSAF: The Republic of Singapore Air Force

RTPI: The Royal Town Planning Institute

RU: Rugby Union

SCE: Superintending Civil Engineers

SEAC: The South East Asia Command was the body set up to be in overall charge of Allied operations in South-East Asia during World War Two.

SE-Asia: South-East Asia

SGH: Singapore General Hospital

SIS, SIV and SISV: The Singapore Institute of Surveyors, the Singapore Institute of Valuers and the Singapore Institute of Surveyors and Valuers.

SIT: The Singapore Improvement Trust

SMM: The Standard Method of Measurement is published by the RICS. It was first published in 1922.

SS: Steamship (prefix for the name of a merchant ship)

SSSI: The Surveying & Spatial Sciences Institute

SVC: The Singapore Volunteer Corps

TAS: The Telecommunication Authority of Singapore

UK: The United Kingdom of Great Britain and Northern Ireland

UN: The United Nations

UNESCO: The United Nations Educational, Scientific and Cultural Organization

UOB: United Overseas Bank Limited, a multinational banking corporation headquartered in Singapore.

URA: Urban Redevelopment Authority of Singapore

WWI and WWII: World War One and World War Two.

YMCA: Young Men's Christian Association

Contents

Preface v
Foreword by Clement Lau, FRICS vii
Foreword by Seah Choo Meng ix
About the Author xi
Reader's Guide xiii
List of Abbreviations xv

Part I: Surveying in Pre-Modern and 19th-Century Singapore

1. Early Singapore and the Role of Surveyors 3
2. 1868–1900 — The First Chartered Surveyors in Singapore: Lermit, Chevallier and Coghlan 31
3. The Story of Regent Bidwell 61
4. The First Eurasian Chartered Surveyors in Singapore 95

Part II: The Work of Surveyors in the Early 20th Century

5. 1900–29: The Story of Other Chartered Surveyors of This Period 109
6. Langdon Williams and his Work with the Singapore Improvement Trust 145
7. 1930–39: Quantity Surveyors Emerge in Singapore 161
8. The Story of Seah Mong Hee, the First Asian Chartered Quantity Surveyor 193
9. 1941–46: The War Years and Immediate Aftermath 201

Part III: The Work of Surveyors Post-World War II

10. 1947–51: The Post-War Years 215

11. The Road Towards Malayanisation 239

12. The Critical Transition Years: 1957 and 1958 253

13. The Separation of R.I.C.S. Malaya and R.I.C.S. Singapore, 1958–61 275

14. The Work of Chartered Surveyors in the Singapore Improvement Trust, 1952–59 281

Part IV: Surveyors in the Pre and Post-Independence Eras

15. 1959–65: Singapore Independence and the Transition to Local Professional Bodies 305

16. R.I.C.S. President's Visit to Singapore in 1966 339

17. 1966–68: The Immediate Years After Singapore's Independence 351

18. Singapore's First Female Chartered Surveyor 379

19. 1950–68: Local Education and the Influence of Chartered Surveyors 383

20. John Aspinall and his Role in Singapore's Post-War Property Tax System 393

21. Individual Stories from the 1960s 403

Acknowledgements 417

Endnotes 423

Bibliography 481

Image Credits 485

Index 489

Part I

Surveying in Pre-Modern and 19th-Century Singapore

Chapter

1

Early Singapore and the Role of Surveyors

The early history of Singapore, from the 2nd century until the arrival of Sir Stamford Raffles in 1819, has been documented in numerous books, publications, journals and extensive research. It is well documented that Singapore, under a variety of different names, was a notable trading port/city of significant size at various points in its history from the 2nd century CE onwards.

Sandstone foundations dating back to the 14th century marking a large palatial building along the slope of Fort Canning — and the *keramat* on the site of the purported resting place of the last king, Iskandar Shah — provide evidence of building structures during these times of great prosperity for Singapore. Indeed, research resulting from (and supported by) archeological excavation work carried out since 1984 have told us much about the early history of Singapore based on many hundreds of thousands of fragments of jewellery, pottery and numerous other artefacts found at Fort Canning and elsewhere.

The 1984 Fort Canning digs led by John Miksic (continuing today under the Archaeological Unit of Nalanda-Sriwijaya Centre) and later archaeological investigations (for example, Miksic's digs along the Singapore river from 1987 onwards) have provided more evidence to support the stories in the Malay Annals, confirming that the built environment in pre-modern Singapore mirrored the development of other maritime cities across South-East Asia. Through others, including Peter Borschberg, research using other data sources, principally texts, maps, charts and scientific data, has provided further proof of the true size and richness of Singapore's pre-modern history. As with archaeology, the search in letters and maps is long and difficult, with clues often not in obvious places. Over the coming years, more evidence is sure to be found to deepen our knowledge of Singapore's rich pre-modern history, and it is hoped that this will include evidence of surveying skills in the evolution of pre-modern Singapore.

Imran Tajudeen's 2005 paper provided excellent drawings to illustrate the common morphology in pre-modern South-East Asian port cities, including Singapore and, in particular, the morphology of the fortress (Citadel) and its environs using Kampong Gelam (Glam) as the model. The results of his efforts mirror what the Fort Canning version looked like some 300 years earlier. More recently, Kwa Chong Guan, Derek Heng, Peter Borschberg and Tan Tai Yong defined the link between pre-modern Singapore and modern Singapore as the huge amount of maritime activity in the region and resulting trade occurring on the "port-city" island with a series of confluences and conflicts (as per Borschberg), defining the peaks and troughs of its journey through the centuries.

The Singapore Stone might have explained more — it was about 10 feet in height and 9–10 feet in length and located at the mouth of the Singapore River. The stone was believed to tell the story of Singapore from the 10th to 14th centuries and may have been inscribed at a time when Singapura/Temasek was under the control of the Majapahit kingdom. John Watson Laidlay (1808–85), an India-based merchant with expertise in ancient languages, concluded that the inscription was written in Kawi script, but he was unable to decipher it.

Later, Dr. J.G. Casparis deciphered one or two words, which he attributed to the Old Javanese language, while Prof. Drs. Boechari suggested that the inscription was likely the Sanskrit language as the writing style appeared more like the Sumatran than Javanese writing style, with Sanskrit being used in Sumatra at that time. However, nobody has been able to fully identify or decipher the existing fragment of the inscription on the Singapore stone that lies in the National Museum of Singapore. The stone was blown up in 1843 to build the Commander's living quarters. Arnold Wright states that George Coleman (more on him later) was responsible, but other sources tell us that Coleman was on leave at the time. Colonel James Low, a British Army officer, salvaged several slabs of the stone after it had been blown up, and sent three to Calcutta, where they arrived in June 1848. Another fragment, found by Colonel Butterworth (Governor from 1843 to 1855), was also sent to Calcutta. One fragment was returned to the Raffles Museum and Library (now the National Museum of Singapore) in 1919. This piece is still displayed there, but at present, there is no definite location for the other surviving fragments.

The Malay Annals (Sejarah Melayu) provide some insights into what the built environment looked like from the time of the founding of Singhapura by Sang Nila Utama in 1299, during the evolution of the Melakan Sultanate and ending with the Portuguese attack and colonisation of Melaka in 1511. Sang Nila Utama's reign and that of his immediate descendants was a time of prosperity for the island, and the Annals confirm that it was frequented by merchants innumerable from every quarter and its ports were "very populous". A fortress was constructed with an inner and outer gate surrounded by a moat, with the palaces of the Royal family at the centre (along with a large hall where they would dine, and halls of state, segregated into different galleries, in accordance with the status of those summoned to attend by the Royal family) as well as the homes, shops and marketplace of the "people of the raja" located between the inner and outer walls of the fortress.

The Annals tell of the existence of "artifices" (skilled workers) in Singhapura during these times, which points towards skilled people, including surveyors, capable of building and maintaining the fortress and all buildings inside it; not just the large palatial buildings but also the wood and attap homes (for artisans and workers) as well as shops and other buildings. The construction of sea vessels is noted in the Annals, necessitating a shipyard along with facilities to repair sailing vessels too, given the large volume of merchant vessels coming in and out of the island for trading purposes. A rice granary is also mentioned in the Annals, no doubt to feed the Royal family, visitors and all those working inside the fortress, as well as troops employed from time to time to defend Singhapura. This defence would take place at sea just off the southern shores of the island, but on the occasion of the second Majapahit attack, Sang Ranjuna Tapa opened the gates of the fortress to let the aggressors inside. On that occasion, the attack comprised 300 vessels and 200,000 Javan troops, which indicates the size of Singhapura, such was the force needed to overcome it. In summary, the evidence gathered since 1984 substantiates the assertion by Arnold Wright in *Twentieth Century Impressions of British Malaya: Its History, People, Commerce, Industries, and Resources* that six centuries before Raffles arrived, Singapore was the "Constantinople of these Eastern seas, the seat of Malay learning and commerce, the focus of the commerce of two oceans and part of two continents".

The history of surveying in modern Singapore dates to the Ross survey and map produced in 1819. Several surveys and maps of Singapore had been produced since the 17th century, but the first one bearing the island's name spelt as "Singapore" was a survey conducted by Captain Daniel Ross in 1819 on the behalf of Raffles. This was the "Plan of Singapore Harbour February 1819", first published in the *Calcutta Journal* on 1 May 1819. This plan identified only two villages along the entire coastline from Tanjong Pagar to Tanah Merah — the Temenggong's village near the mouth of the Singapore river and a village near Kampong Glam. Other settlements (Malay and *orang laut*) were discovered around the island, but apart from these and the 20 gambier plantations close to the future town, the island was largely jungle, with around 1,000 inhabitants who subsisted by fishing and trading.

The First Resident of Singapore, Major William Farquhar, and his assistant engineer (Lieutenant Henry Ralfe) conducted land surveys in Singapore, as did Captain James Franklin of the Bengal Cavalry in 1822, and his survey — "Sketch of the Island of Singapore" — was the first map that captured the full outline of the Singapore island. Lieutenant Philip Jackson, Garrison Engineer of the Bengal Artillery arrived in 1822 and was appointed the Surveyor and Registrar of Land on 1 February 1826 with a monthly salary of 300 rupees. His responsibilities included surveying lands and registering grants and transfers. A surviving sketch of the town by him from 1823 shows only a collection of "temporary" Anglo-Indian style bungalows built of timber with pyramid-shaped or hipped roofs, thatched with *attap* and with verandas down the side — the sketch included the home of Francis Bernard, the Police Superintendent. At this time, Police costs were funded by taxes on opium and liquor, whilst gaming revenue was used to keep the streets clean.

Jackson's sketch is the earliest evidence of the "European town" having developed along the shoreline on the north of the Singapore river mouth, and the style of house is to be expected, given that most of the Europeans then in Singapore had come across from Bengal in the service of the East India Company. The Anglo-Malay style of houses (placed on stilts, with a central entrance porch projecting from the front of the house) had not yet come into prominence.

Raffles formed a Town Planning Committee in November 1822, through which he segregated the various races and nationalities into specific locations and specified the area for Government buildings, religious worship, commerce and education and included details such as house and street sizes and the type of materials to be used. He commissioned Jackson to produce a plan (the "1822 Jackson Plan") for the Town of Singapore. This Plan left a lasting mark on the Singapore urban landscape, e.g., the Commercial area (Commercial Square) is now Raffles Place, and areas for religious worship in the city largely remain as per the Plan, as does the proposed site of the Botanic Garden.

During the period from 1819 to 1860, as Singapore's position in the region became well established, the short-term houses and commercial buildings were replaced with permanent structures. From 1823, land could be purchased at an auction on a permanent lease, resulting in plot purchasers willing to spend money on more permanent structures to replace temporary ones. Raffles regulated the size of house plots, road widths and building heights (no more than three storeys), and commercial buildings had to be of bricks with tiled roofs and a solid brick or stone foundation. Indeed, the price of land was also determined at this time. In these early times, verbal licences to occupy were granted free of rent with a permit to apply for a regular lease after two years, but by 1826, only 10% of the island was occupied, when the earliest 999-year leases were sold/issued and recorded. These early leases did have land diagrams to show the land by reference to roads, which defined the north, south, east and west boundaries of the land within the lease. These "pioneer" leases also included a rough written description of the land and were subject to an annual quit rent. Fifteen-year leases also existed around this time, again subject to a quit rent and conditions relating to land clearance, with options for further 15-year terms and an option for a 999-year lease. Although these leases were prohibited in 1831, some 415 acres of land were mostly still held on these initial 15-year leases till 1833.

There was still a great deal of uncertainty around Singapore's future — the political situation between the East India Company, the British Government and the Dutch Government was tense, and relations were not good with the Sultan and Temenggong (a member of the nobility as well as being the Sultan's most senior adviser and protector) despite the financial agreement reached by Raffles in 1819. Despite this, through Raffles and those who governed Singapore in his absence, the price was 3,000 Spanish Dollars for plots that were 150 feet deep and had a 50-feet frontage on the river, along with an annual quit rent of 38 Spanish dollars. Residential plots with an area of 1,200 square yards fetched 400 Spanish dollars and a quit rent of 28 Spanish dollars a year — a far cry from current property values along the Singapore River but still a huge price for a permanent lease of land in Singapore's very early days! In 1838–39, the 99-year lease started to emerge as a catalyst for ensuring that the impenetrable areas of jungle close to the town were cleared and built upon, i.e., the lease placed conditions on them relating to the construction of tiled roof houses (thus aligned with Raffles' earlier regulations).

The Government failed to achieve the same level of price for the 99-year leases as they had for the 999-year leases due to the shorter tenure.

For agriculture, it took until 1840 for legislation to govern land lease grants — leases were fixed at 20 years with an option to renew for 30 years and were free from rent for the first 2 years. After that, rent was only 4 annas per acre for 3 years, 8 annas for the next 5 years and 1 rupee per year after that (1 anna was equal to 1/16 of a rupee). This changed to allow Indentures (or freehold title) for land outside the Town to encourage agriculture, but not long afterward, the decline in Singapore agriculture began.

The word "Government" is mentioned above, but titles to all land from 1819 onwards were, in fact, derived from the East India Company. However, in 1858, the British Crown began to sell land on leases (and some fee simple sales) under the royal prerogative. Then, in 1867, all control vested in the Colonial office, after which other forms of tenure prevalent ranged from 30, 60, 99 to 999 years through to perpetual leases, sold on a premium, the one-off price paid on taking the lease, and thereafter, an annual ground rent known as a quit-rent.

To keep pace with all this leasing activity, the island was gradually mapped and surveyed, resulting in several key infrastructure works, including bridges, sewers and roads. At this point, some early buildings of significance began to emerge. Fortunately, some of that early built infrastructure in Singapore has survived to the present day.

In creating a sustainable built environment in Singapore during this period, several architects, engineers and surveyors (mostly in the Government service) played a prominent role in the development. A number of these individuals had training as surveyors in addition to their architectural and engineering training — and during the first 30 years of modern Singapore, George Coleman and John Thomson were the most prominent among them. The impact of Coleman and Thomson on Singapore is still felt today.

The careers of Coleman and Thomson preceded the formation of the Institute of Surveyors in London in 1868, but their surveying training was the foundation of their groundbreaking work on the roads, bridges, sewers, general infrastructure and buildings of Singapore between 1819 and 1855. Their surveying abilities, combined with their engineering and architectural skills, were a major impetus for the development of early Singapore.

George Coleman

George Drumgold Coleman (1794/95–1843) was an architect born in Drogheda, Republic of Ireland. He was the son of James Coleman, a merchant of Drogheda who had married into the Co. Louth merchant family of Dromgold. Records for this time are scarce, both from a Drogheda perspective and from a national perspective (many public records were lost to fire in 1922 during the Civil War), but Coleman (and his siblings) almost certainly went to the Protestant School in Drogheda, and he may also have been articled as an architect and surveyor to Francis Johnston. Again, no record exists of Coleman's training (there is no record of any certificate from the Dublin Society or the Royal Academy School), but the book by T.H.H. Hancock entitled *Coleman's Singapore* suggests that the only plausible explanation is that he was articled to and trained by a Dublin architect, and the fact that the renowned architect Francis Johnston was known to John Palmer of Calcutta leads to a strong assumption that Coleman was articled to Johnston. George travelled across Europe to learn all about different European styles of architecture and spent time working in Calcutta (1815–19), where he developed a strong reputation as a skilled architect, project overseer and contractor.

He then went to Batavia (now Jakarta) from 1819 to 1821/22 and further enhanced his reputation, and there he first came to the attention of Raffles. Coleman arrived in Singapore in June 1822 but then returned to Java for 2.5 years before finally coming back to Singapore, where he remained right through to 1841. He was Singapore's first officially appointed Government surveyor, and in August 1833, he was made Superintendent of Public Works, Overseer of Convict Labour and Land Surveyor. During his time in Singapore, following the plans approved by Raffles, he constructed many public roads and bridges, forming

Singapore's very first town infrastructure. Coleman was a proponent of neo-Palladianism, and all his building designs take their inspiration from Palladian-inspired buildings, and he took inspiration from the buildings and publications of the architect Colin Campbell (1667–1729).

The book by Jane Beamish and Jane Ferguson — *A History Of Singapore Architecture: The Making Of A City* — is worth reading to understand Coleman; they provide detailed information on buildings in India, Europe and London, which provided Coleman's inspiration, along with a photographic comparison between buildings that inspired Coleman and the ones he designed and built in Singapore. At that time, Singapore suffered from a shortage of building materials, skilled labour and building funds. Beamish and Ferguson's book gives an enlightening (if not shocking) overview of solutions adopted by Coleman (and others) in these early years, including the use of convicts in the construction process. Coleman stopped working due to ill health in June 1841, at which point he left Singapore. He sold the house he had built in 1829 (No. 3 Coleman Street), and it became a hotel under various names, later being subdivided into shops and a home for what is believed to have been hundreds of slum dwellers. Having been neglected for many years, it was demolished in 1965. His dwelling was on a piece of land of significant size — indeed, in 1970–71, the present-day Peninsula Hotel and Shopping Centre was constructed on its site. Coleman returned in 1843 one final time and died a few months later. He was buried in a mausoleum in Fort Canning Cemetery (now Fort Canning Park), the Christian Cemetery where Coleman had earlier constructed two Cupolas as shelter for visitors to the Cemetery. Coleman's Cupolas survive to this day inside the former Cemetery — and part of Coleman's tombstone also survives inside the former Cemetery, along with a plaque commemorating his life (see Figure 1.2(b)).

As well as being the Government surveyor, Coleman had his own practice as an architect-contractor and was highly influential in the creation of many of Singapore's first significant building structures. Coleman was responsible for the design and construction of:

- the **Armenian Church** (still standing today)
- the original **St. Andrew's Church** (later demolished and rebuilt)
- **Telok Ayer Market** — the second market on the site. His structure survived from 1836/38 until the current structure, now called Lau Pa Sat, was built in 1894.
- **Raffles Residence** (long since demolished) on Fort Canning Hill. On its site is a brick and tile property constructed in 2003 called Raffles House.
- The original **Singapore Institution** (now known as Raffles Institution) building on the corner of Bras Basah Road and Beach Road. This magnificent building was demolished to make way for Raffles City Shopping Centre, including Fairmont Singapore Hotel and Swissotel The Stamford Hotel.
- For John Argyle Maxwell, he designed and built one of the first permanent residential buildings in Singapore. It was in the Palladian manner, adapted to the tropical climate by incorporating a verandah and overhanging eaves to provide shade. This house, just off High Street, was later leased to the Government and was called (Old) Parliament House and now **The Arts House**, although very little remains of the Coleman structure. Its original use was a court house and Government offices, making it the first office building in modern Singapore (see Figure 1.1(a)).
- He also constructed the first (brick-built) version of the bridge that links Hill Street with New Bridge Road over the Singapore River. This bridge is long gone — the bridge there today in the same position is known as **Coleman Bridge**.
- It is possible that he designed and built the **Istana Kampung Glam** (still standing) and the now-demolished **Reception Hall** for the Temenggong's Istana at Telok Blangah Road (later known as the State of Johor Mosque), but there is no firm evidence to prove this.

FIGURE 1.1. (a) Old Parliament House, now the Arts House, off High Street, Singapore: The design and construction of the original main structure were by George Coleman, although little remains of his original building structure. (b) Caldwell House: Designed and built by Coleman in 1841 as a residence for Mr. Caldwell, the Senior Magistrate's clerk. It is now integrated into the CHIJMES Historic Building Complex in Victoria Street, Singapore.

- **Mr. Caldwell's residence**, Victoria Road, Singapore (see Figure 1.1(b))

Fortunately, a handful of Coleman's buildings have survived to this day, so his legacy lives on in these buildings as well as in central Singapore's road infrastructure and the roads that bear his name — all of which have somehow survived the many changes witnessed by Singapore since 1819.

FIGURE 1.2. (a) Naraina Pillai's statue, erected on the Singapore River in 2018. He was Singapore's first known building contractor. (b) George Coleman's gravestone in the former Fort Canning Cemetery (now Fort Canning Park).

According to Robert Ibbeston, the resident (1830–32), Coleman was paid 15% fees as an architect plus his contractor profits — his wealth later enabled him to purchase a lease of land on the streets bearing his name — Coleman Lane and Coleman Street. So, as well as being a trained architect, surveyor and quantity surveyor, Coleman owned one of Singapore's first building and construction firms. He was not the first-ever building contractor — that was Naraina Pillai (see Figure 1.2(a)), a merchant who came to Singapore on Raffles' second visit. Pillai opened a brick kiln in Tanjong Pagar and set himself up as Singapore's very first building contractor after seeing the rapid rate of house building. He got help from his friends in Penang as bricklayers and carpenters.

John Thomson

John Turnbull Thomson arrived in 1841. Having studied at Duns Academy, Berwickshire, and completing his education with a Mathematics course at Marischal College, Aberdeen, Scotland (now part of Aberdeen University), he was appointed as the first Government Surveyor of the Eastern Settlements covering the three Straits Settlements. His focus was

Singapore — later, separate Government Surveyors were appointed for Penang and Malacca. From 1841 to 1853 (when he took an absence of leave due to ill health), and for a short time from 1854 to 1855, Thomson served not only as a surveyor but also as an architect and engineer. Thomson Road is named after him. Aside from being a well-respected artist who captured the essence of mid-19th-century Singapore in his paintings, he was also a prolific mapmaker whose maps show the transformation of Singapore from the early days of Singapore town through to the start of the development of an infrastructure for the interior of the island.

The highlights of his survey and mapping work are:

- Plan of the Town of Singapore, 1843.
- Plan of Singapore Town and Adjoining Districts, 1844 and 1846 — the most detailed map of Singapore from that time that is known to exist
- A chart showing the Survey of the Straits of Singapore, 1846 (revised in 1855) — a topographical survey of the island of Singapore and its dependencies, the marine survey of the Straits of Singapore and east coasts of Johor and Penang, and the Horsburgh Lighthouse on Pedra Branca.
- Map of Singapore Island and its Dependencies, 1852.

John Thomson was responsible for the design and construction of many significant engineering works, including bridges, major new roads inside Singapore and hospitals. His designs and town water projects led directly to the creation of MacRitchie Reservoir. The notable buildings and structures he designed and built were:

- **Horsburgh Lighthouse** on Pedra Branca (still there today)
- **Ellenborough buildings** (possibly the very first purpose-built row of shops, demolished in the 1970s — now where Central Mall and Clarke Quay MRT stands)

- Tower and spire of the original **St. Andrew's Church/Cathedral**
- The first bridge across the **Kallang River**
- **Thomson Bridge** (Elgin Bridge was later constructed in its place)
- **Hajjah Fatimah Mosque** on Beach Road, Kampong Glam end (still there today)
- **European Seaman's Hospital**
- **Original Chinese Paupers' Hospital Building** — later to become the Tan Tock Seng Hospital
- **Dalhousie Obelisk, a monument that now stands by the Esplanade** (see Figure 1.3(a)). The monument was designed by Thomson and constructed in honour of the Marquis of Dalhousie, James Andrew Broun-Ramsay, the Governor-General of British India, who helped the development of Singapore and was also a generous benefactor. Thomson's design drew inspiration from the world-famous Cleopatra's Needle by the River Thames in London, and
- **Cavenagh Bridge** (named in honour of Major General Orfeur Cavenagh, Governor of the Straits Settlements from 1859 to 1867) was designed by Colonel G.C. Collyer, chief engineer of the Straits Settlements, with

FIGURE 1.3. (a) Dalhousie Obelisk, located between the Esplanade and Cavenagh Bridge. Both were designed and built during John Thomson's time in charge of the Public Works Department. (b) Cavenagh Bridge, across the Singapore River (located between Esplanade and Fullerton Hotel).

> R.M. Ordish of the Public Works Department (PWD), which was at that time under Thomson's management. Construction, importation from Scotland and assembly of the Bridge in Singapore took place after Thomson's departure. See Figure 1.3(b).

Thomson's workload took its toll, especially during the construction of Horsburgh Lighthouse, where he was based for many months. He and his family moved to New Zealand, where the climate was better suited to him and helped him recover his health. After retiring, he died in Invercargill, New Zealand, on 16 October 1884, aged 63, and is buried in the John Turnbull Thomson Mausoleum and Family Plot there.

Coleman and Thomson are worthy of recognition as surveyors of great influence. It is hoped that this book offers recognition for their contributions to modern Singapore's built environment. There is no doubt that they laid solid and sustainable foundations for the development of Singapore, which, in turn, allowed Chartered Surveyors to operate in Singapore from the 1880s onwards.

A succession of Surveyor-General and Engineer roles followed Thomson over a short period of time. Originally, these roles were under the British East India Company. However, on 28 June 1858, after the Indian Rebellion of 1857, the rule of the British East India Company was transferred to the Crown in the person of Queen Victoria. Then on 1 April 1867, the Straits Settlements (at this stage comprising the Malay Peninsula and Singapore) became a British Crown colony, making the Settlements answerable directly to the Colonial Office in London instead of the government of British India (representing the Crown) based in Calcutta. As a result of all these changes, the Survey Department was, by 1869, under the direct control of the PWD, which had been created around 1867. Furthermore, the Survey Department (also known as the Survey Branch) was managed by the Colonial Engineer, later known as the Colonial Engineer and Surveyor-General of the Straits Settlements.

The PWD was set up to take responsibility for all works of construction and maintenance undertaken by the Government of the

Straits Settlements. In 1868, when John McNair took over, its provision for public works was $148,000. So, the Colonial Office shifted its focus to roads, drainage, infrastructure works and public buildings, and less on actual survey work. Arnold Wright's *Twentieth Century Impressions of British Malay* tells us that by 1908, the PWD's remit included maintenance of inland navigation, repairs and additions to public buildings and bridges, the acquisition of land and buildings for public purposes, and the repair and maintenance of roads. In 1908, the total length of roads maintained by the PWD in the Straits Settlements was 591 miles, of which 90 miles were in Singapore.

The various Colonial Engineers and Surveyor-Generals of the Straits Settlements over this period are set out below — it should be noted that from 1867 onwards, most of those working in public service in the Straits Settlements were either former Royal Engineers or former officers of the Madras Army, which was from 1858 part of the British Army, having earlier been created by the East India Company:

- **Jules M. Moniot** took over in 1855 as the Surveyor-General for the entire Straits Settlements (not just Singapore) and moved from Penang to Singapore in 1857 in this role, but had to retire in 1862 because of ill health. (He died in 1863.) His role was taken over by **Daniel Quinton**, who was promoted from the Malacca surveyor to Moniot's job and moved to Singapore in 1867. He, too, took ill and retired in 1869.
- **Major John Frederick Adolphus McNair (1828–1910)** then became Colonial Engineer from 1868 until 1880/1, when he retired. He oversaw the construction of St. Andrew's Cathedral, designed the former Empress Place Building (Asian Civilisations Museum) and Government House (Istana), and made alterations to the Old Parliament House (The Arts House). He was also on the Singapore Institution Committee that discussed the formation of the Museum and Library (now the National Museum).

FIGURE 1.4: 1908 staff photographs of the Engineer and Surveyor-General's office showing: (a) F.J. Pigott, (b) Robert Young and (c) John Van Cuylenberg.

- **Captain Henry E. McCallum,** later **Sir Henry McCallum** (who delivered the 1881 Map of Singapore town), was the Colonial Engineer after McNair through to 1896 when he left Singapore. In May 1882, McCallum submitted plans for the development of the **Raffles Museum and Library** in **Stamford Road** — it is now the **National Museum**. Work started in 1884 and finished in 1887.
- **Colonel A.C. Alexander** was appointed Colonial Engineer and Surveyor-General of the Straits Settlements in 1897. He died in 1898 and was replaced by **Major Alexander Murray** that same year — Murray retired in 1909. Finally, the Deputy Colonial Engineer **F.J. Pigott** (see Figure 1.4(a)) took over, and he was the last Colonial Engineer and Surveyor-General.

In Wright's *Twentieth Century Impressions of British Malaya*, he allowed F.J. Pigott, MICE (then the Acting Surveyor-General, Straits Settlements for the PWD) the space to explain Singapore's land survey and revenue work. Pigott's section of the book provides a unique insight into how land surveys and land revenue operated in Singapore from 1819 until 1908. Below is a summary of what Pigott said:

- Since the early days of Singapore, leases of 999, 99, 60 and 30 years had been given out.
- Land in Singapore was held from the Crown by grant or lease, with the conditions of tenure differing depending on the Government policy at the time. There were at that time 18 different forms of title in Singapore.
- The early land surveys from 1827 onwards were only rough sketches and not representative of what was on the ground.
- The 1839 Act XVI enabled rent collection in the Straits Settlements (see my note in the Endnotes on this).
- Land tax revenues (and land rent revenues) enabled the creation of the (Land) Survey Department, which by 1906, was also referred to as the Revenue Survey Department(s) of the Federated Malay States, responsible for the primary survey of the Crown estate and all surveys affecting title to the land.
- The Colonial Engineer was responsible for both the PWD and the Revenue Survey Departments.
- As of 1906, land revenue from rents in Singapore amounted to $104,482 for the year and totalled $377,972 for the Straits Settlements, whereas the cost of running the Land Office was $25,775, which was $6,613 less than in 1905.
- Thomson, Quinton and Moniot (previously referred to) carried out the first reliable land surveys.
- From the 1870s onwards, with the development of trade and agriculture in the Straits Settlements and the increased importance of the Survey Branch, their work focused on land applications, settlement of encroachment cases, building site surveys and engineering purposes. Priority was given by the Survey Branch to applications for registration of a title deed — once an application was lodged, the Survey Branch would be notified, and they would check the application details to see if a fresh survey plan is required.

- In 1870, a Legislative Council report led to the Surveyor-General being made responsible for maintaining municipal property, including land revenue.
- From 1881 to 1896, more systematic and scientific methods were adopted, including triangulation, the use of theodolites and 100-feet tapes. Indeed, in 1884, the Council supported the recommendation of a trigonometrical survey of Penang & Province Wellesley, Singapore and Malacca. Singapore was then 143,947 acres in size.
- For surveys, permanent granite centre marks (known to the Survey Branch as "Traverse and Referring Stations") would be placed, and bearings would be observed with a 7-inch theodolite of Everest's pattern divided into 10 seconds of an arc, and a 5-inch transit theodolite graduated to read to 20 seconds of arc, depending on whether the work is in the town or the country.
- Ground marking was done by way of granite boundary stones, 4 × 3 × 3 feet with centre marks.
- A Survey School was set up during this time, running a 3-year surveying course (more on this below).
- The damage or removal of Boundary stones and markings was an offence, first recorded under the Boundaries Ordinance 1884 (this followed the 1884 Boundaries Bill, instigated by the Surveyor-General, supported by Seah Liang Seah, then-acting Colonial Engineer and a big supporter of the land survey and revenue work (and after whom Liang Seah Street is named) and George T. Addis, Commissioner of Land Titles). A "Survey of the Colony" followed, along with the creation of a new position, known as a Demarkation Officer. Failure to comply with a notice from the Collector of Land Revenue to keep the boundary defined would result in the boundary being defined by the department at the expense of the guilty party.
- Surveys for land registration purposes were a priority for the Survey Department, as reinforced by the Registration of Deeds Amendment

Ordinance of 1886, which was brought into force in Singapore on 1 October 1907.

- Scale maps were introduced: a scale of 40 feet and 50 links to the inch in the town, 1 chain to the inch in the suburbs and 4 chains to the inch in the country.
- A compilation of old leases and grants was conducted to identify occupied land without title.

Singapore Property Tax System 1825–87

The origins of the Singapore property tax system were in the British rating system and the "annual value" concept. Excise taxes were first levied in 1820, but it was not until 1825 that houses were first assessed. These revenues — called property tax or rates — were used to pay for municipal workers. In 1832, a 5% rate was levied on all houses in the town, increasing to 8.5% in 1840 for town houses, with a separate levy for those out of town. The rate was raised to 10% in 1848 (5% for out-of-town, i.e., "country" houses).

The 1839 Act XII established the basis of the "Assessment" or "Tax" — it was based on the real annual value (or not exceeding 10% of the value) of every dwelling house and other properties in Singapore, Malacca and George Town, with the annual rate levy applied to the Assessment being fixed at no more than 5%. The Act empowered the Chief Civil Officer (under instructions from the Governor of Bengal) to collect the Assessment and Taxes leviable under the Act in the same way as if they were employed in the collection of Land Revenue. The word "Taxes" referred more to the carriages, carts and other vehicles that were covered by this Act.

The "rates" system was British in origin but arrived in Singapore via India, as Singapore was during the 1830s and 1840s, governed by the British East India Company in India. The series of key legislation in those years is as follows:

- Agitation among mainly European residents (anxious to foster local autonomy and control over urban affairs) was exacerbated by worsening sanitary conditions. This resulted in the passing of Act XII of 1839, which allowed for the establishment of an assessment fund to be used for municipal purposes. The fund was disbursed by executive officers of the Government and used mainly for the upkeep of a small Police force.
- Sanitary ills and governance concerns continued, and the European residents — as "ratepayers" — demanded change. Act IX of 1848 empowered a Municipal Committee of five assessors (2 officials, 3 non-officials) to administer the fund.
- The 1853 Act conferred powers on the Municipal Committees of the "stations" in the Straits, i.e., Malacca, Penang and Singapore. These powers included the assessment, levying and collection of rates/taxes and the appointment of (inter alia) surveyors to oversee public works.
- Act XXVIII of 1856 provided for the establishment of a Municipal Board comprising five members — the Resident Councillor (and later, the Colonial Secretary after the 1867 transfer) as President, one other official and three representatives chosen by ratepayers paying annual rates of over 25 rupees. Complementary Acts in 1856 directed the municipal commissioners to apply the funds to conservancy and improvement. Again, the focus was on improving and maintaining the infrastructure, including the Police force (albeit operated independently under Act XIII of 1856), burial grounds and to "pull down ruinous houses".

The Indian administration of Singapore ceased in 1867 when Singapore became a Crown Colony administered directly by the Colonial Office in London, and under the Municipal Ordinance of 1887, rates were levied on officially assessed Annual Values. A rural Board administered non-urban parts, and rubber estates were brought into assessment in 1937. Steps were taken to ensure revenues were used effectively on a value-for-money basis and that the administration of those funds was transparent through a part-elected municipal committee established in 1856.

Over the next 20 years, the position of successive Governors differed on the involvement of Government officials, (British) Government control and public/ratepayer involvement in the administration of local affairs, but in 1886, the departing Governor (Weld) set in motion a bill that was the first strand of legislation passed in the Crown Colony to control the built environment, triggering wider interest in the running of municipal affairs.

There will be more in later Chapters of this Book on the prominent role in relation to the assessment of rates played by Chartered Surveyors such as Harold Carpmael and Fred Snell in the pre-World War II years and John Aspinall in the post-World War II years.

Survey School

The Survey School of the Revenue Survey Department in Singapore was the first School of its kind in the region — it also helped to develop the careers of Land Surveyors, at least one of whom became a Chartered Land Surveyor. The original School Instructor was A. Sutherland (late of the Royal Engineers), who was there from when it first opened in 1891 until it closed in 1894. However, it re-opened in 1902 and Wright's book tells us that the Instructor at the School upon re-opening was John van Cuylenburg (see Figure 1.4(c)).

John van Cuylenburg (1862–1940) was a well-known and well-regarded draughtsman and surveyor. He was not a Chartered Surveyor, but he was influential in the emergence of Land Surveyors born and trained locally. Born in Colombo, Sri Lanka, on 3 June 1862, he was educated at Colombo Royal Academy, and on the recommendation of the Ceylon Government, he was, in 1881, appointed as the Singapore Plotter and Computer. He later transferred to Malacca, returning to Singapore in May 1885 as a draughtsman. In 1888, he became District Surveyor, and in addition to his many surveying duties, he was also commissioned by the Royal Asiatic Society to prepare a new map of the Malay Peninsula. After a short surveying stint on Christmas Island, he was appointed as

Survey School Instructor in 1902, again in addition to his other duties. His Map of the island of Singapore and its dependencies still exists today. John was married to Arabella Smith from Sarawak, and he died in Kuala Lumpur on 10 September 1940. His descendants still live across Singapore and South-East Asia.

Going back to 1889, the idea of a Survey School was first mooted and gained the support of Henry McCallum (the Colonial Engineer in Singapore mentioned earlier in this Chapter) and Captain Cameron, the Colonial Engineer based in Penang. The idea was to set up regular training schools for surveyors in Singapore, Penang and Malacca, with teachers brought over from England for this very purpose. *The Straits Times* reports that the demand for surveyors in both Government service and private practice was strong, and it made economic sense for the State to train up surveyors born in the Straits Settlements rather than using imported surveyors — the point was also made that local surveyors would be more "content with their lot" than surveyors imported from overseas. By September 1890, more details of the first School (in Singapore) emerged, with the intention of starting with 20 students to be selected on passing a general knowledge examination set by the Surveyor-General and trained at the Government's expense. The allowance was $12 a month, rising to $20 a month after half a year and to $40 a month on passing an examination to become 5th Grade surveyors. The Government's objectives were clear — they wanted to give "*better-educated lads of the Eurasian population the opportunity of obtaining a skilled training without cost and with the prospect of an assured and improved livelihood*". The preference towards Eurasians was said to be because there was a much greater chance of them staying in Government service for their entire career, as opposed to someone from England who may not stay in the Straits Settlements for long. The intention was to create another School in Penang in due course.

The Singapore Survey School's first day was 16 October 1890, with Sutherland as the School Instructor. The school was situated in a house on Beach Road facing the sea — this house was also the home of Sutherland

and his family. In April 1893, the (School) house was struck by lightning in a violent thunderstorm; fortunately, nobody was badly injured, although Mrs. Sutherland and her children were hit by falling plaster.

In August and September 1891, the Colonial Secretary J.F. Dickson (after whom Port Dickson is named) placed advertisements in the local Press, inviting applications for the second intake in Singapore to be submitted to the Survey Department. The criteria included an age preference of between 17 and 21 years old, a medical examination and a 6th Standard Certificate, with additional passes such as Algebra and (Euclidian) Geometry being preferred. In addition to the September 1890 announcement, the Colonial Secretary's advertisement mentioned a 5-year Bond to work in Government service after passing the school's surveying examinations. Daily classes (apart from Saturday and Sunday) would be held, and after the first three months, field work and plotting would be added to the curriculum. Time off for sickness would only be allowed with the production of a Medical Certificate from the Government Medical Officer. After six months, the students would be split into field and office work, depending on which they were most suited to. There is an 1891 *Daily Advertiser* article on the first examination results from the Survey School, congratulating Sutherland (then still a teenager) on his efforts, particularly as he had given up a job in the Square on a higher salary to become the Instructor. The results were deemed "satisfactory" with Adolphus Norris and Loftus Beins scoring 311 and 303, respectively, out of 429 — their reward was a higher monthly allowance of $20 (the monthly allowance was $12 a month during the first six months). After 12 months, the students could present themselves for the second and final examination, and if they passed, they would be appointed surveyors of the 5th Grade, which commanded a salary of $480 per annum.

In June and September 1891, both Beins and Norris were members of the Survey School Club's football team. The teams are listed below — most, if not all, the team are from well-known Eurasian families in Singapore:

> June 1891, playing against the Chinese Club with J.J. Oliviero in goal; A. Norris and L. Beins at the back; A. Hansen, R. Tessensohn and F. Minjoot the half backs; E. W. Klyne, Chas. Mosbergen, S. Quays, A. Read and C.F. Oliviero the forwards.
>
> September 1891: H.E. Valberg in goal; S. Quays, G. Ahin, C. Oliviro, W. Cowie and A. Norris at the back; with L. Beins and E. W. Klyne on the right wing, F. Minjoot in the centre with A. Read and C. Mosbergen on the left wing.

The annual report for the year 1892 from now-Major Henry McCallum, published in May 1893, stated that the school was showing improvement, with most students having passed their examinations and some finding employment at the end of their Final Examinations — Adolphus Norris and Edward Klyne found surveying positions in Perak and Loftus Beins found work in Malacca. The school students also did some work supporting the Survey Department in 1892, including map tracings, and a portion of the new Town Map was prepared by way of a practice exercise. The school's expenditure in 1892 was $3,922.82

As of July 1893, the School's examinations saw seven out of eight students pass. There were falling numbers of staff in the Survey Department, and a further problem was that the students (the vast majority of whom were Eurasians) could not get suitable employment at the end of the examinations because there was a preference for "highly paid outsiders". As a result, a decision was made to close the school in 1894, at which point Mr. Fry had taken over from Sutherland, who moved to Penang to replace Survey Department staff who were leaving due to ill health or other reasons. The changes saw Robert Young (Surveyor 1st Grade) (see Figure 1.4(b)) leave Malacca to be replaced by **John Pennefather** (more on him in later Chapters). The newspaper article does not explain the preference for "highly paid outsiders", but the appointment of Pennefather (from a prominent Eurasian family) to the highest surveying role in the Survey Department in Malacca confirms that either change was about to happen or discrimination was not as widespread as the *Daily Advertiser* was inferring.

Notwithstanding the issues of alleged discrimination, due to growing demand and the recognition of its success, the school was re-opened at the end of 1901, at which point we start to see a more diverse range of students, not just those from Eurasian families.

The Survey School was an initiative led by the PWD (Land/Revenue Survey Department), with links to the Royal Engineers, and in line with the Government's objectives — the vast majority of the Survey School students came from Eurasian and Anglo-Indian families. Examples are Norris (born in 1873 in Singapore to a well-known Eurasian family, who later worked for Lermit & Westerhout alongside Arthur Lermit — more on his father, Alfred Lermit, in Chapter 2); Klyne, later a surveyor with G.F. Fernandez & Company; and Beins, who worked in Government service in Malacca and elsewhere.

It is no coincidence that during this time, a small number of Eurasian surveyors became Chartered Surveyors. There is no doubt that the school played a role in helping them achieve their professional goals, given that travelling to London to study and work full time to get a professional surveying qualification was impossible.

Further evidence of the Survey School's link to the Eurasian community came in April 1912 when the School's examinations (with Mr. Towner and Robert Young as the examiners) saw in second place **George Pennefather, son of John Pennefather**, a Chartered Surveyor who features in Chapter 4. Earlier, in 1894, **Wilfred Nicholas**, who later became a Chartered Surveyor, and who is also featured in Chapter 4, studied at the Survey School. His Certificate was signed by a Colonel of the Royal Engineers, evidence that whilst the school was PWD-led, the Royal Engineers were closely involved.

Later, in 1917, the Survey School sought to recruit four students, "boys of good character between 16 and 19 years of age, of good health (certified by a Government Doctor) and preferably born in the Straits Settlements". They must have taken the Junior Cambridge Examination or other examination deemed acceptable to the Surveyor-General. The 3-year course started with an allowance of $300 per year, rising to $420 and then

$540 for the final year, subject to a surety sum of $500 and an undertaking to complete and, if required, serve a 5-year post-completion employment bond with the Government. Applications had to be accompanied by appropriate documentation as well as references and submitted to the PWD Executive Engineer.

The PWD set up other schools in Malaya and other locations (e.g., India), but Singapore was the only one wholly dedicated to surveying work. The closest to it in the region was the Treacher Technical School in Kuala Lumpur (named after Sir William Hood Treacher, KCMG, a former Resident-General of the Federated Malay States), later to become the Kuala Lumpur Technical College and now a University (UTM). The school was set up in 1904–06 to provide specialised instruction for the apprentices in the Public Works and Railway Departments focused on the PWD Subordinate Examination. It had Junior and Senior Divisions, and school subjects as of 1906 were algebra, trigonometry, elementary statistics, building materials, building construction, surveying, earthwork and roads, bridges and applied mechanics, plan drawing, and estimating. The first headmaster was D.H. Laidlaw, a PWD engineer, and one of its first trainers was K.H. Maule-ffinch, who is mentioned further in Chapter 4.

Singapore's First Surveyors

Building a Name: The History of the Royal Institution of Chartered Surveyors by Jennifer North describes the role of a surveyor as being one who oversees "land, buildings, construction and the environment. They measure, manage, value, develop and protect the most fundamental assets of the world". Given the small size of Singapore and the perspective that land was (and is) a precious commodity, it is evident that these surveyors took their role as overseers of land seriously. They built some of Singapore's most important structures, such as the National Museum, and through their survey work, they laid out the island's infrastructure. They even set up a Survey School so that they could pass on their land-surveying knowledge — this School was the first of its kind in South-East Asia and produced at least one (if not more) of the first locally-trained Chartered Land Surveyors.

These surveyors captured the size, extent and use of the island's land and property assets, and in so doing so, laid the core foundations for the development of Singapore's built environment by future generations.

Although the first surveyors were not Chartered, their training and dedication to public service were wholly consistent with these words from the RICS Royal Charter: *to secure the advancement and facilitate the acquisition of that knowledge which constitutes the profession of a surveyor... and to maintain and promote the usefulness of the profession for the public advantage in the United Kingdom and in any other part of the world.*

Chapter 2

1868–1900 — The First Chartered Surveyors in Singapore

Lermit, Chevallier and Coghlan

Several decisions around the Singapore Municipal and Government affairs between 1819 and 1887 led to an increase in trade, population growth and a significant increase in building construction along with infrastructure to meet the growing needs and importance of Singapore.

Meanwhile, in London, in 1868, the Institution of Surveyors was formed. By the late 1870s, members of the Institution of Surveyors had already started to seek employment opportunities around the world. The first Chartered Surveyors arrived in Singapore in the early 1880s as a result of Singapore's growing importance, resulting in the Straits Settlements moving from the control of the British East India Company and becoming a British Crown Colony in 1867.

Singapore Commerce, 1868–1900, and the Role of Surveyors

In the early decades of post-Raffles Singapore, the role of administering and regulating land and property in Singapore was dominated by other professions (and by clever, skilled entrepreneurs with no professional training), but the legislation and administrative decisions in question laid the ground for Chartered Surveyors to find opportunities in Singapore by the end of the century.

Through the British India administration, the English doctrine of estates and tenures operated in Singapore (as well as Malacca and Penang) so that all land in theory was vested first in the East India Company and, later (after 1867), the Crown. This was the case from when Raffles arrived but was reinforced by the Second Charter of Justice, English statutes that came in force on 26 November 1826. As a result, from almost the moment that Raffles landed in Singapore, land and property transactions took place.

For at least the first 20–30 years, many transactions took place under the auspices of A.L. Johnston & Co. (1820–92), a firm set up by Alexander Laurie Johnston, one of the first settlers in Singapore and a friend of Raffles. Johnston was later joined in the business by the Read family, also early Singapore residents. Their business was much wider than just the sale of land and property — they were amongst the most prominent businesspeople in the town, and they traded in the entire range of goods and commodities that came into the port of Singapore. Johnston was a founding member of the Singapore Chamber of Commerce in 1837 and was elected its first chairman. Johnston's Pier at Collyer Quay (demolished in 1933 and replaced later with Clifford Pier) was named after him.

Another active firm in the land and property market at this time was Martin, Dyce and Company — a trading firm of merchants and agents. From 1842 to 1847, its Singapore office was headed by Charles Alexander Dyce (sometimes referred to as Charles Andrew Dyce), a Scotsman of many talents, who, in 1847, added the impressive tower and spire to the Cathedral of the Good Shepherd in Bras Basah Road. One notable land sale that the firm was involved in was a large land area owned by C.R. Prinsep Esq. between Orchard Road and Bukit Timah Road, offered for sale in 1859.

Against the backdrop of this growing enterprise in land and property transactions, it was in the 1880s that the first three Chartered Surveyors arrived in Singapore. They became involved in the sales of land and property at auctions and private transactions through auctioneers and estate agents such as Powell & Co. and Coghlan & Co. (more on them later).

Powell & Co. was established in 1863 by H.T. Powell, then-Secretary of the Singapore Exchange. As well as real estate auctioneers and valuers, they were also general traders; for example, they sold billiard tables. Based at 20–22 Raffles Place, Powell & Co. conducted land sales on behalf of the Colonial Government, arranged mortgages, valued property and were rent collectors. When Powell retired, J.T. Lloyd and C. Dunlop took over the business, and by 1908, Lloyd was the sole proprietor — he was also a long-time Singapore resident, having arrived in 1870.

New legislation brought in by the administrators of the Straits Settlements (under a governor based in Singapore and his executive and legislative councils) eventually brought about better regulation and structure to property transactions, ownership and management and enabled greater transparency, but it took a long time for change to happen. However, prior to this, the situation was chaotic. The entrance of professionally qualified surveyors, architects, property valuers and property auctioneers into the Singapore business community from 1883 onwards coincided with this improved regulation and important legislation. Key examples of the important legislative changes are:

- Boundaries and Surveys Ordinance of 1884 — Ordinance No. VIII
- Crown Lands Ordinance of 1886 — Ordinance No. II (followed by the Crown Lands [Grants in Fee Simple] Ordinance 1902)
- Land Revenue Collection Ordinance of 1886 — Ordinance No. IV
- Conveyancing and Law of Property Ordinance 1886 — Ordinance No. VI
- Conveyancing and Law of Property Act (Cap 268)
- Ordinance 6 of 1894 — Bills of Sale Ordinance 1894 (preceded by the 1886 Ordinance, No. XII
- Bills of Sale Ordinance of 1886 — Ordinance No. XII
- Registration of Deeds Ordinance of 1886 — Ordinance No. XIII
- Ordinance 21 of 1906 — Auctioneers Licensing Ordinance 1906 Amendment Ordinance 1906
- The Foreshore and Sea-Bed Ordinance I of 1901 (the leasing of foreshore and seabed land for periods not exceeding 100 years)

Prior to 1858–67, there was a range of licences, permits, leases, grants and arrangements with varying detail but little in the way of proper boundary definition. Furthermore, there was no registration and no process for tracing the real owners or leaseholders in the event they did not pay their quit-rent or if a boundary issue arose. So, much confusion about the Colony's land system existed in the 1850s and 1860s, exacerbated by the fact that legislators and townsfolk had differing opinions as to the direction it should take, creating more uncertainty.

Consequently, the legislation from 1883 to 1906, coupled with the extensive land survey work done from 1881 onwards (as mentioned in the Chapter 1), was a major factor in enabling Alfred Lermit, Harvey Chevallier and Henry Lake Coghlan to enter the market with success.

Alfred Lermit

Alfred William Lermit (1850–1929) (see Figure 2.1(b)) was born in Lexden, near Colchester, Essex, England, the son of Gerald Thompson Lermit (1825–94) and Elizabeth Henrietta Downes (1825–92). He married Emily Jervis in Hampstead, England, in 1879. They are seen together in Figure 2.1(a).

He qualified as a Professional Associate of the Surveyors' Institution in 1878.

Prior to reaching Singapore, he had lived in Hampstead, London — first at 1 Stow House, North End, and then at 1 South Hill Park Gardens — both substantial residences that are still standing today. Lermit was by then practising in Chancery Lane, London, and one of his trainees there (who was also a lodger at the home of Alfred and Emily Lermit at the same time) was a young man by the name of Harvey Chevallier. There is more on Chevallier later in this Chapter. Lermit already had links to the region through his grandfather, Alfred Lermit, who was a Military Officer with the East India Company.

Alfred and Emily's arrival in Singapore was announced in *The Straits Times* on 11 June 1883, where it was stated that owing to Alfred's arrival, the business activities of Messrs. Crane Brothers' business at 7 Raffles Place would henceforth be extended beyond land agency, valuation and auctioneering to surveying, levelling, preparation of designs for buildings of all descriptions as well as the demarcation of boundaries, laying out of building estates, roads, plantations or other works.

Arnold Wright's *Twentieth Century Impressions of British Malaya: Its History, People, Commerce, Industries, and Resources* tells us that:

"Mr. Alfred William Lermit, of Messrs. Tomlinson & Lermit, was born at Colchester in 1850. After being trained as a land agent and surveyor, he commenced practice on his own account in London in 1871, and during the next ten years, built up a good connection.

In 1882, he qualified as a fellow of the Surveyors' Institute. He came out to Singapore in the following year, and practised in partnership with Messrs. Crane Bros., and others, for some time. Mr. Lermit has carried out surveys for the Government in Province Wellesley, in the island of Singapore, and in Johore.... For the past few years, he has been in partnership with Mr. Tomlinson."

FIGURE 2.1. (a) Alfred and Emily Lermit. (b) A portrait of Alfred Lermit. The first photo was taken during Lermit's early days in Singapore while the second photo was taken in the 1890s at the height of his Singapore career.

By September 1883, he had moved on and set up a practice with the well-known and highly respected (PWD) Government surveyor, Arumugam Annamalai ("Namly") Pillai at 29 Malacca Street in a new architectural and surveying firm. Pillai came from a wealthy Hindu family in Ceylon (now Sri Lanka) and later became a major landowner in Singapore — indeed, roads on his vast Bukit Timah estate were named after him: Namly Road, Namly Road, Namly Crescent, etc. Their business was modest and only lasted a period of around three years, but this partnership between an Asian and a European was one of the first of its kind in Singapore and a first for surveying and architecture in Singapore itself.

Having established himself in Singapore over a period of four years, Lermit made another career change, one that would have a significant impact on the built environment in Singapore — the formation of Swan & Lermit, which evolved a few years later into Swan & Maclaren. In early 1888, Lermit formed Swan & Lermit with an engineer by the name of **Archibald Alexander Swan (1857–1911)**, who had come to the region in the 1880s from Scotland to manage survey and construction works for new roads, railways, buildings, and piers throughout the Malay Peninsula and in Johor, Borneo and Sumatra. Swan was also involved with oilfields and mines across the Malay Peninsula. Swan was an Associate Member of The Institution of Civil Engineers by 1885 and a Member by 1896.

Both Swan and Lermit were well placed to offer their services locally — indeed, as of 1889, both Swan and Lermit were two of only four surveyors in Singapore recognised to work for the PWD — Pillai was one of the other remaining two. Indeed, as evidenced by Julian Davison's *Swan & Maclaren — A Story of Singapore Architecture*, the work of Swan & Lermit was predominantly survey work supporting the construction of railways and waterworks across the entire length of the Malay Peninsula. Their offices were in Flint's Building, which was located on the corner of Battery Road — this building was later replaced by the Laidlaw Building and is now the site of the Maybank office tower.

During the short commercial life of Swan & Lermit (1888–92), additional engineering and surveying staff came into the practice at the same time as Harvey Chevallier (see later in this Chapter) moved on. These additional recruits were from Scotland and comprised Civil Engineers and an architect, which reflected the nature of the services performed by the practice, predominantly survey work across the Malay Peninsula and Borneo with a limited portfolio of architectural design work in Singapore.

Despite the predominance of survey work, Swan & Lermit are also attributed to the design of several landmark buildings in Singapore between 1888 and 1892. Two of these buildings are still standing today — one was a godown built for Messrs. Paterson, Simons & Co. that now faces onto Read Bridge, in what is now Clarke Quay and the other was Our Lady of Lourdes Church in Ophir Road, which is now a National Monument (see Figure 2.2(a)). Some historical sources state that the plans for the Our Lady of Lourdes Church were prepared by Swan & McLaren. However, the land was dedicated to the Church in 1885, the foundation stone was laid in 1886, and construction was completed in 1888 — yet Swan & McLaren did not exist until 1892. However, what is certain is that the plans for the Church were initiated by Father Meneuvrier in Singapore. He worked with Alexandre Izambert of Paris, who supplied the foundation plan, the cast iron columns and metal frameworks from his Paris iron foundry. Indeed, at the same time, Izambert was also involved in a very similar "iron church" project at Fernan-Vaz in Gabon. So, the exact involvement of Swan & Maclaren or Swan & Lermit is unclear. Izambert had been born in 1850 and owned an Iron Foundry at Boulevard Diderot in Paris. He won the only Gold Medal at the third Paris World's Fair (the "Exposition Universelle") in 1878 and was highly regarded for his iron frame constructions. He was also involved in the railway construction industry in Africa, so it is possible that he knew Swan through their railway industry links, and through that connection, Swan & Lermit became involved in the construction of the Church.

FIGURE 2.2. (a) Our Lady of Lourdes Church and (b) Singapore Philatelic Museum (former Anglo-Chinese School) on Coleman Street.

In the late 1880s and early 1890s, Alfred Lermit was also the Secretary of both the Tanjong Pagar Land Company and the Singapore Land Company Limited. The Tanjong Pagar Land Company was responsible for many land reclamation works in the 1890s, along with strategic acquisitions for reclamation works at the Anson Road end of Tanjong Pagar; today, this land is an integral part of the Tanjong Pagar Terminal. Due to his expert knowledge of Tanjong Pagar and his growing expertise as an architect, Lermit was commissioned to be the architect for the Boustead Institute, the full cost of which Edward Boustead had set aside to be constructed after his death. The Boustead Institute opened in June 1892 at the junction of Tanjong Pagar Road and Anson Road, and became a place of free lodging for destitute, sick and homeless sailors. Unfortunately, this magnificent landmark building (see Figure 2.3) had to be demolished in the 1970s to facilitate the port expansion.

In 1890, Swan & Lermit was commissioned to redesign the road junction of Battery Road, Bonham Street and Commercial Square (now Raffles Place), with the demolition of some old godowns, the widening of Battery Road and the erection of new buildings set back along the new line of Battery Road. The current road layout at this junction is almost identical to how it was when Swan & Lermit's project was completed. In the same year, Lermit delivered another landmark on Battery Road — the Dispensary Building.

FIGURE 2.3. Boustead Institute, Anson Road/Tanjong Pagar Road, Singapore, taken from a pre-WWI postcard issued by Wilson & Co. of Singapore.

In 1892, Lermit was on the move again, and he left Swan & Lermit, at which point the company became known as Swan & Maclaren. Until that point, Maclaren (full name James Waddell Boyd Maclaren) had been an assistant to Swan in the practice. Julian Davison's excellent *Swan & Maclaren* details Lermit's achievements during the transition from Swan & Lermit to Swan & Maclaren during this period and the success of the company in subsequent years, a success which was in no small part attributable to the Chartered Surveyors who worked for Swan & Maclaren in Singapore from the 1890s onwards.

From 1892, Lermit practised on his own account as a property agent (from 3 Raffles Place), although he still retained his links to Crane Brothers as the Company Secretary. In 1897, he formed Lermit & Westerhout, which was responsible for the design of several buildings in Singapore. In 1900, Lermit & Westerhout were involved in a case of negligence called upon by Goh Eh Choon, Chop E. Seng Hong and Syed bin Mohammed bin Ahmed Alsagoff against Manasseh Meyer of Meyer Bros., concerning the collapse of two houses at 62 and 63 Boat Quay. The claim from the plaintiffs was that the dependent did not take proper precautions whilst repairs were being done to the houses in October 1899. Lermit & Westerhout supported the claim and gave evidence on behalf of the plaintiffs, who won the case.

FIGURE 2.4. 38 Oxley Road: A famous building designed by Lermit & Westerhout and the former home of the Founding Prime Minister of Singapore, Lee Kuan Yew. The building plot was laid out for a twin pair of bungalows, Nos. 38 and 40 Oxley Road, built in 1898 and originally named "Castor" (No. 38) and "Pollux" (No. 40). Only No. 38 remains.

Lermit & Westerhout became **Tomlinson & Lermit** in May 1904 when Sam Tomlinson entered the partnership. The May 1904 partnership resulted from a merger between Lermit & Westerhout and Tomlinson & Chye Tian Fook, and it created a partnership that could offer a whole range of services, including civil engineering, architecture and valuation as well as being licensed surveyors. Their offices were at 1 Raffles Place until 1907 when they moved to 23 Raffles Place and later to 17 Armenian Street in Singapore.

Tomlinson and Lermit worked together on the design of many Singapore commercial buildings and houses, some of which survive today, including the Singapore Philatelic Museum on Coleman Street (see

FIGURE 2.5. Nos. 7, 8 and 9 Bras Basah Road Singapore, designed in 1903 by Tomlinson & Lermit.

Figure 2.2(b)). The Museum was originally part of the Anglo-Chinese School on Coleman Street, for which the Crane Brothers designed the first School building on Coleman Street in 1893 (with the construction project managed by Lermit). Tomlinson & Lermit designed the building extension, which was commissioned by the Trustees of the School in 1906.

Westerhout left the practice around 1911/12, but the Lermit–Tomlinson partnership stayed together until 1914 when they separated too, with Tomlinson setting up in Robinson Road and Lermit practising independently from Room No. 20, Hotel de la Paix. Not long after this, Lermit retired from full-time practice in Singapore.

Another wonderful example of Tomlinson & Lermit's design is the Parsonage that lies next to the Armenian Church (see Chapter 1 in the context of Coleman) at the junction of Hill Street and Coleman Street. Built in 1905 and ready for use by October 1905 at a cost of $7,000, it was funded by Nanajan Sarkies in memory of her husband, John Shanazar Sarkies. The first occupier was Monseigneur Iwathian, the Armenian Archbishop. It is an excellent example of a Singapore two-storey "black and white", including the projecting porch to protect visitors against the weather extremes, and its louvred windows. At the rear are the original servants' quarters, separated from the main house. It lies on the grounds of the Armenian Church and is now the home of the Armenian Heritage Gallery. See Figure 2.6.

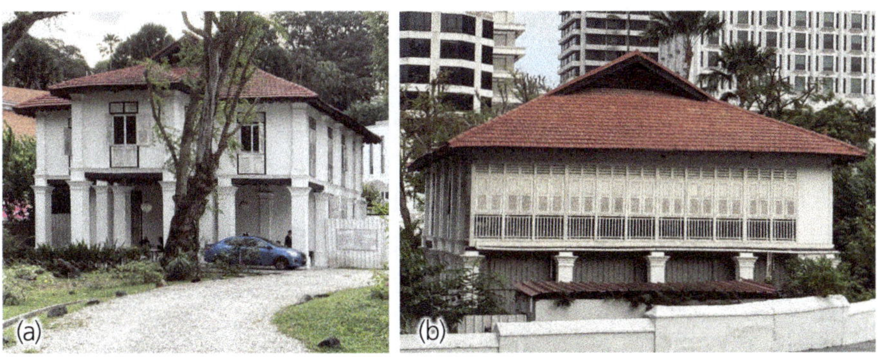

FIGURE 2.6. (a) View taken from Hill Street of the covered front porch, former Parsonage next to the Armenian Church, and (b) its rear view. The photographs show the building during major renovations undertaken from 2020 to 2022.

Alfred Lermit was passionate about the protection of the public when it came to matters of building safety. Indeed, a letter of his (signed off as a Fellow of the Surveyors' Institution) was published in *The Straits Times* on 6 May 1903, following a building collapse on Bras Basah Road on 14 April. He expressed the opinion that until personal liability (coupled with the risk of being barred from architectural practice) was introduced, the public would not be safe from building collapses. He said that had the collapse occurred half an hour earlier or later, it might have had catastrophic effects, given that children from the nearby Raffles School walked home this way after school. He urged that the Superintendence of Buildings in the Municipality of Singapore should fall under a suitably qualified — and licensed — professional similar to the District Surveyor in the UK at that time, someone devoted to public building matters, and that the cost should fall on building owners and not the ratepayers.

The incident in question resulted in the building owner (Syed Mohamed Alsagoff) being fined (or "mulcted", as *The Straits Times* described it) $250 for deviation from the building plans that had been agreed upon by the Municipal Commissioners. The building in question was the Bras Basah wing of the Raffles Hotel, which Alsagoff leased to the Sarkies Brothers. The new wing was finally completed in 1904.

Mr. Pierce, the Municipal Engineer, was called upon to explain to the Court from a technical perspective why the deviation from the plan was of a serious nature. He explained that the deviation involved the construction of a wide single arch (instead of two arches in the original approved building plans) on the first floor of the new Bras Basah wing of the Raffles Hotel, and the width of this arch was such that the end wall was not protected from the thrust of the arches. As rooms in this wing were to be used by the Hotel's guests, the architects' (Swan & Maclaren) intention for this deviation from the agreed building plans was to let more light into the rooms.

Swan & Maclaren had submitted plans to the Municipal Architect, but they had not been approved — the newspaper article strongly inferred from the evidence given in Court that the building owner had assumed they would get approved and hence told his contractors to go ahead. The newspaper added that "nothing was said about alterations on the upper floors", and whilst conversations clearly took place, the changes were not approved, and the plans showing the alterations had not been initialled by David Craik.* The implication in the newspaper articles of the time, without being specific on the point, was that the Building Inspector (Mr. Scott) verbally agreed to ground floor alterations but not to the first floor alterations and that when it was finally realised (by Scott or by Swan & Maclaren, or both) that there was a serious flaw, attempts were made to remedy the situation but it was too late, and the building collapse happened.

There was clearly a breakdown in internal communication within Craik's team (the Municipal Architects, Municipal Commission) and between Craik's Building Inspector and the contractors doing the building work for the owner. This was made worse by the fact that Swan & Maclaren were only consulted "as and when" during the project. Both sides also stated, as witnesses, that the Building Inspector had seen the work and had assumed that the re-design and construction were sound as Swan & Maclaren was a renowned firm in Singapore. The newspaper article does not say if he talked to the Municipal Architect (or the owner's architects Swan & Maclaren) about the change or if he noticed (or failed to notice) that there was a deviation from the approved plans, but the building owner did say under oath that he did not blame the Building Inspector for not spotting the change as the Inspector did not have time to examine the details.

Back to 1905, an unfortunate incident involving the Lermit household took place on 20 December when a 28-year-old New Zealander named Chapman, who was staying with the McGill family at their residence known as Ayer Jerneh on Serangoon Road Singapore, accidentally shot dead Lermit's gardener's Javanese wife as she fed her baby. The Lermit family lived next door to a property named Goldburn. Several bullocks

* David McLeod Craik (1873–1938), FRIBA, was a Municipal architect and a long-term resident of Singapore and Malaya. Ten years after the abovementioned incident, he was invited to become a partner with Swan & Maclaren, where he remained for the rest of his life.

had trespassed onto Ayer Jerneh, and Chapman attempted to resolve the matter by shooting at the bullocks, which were 40–50 yards from him. Unfortunately, he missed thrice, and his last shot entered the barn at Goldburn, injuring and killing the poor lady. Chapman was charged the next day with culpable homicide not amounting to murder. In January 1906, Chief Justice Sir Lionel Cox decided that there would be no good purpose served by giving Chapman a jail sentence, so he fined him $500 instead. The Court was told that the widowed husband was satisfied with the way that Chapman had conducted himself in the aftermath of the shooting, with the *Eastern Daily Mail and Straits Morning Advertiser* reporting that he would be using the money to take his family to Mecca to obtain a special benediction for his wife in the Great Beyond and to set up his own business.

Sadly, the verdict was met with applause from Chapman's supporters in Court, which is a sad reflection of the times, when the lives of white Europeans were clearly of much greater value to the colonial rulers than the lives of South-East Asians.

As well as being involved in the design and construction of many buildings in Singapore, Lermit owned many parcels of land. Indeed, at a Powell & Co. auction in March 1901, Lermit purchased a 65,273-square-feet building lot, near the 7th milestone at Bukit Timah Road, on a 999-year lease for $700, along with six smaller building lots totalling 57,054 square feet for an additional $500. The land in question is approximately where the Bukit Timah Market & Food Centre now stands. Lermit was regarded as an expert in property and land values. For example, he gave expert evidence of land values in the important Telok Blangah land dispute case of 1906. This case was brought by the executors of H.H. the Tummungong Abu Bakar (who took the title of Sultan in 1885) in relation to two pieces of land in Telok Blangah. The origins of the dispute go back to a parcel of 500 acres of land that had been conveyed back to the Tummungong by the Crown in 1863 — it had been taken by the Crown in 1823 as part of the cession of the Singapore island by the Sultan. The case was rather complicated and involved a "long possession" argument in relation to land, part of which was alleged to be part of the 500-acre parcel. It also involved the occupation of land by the Sultan, his family and the *orang rajah* since 1823. Lermit gave evidence to the Court as a witness for the plaintiff in relation to the value of the disputed areas.

Lermit also surveyed and measured another parcel of land in Telok Blangah that was subject to a claim by the same executor against George D'Almeida, who was also claiming "long possession" of a parcel of land that he purchased for $400 from a Bugis man who had previously been in possession of the land for 25 years, with a Statutory Declaration to support this assertion. D'Almeida put forward the argument that he had paid rates on the land to the Municipality for a certain period of years, as had the previous owner, and indeed one witness called upon was an employee of the Municipal Assessor's office, who confirmed that this was correct. Lermit's earlier survey of the land in question was subjected to much questioning during the Court case, as was his valuation of $2,000 per acre.

1906 was a busy year for Lermit — he was also called as an expert witness in several land dispute cases, and in each case, the greatest reliance could be placed upon his evidence. One prominent Supreme Court case related to land on Orchard Road, with Lermit's evidence not only being significant because of his involvement in surveys of the land dating back to 1884, but also because his involvement was in the capacity of a surveyor employed to carry out surveys, and not as a landowner, meaning that he had no vested interest in the outcome of the dispute. Lermit's evidence led to the decision by Justice Law (Sir Archibald Fitzgerald Law) that encroachment had taken place and that the land was not "no man's land" as the defendant was claiming. Another case in 1906 involved building encroachment where he was called upon to confirm that the pillars supporting the building structure had not been built in accordance with the official plan.

After he became a sole practitioner in 1914 (upon the break-up of Tomlinson & Lermit), Lermit continued to practice as a registered surveyor in Singapore until around 1923, at which point he retired to live with his son in Malaya.

He died on 15 December 1929, after a short illness, in the Bukit Dinding Estate, Bentong, Pahang, at the home of his son, Ernest Vincent Lermit. He was 80 years old. His funeral took place in Kuala Lumpur at Cheras Road Cemetery, Cheras, where he was buried.

The Straits Times, on 19 December 1929, paid tribute to him as a well-known personality in Singapore with many friends. He was described as having a genial and kindly manner and being a man with wide classical and scientific reading. He was also described as the "Nestor of Singapore architects", which meant that he was much respected for his seniority and experience, and though he was inclined to offer advice that was not always taken, he possessed a great sense of humour and enough intelligence to be unconcerned when his counsel was not always followed.

His wife, Emily Frances Lermit (born Emily Frances Jervis), died in Rous Lench, Worcestershire, England, on 21 December 1938, aged 84. She is buried in the Rous Lench Churchyard, and her gravestone states that she was the wife of "the late A.W. Lermit of Singapore". Their youngest son Charles Rowland (b. 28 February 1887, Singapore), died in November 1909, aged 22, in a bicycle accident in Evesham, England. He, too, is buried in Rous Lench.

Another son, Major (later Lieutenant-Colonel) Arthur Alfred Lermit (b. 1881, Hampstead), OBE, was also a Chartered Surveyor (a Colonial Fellow) based in Singapore in his early career as a surveyor with Swan & Maclaren from 1911 to 1914. His photograph is shown in Figure 2.7. Returning from Europe after active service in World War I,

FIGURE 2.7. Arthur Alfred Lermit, son of Alfred and Emily Lermit.

he then spent his entire post-War career working and living in Johor and Malacca. He was Commanding Officer of the Malacca Volunteer Corps for many years until retiring to England in 1936, where he died in 1975. There will be more stories about him in a future book about Chartered Surveyors and R.I.C.S. in Malaysia.

Aside from Arthur Alfred Lermit, Alfred and Emily had five other children: Jessie Frances (1880–1961), Ernest Vincent (1883–1964), Charles Rowland (1887–1909, mentioned earlier), Olive Marguerita (1889–1966) and Henrietta (1890–1901).

Lermit's Partners: Sam Tomlinson and Johannes Westerhout

Arnold Wright's *Twentieth Century Impressions of British Malaya* gives us the background of **Sam Tomlinson (1859–1935)**. His influence and legacy as a surveyor in Singapore are worthy of mention in this book, even though he was not a Chartered Surveyor. He was one of the best-known professionals in Singapore, heavily involved in all local affairs for many years, including his time as the Municipal Engineer of Singapore. He arrived in Singapore in 1896 and retired to the UK in the 1920s, passing away in 1935 in Grange-over-Sands, England. Born in Yorkshire, England, in 1859, he was educated at Morton National School, Keighley Trade School and the Normal School of Science, South Kensington. At the examination of the City and Guilds of London, he won the surveying medal and came in first in mine surveying in 1882. Having done his professional training at Bradford Corporation Waterworks, he arrived in Bombay (Mumbai) in 1886 and spent ten years at the Bombay Waterworks. He reached Singapore in 1896 as a municipal engineer. Although his training was in surveying, he did not belong to any surveying association but was a member of the Institution of Civil Engineers, Royal Meteorological Society and American Society of Civil Engineers. His major achievements were:

- Plans for the extension of the water supply to the Tanglin and Keppel Harbour districts
- The reservoir at Pearl's Hill and the improvement of the Stamford Canal
- Designed the Jinricksha Department's headquarters and Adis Lodge — the Adis family residence on Mount Sophia, known as "one of the most magnificent mansions east of Suez"
- Commissioned the design (by Swan & Maclaren) of the magnificent "Grand Hotel de L'Europe", a beautifully designed Hotel that was only demolished in the 1930s because of its location, as the site was required for the construction of the new Supreme Court
- A prominent role in the arbitration case for the Government acquisition of the land and buildings for the Tanjong Pagar Dock Company's undertaking of lands for railway construction and water supply purposes and for the land acquisition in Johor for the construction of the State railway through Johor.

Johannes Bartholemew ("Birch") Westerhout (1871–1937) was a Eurasian from Malacca — his grandfather came from Utrecht and settled in Malacca, marrying into the local Eurasian community during the period when Malacca was a Dutch colony controlled by the Dutch East India Company or *Verenigde Oostindische Compagnie* (VOC). Having worked for the PWD in Malacca, Birch moved to Singapore in the 1890s and joined Crane Bros. Westerhout's skills as an architect were instrumental in the success of Lermit & Westerhout (and Tomlinson & Lermit) from 1897 until he broke away to form his own practice in 1911/12. Like Tomlinson, Westerhout was not a Chartered Surveyor, but his career was intertwined with Lermit, Chevallier and Coghlan in Singapore.

One final note: Lermit, Tomlinson and Westerhout all have Singapore roads named after them: Lermit Road is next to the Singapore Botanical Gardens, Tomlinson Road is off Orchard Boulevard, close to Wheelock Place, and Westerhout Road is between Geylang Road and Guillemard Road.

Harvey Chevallier

Harvey Chevallier (1861–1923) was the second qualified Chartered Surveyor in Singapore — he qualified in 1883 and, as mentioned earlier in this Chapter, he trained as a surveyor under Lermit in London and later followed him to Singapore to work for him.

Having qualified as a member of the Surveyors' Institution by the time that he arrived in Singapore in 1886, Chevallier worked for Lermit as a surveyor and assisted him at Swan & Lermit, although by 1888 (not long after the formation of Swan & Lermit), Chevallier was working for the Eastern Extension Telegraph Company in Singapore.

Chevallier was born in 1861 in Suffolk, England. The Chevalliers, originally from Jersey, lived at Aspall, Suffolk, and can trace their ancestral roots back to English Royalty (to Edward I). A newly discovered strain of fine malting Barley was named after his grandfather — indeed, Chevallier barley was renowned in England for the quality of the beers it produced. The Chevallier family is connected by marriage to the Cobbold family, a prominent family of landowners, politicians and brewers in Ipswich, Suffolk, who have also provided five Chairmen for Ipswich Town Football Club.

He was a District Officer in Kuala Pilah (a town halfway between Kuala Lumpur and Malacca) from 1891, and this is the most likely reason why he did not continue as a member of the Institution and ceased to be a Chartered Surveyor. At times, he covered the duties of the (Acting) District Officer for Singapore, so his time in Singapore from 1891 onwards was limited. A District Officer was essentially someone who administered Government business in a specific town or place in the Straits Settlements. He played an active and progressive role in his time in Kuala Pilah.

By 1900, he was in Government Service at Sungei Ujong, and then as District Officer in Perak (where he was held in high regard by all those who came into contact with him, including the Sultan of Perak), he became the British Government's Administrator to British Brunei from May to December 1907 in an acting capacity, and then permanently (Brunei and Labuan) from November 1909 to November 1913.

Towards the end of his career in Brunei, his surveying and legal skills, combined with his administrative skills, were put to use in November 1913 when he was involved in a disputed river rights case in Brunei relating to the Sibuti river and a claim for Perpetual Cessation Money deriving from the river rights, amounting to $500 per year. The defendants appealed against his decision in favour of the claimants, and the Supreme Court decided in favour of the appellant, which would have disappointed Chevallier, who held his own during a long cross-examination period during the appeal. It was a fascinating case pertaining not just to land law but to the rights of inheritance of "*anak gundik*", i.e., the morganatic or illegitimate children of a Sultan.

His surveying knowledge did receive some criticism in *The Straits Times* on 28 July 1913, when he issued a Government report on the island of Labuan. The report was described as a simple array of facts and figures, missing the "illuminative, descriptive touch" of reports from earlier years, possibly due to a lack of time on Chevallier's part. The only point of interest in his report was that Labuan was not stagnant, and that the PWD had carried out valuable infrastructure works in the past year for the benefit of the island.

However, his annual Reports during his time in Brunei demonstrated his strong abilities as a Government administrator, and his knowledge of land and property matters enabled him to administer matters such as land tax and the Survey Departments's activities with great skill — he oversaw the transition from poll tax to land rent as well as oversight of the land demarcation effort of the Government Survey Department, which was quite intense due to efforts (driven by Chevallier in partnership with the Sultan and others) to put a zoning structure in place for future land uses such as agriculture, rubber production and mining. He also administered prospecting licences and resulting fees for the likes of British Borneo Petroleum. Indeed, Chevallier said in 1913 that the "prospects of finding oil are favourable, and future developments will be awaited with eagerness".

Upon retiring from his position in November 1913, newspapers of the time covered his farewell (as a "hearty send-off") and departure from Labuan. However, by November 1914, he was back in Labuan, this time as District Judge, where he had previously been District Judge back in 1904. He retired soon after to England and died in Trimley St. Mary (a village between Ipswich and Felixstowe) in Suffolk on 2 September 1923.

He was unmarried.

He was described as a "man of kindly, quiet disposition and yet firm and just in his decisions and tactful in his dealings with the Malays".

Henry Lake Coghlan

Henry ("Harry") Lake Coghlan (1870–1938) (see Figure 2.8) was the third Chartered Surveyor in Singapore, arriving in Singapore in 1890.

Arnold Wright tells us that he was of "Irish parentage" and that H.L. Coghlan & Company Chartered Surveyors and Auctioneers was a highly respected firm of Auctioneers, Surveyors and Estate Agents. Coghlan was born in Cowes, Isle of Wight, England, on 16 April 1870 to Michael Coghlan (also spelt Coghland) of Hampshire, a Coast Guard, and Isabella Maria, daughter of Abel and Mary Lake of Portland, Dorset. He had a sister, Honoria, who later joined him in Singapore — she married and had a family here. Their father, Michael, died young, so Isabella took Henry and Honoria to live with her parents in Wakeham, Portland.

Isabella then married William Henry Pye in Portland in 1875. William was an Army Private (later a Prison Warder), and they all lived with the Lakes. However, Pye was the son of a Lancashire Cotton Manufacturer, and it is possible, although not certain, that Henry Lake Coghlan managed to get a good education due to his stepfather's family wealth or connections — for sure, coming from a poor background and losing his father when still an infant could have led to him (as the eldest child) starting work at a young age (10 or younger) to keep the family going. His grandfather, Abel Lake, was a Navy Pensioner and was unlikely to have been able to fund an education for Henry and Honoria. Coghlan arrived in Singapore in 1889, by which time his mother had died, and his stepfather was still living with the Lake family.

Coghlan worked for Powell & Co. auctioneers as an assistant during his first few years in Singapore. Powell & Co. comprised two separate business lines — one under Charles Dunlop ran the Malayan Mining Concessions Department and the other under John T. Lloyd ran the auctioneer and estate agency side of the business. Coghlan first worked for Dunlop and later for Lloyd. He clearly excelled at both sides of the business, as in later years Coghlan was a leading figure in the Singapore business community in the field of mining concessions and property transactions.

He moved to the General Post Office, where for several more years, he managed the Registered Letter Department. In 1901, having passed the Straits PWD examination to become a Licensed Surveyor, he set up a business as an auctioneer with a local partner. Arnold Wright's *Twentieth Century Impressions of British Malaya* tells us that: "No one is better able to feel the pulse of trade and finance or know the value of property better in a place like Singapore than an old established firm of auctioneers, surveyors and estate agents such as Messrs. H. L. Coghlan & Co. of 5 Raffles Place (see Figure 2.9(b)), who have one of the most extensive connections in the Straits".

FIGURE 2.8. Photographs of Henry Lake Coghlan from the time of his arrival in Singapore through to his retirement.

FIGURE 2.9. (a) The former Singapore Central Fire Station, plans for which were approved in 1907 by the Municipal Commissioner's Board during Henry Lake Coghlan's time as a Board member. (b) H.L. Coghlan & Co.'s premises at 5 Raffles Place, Singapore. This building remained *in situ* until the 1960s. Its site is now occupied by UOB Plaza 1 office tower, where it meets One Raffles Place.

During the arbitration case for the Government acquisition of land and buildings for the Tanjong Pagar Dock Company's undertaking of lands for railway construction and water supply purposes (and of land in Johor for the construction of the State railway through Johor), Coghlan was commended for his role as a land expert and valuer on behalf of the Colonial Government. In 1907, he was also called upon by the Government (alongside A.J.W. Watkins of Swan & Maclaren) to act as an assessor before the Supreme Court in relation to a compensation claim against the Government for land taken for the construction of a new Beri-Beri Hospital in Pasir Panjang. The land parcel was just over 2 acres in size and had been intended for development as a house dwelling in the period prior to the Government notice of acquisition. The Government offer was $4,402.26 plus the "normal" 15% compensation, giving a total payment of $5,062.59 for the land. Both sides called upon witnesses to give evidence for comparable land values, including Sam Tomlinson of Tomlinson & Lermit.

Coghlan's notable achievement was in 1906 when he delivered a revaluation (on the new exchange rate) of several million dollars worth of real estate held in trust by various corporations in Singapore. He served on the Municipal Council as well, representing Kallang, and he was also a Director and Trustee of the Singapore YMCA (along with Tomlinson), a position he held until his death. In 1908, *The Straits Times* announced that Coghlan had been elected a Fellow of the Surveyors' Institution of the United Kingdom, "which qualified him to designate himself a Chartered Surveyor". As this took place during Coghlan's time in Singapore, he was also allowed the designation "Colonial Fellow".

Whilst serving on the Municipal Council, Coghlan was a frequent Singapore Juror and an active member of the Municipal Commissioner's Board meetings in 1907, 1908 and 1909, at which the plans and drawings (prepared by the Municipal Engineer) were approved for the new Singapore Central Fire Station on Hill Street (see Figure 2.9(a)). This included the acceptance of the tender from Messrs. Chia Tien Siew to supply labour and materials (apart from iron and steelwork) to build the new Fire Station for the sum of $64,500. An additional sum of $10,000 to build a Superintendent's house at the back of the Fire Station was rejected as being too expensive, and since the principle sum was agreed, it would be looked at again at a future date. On completion in 1909, the new Fire Station was so large in comparison to the island's needs that it was said that it would serve the people of Singapore "for ever" — sadly, the island's population growth since 1909 has put paid to that assertion. However, the building has stood the test of time, and it is now home to the Civil Defence Heritage Gallery, the official museum of the Singapore Civil Defence Force. The building is now a National Monument.

During this time, Coghlan's firm was heavily involved in the sale of properties at auctions in Singapore, as well as the local management of several well-known rubber companies. His firm also handled the auction of a wide range of commodities and household goods from around Asia. From around 1901 onwards, Coghlan's operation (and that of his former employer and now rival, Powell) were responsible for the sale by auction of hundreds of residential properties across Singapore, as well as significant parcels of commercial land in Orchard Road, Bukit Timah Road and other parts of Singapore. An example of a successful auction sale was the sale at an auction in 1911 of land and property at 2 and 3 Craig Road (now situated in Tanjong Pagar Conservation Area) for $5,600. His firm was also involved in the sale of many shophouses and parcels of land developed for shophouses, such as on 25 November 1907, when they offered for sale the 999-year lease of a shophouse at 17 South Boat Quay.

Another notable sale by H.L. Coghlan & Co. in 1911 was the property known as "Buitenzorg" on Paterson Road, which was sold freehold for $18,500 — Buitenzorg was one of the grandest properties in Singapore, with grounds extending to Paterson Hill and Grange Road, and in the 1930s, it was the home of the Hu family. Buitenzorg has long since been demolished, and the Marq Luxury Condominium development now lies on its site.

At the same time, Coghlan was living nearby at "Burwood" on Grange Road. By 1910, he had decided to retire from Singapore, and sold his firm to Williams, Draper & Steadman (architects) and then set up an office at 39 Lombard Street London, a building often described as one of the most spectacular examples of Victorian Italianate banking architecture in the City of London. From these premises, he ran Eastern Industries Limited, which became one of the biggest firms in the world for managing rubber plantation companies.

In 1919, in the King's Bench Division, the Lord Chief Justice and a special jury tried the case of Coghlan vs. Lintner & Co., the latter being a commercial rubber company. Coghlan was claiming wrongful dismissal from Lintner & Co., where he had been Managing Director from 1914 to 1918. When the previous Managing Director returned from Europe at the end of World War I, he dismissed Coghlan, claiming that he had improperly received commissions. His claim could not be proven, and the case was decided in favour of Coghlan, who was awarded costs. *The Straits Times* covered the story extensively at the time.

In 1924, he wrote a book entitled *Coconut Industry in Malaya*, and he was also a respected photographer.

He was also one of the founders of Freemasons Lodge Malaya No. 5213 in London, a Lodge that still exists today.

He died in Worthing, Sussex, England, on 13 January 1938, aged 67. His obituary in the *Singapore Free Press and Mercantile Advertiser* confirms how he became involved in the rubber industry. Indeed, his auctioneering firm not only sold land and property, but Coghlan was also the first auctioneer in Singapore to sell plantation rubber, for which demand grew massively with the advent of the motor car (and pneumatic rubber tyres).

He was reputed to have been the driving force behind the use of solid rubber tyres for horse carriages and private rickshas, and his influence in Singapore led to the strong preference for plantation rubber over the hard Pará rubber trees that originated from Brazil. Indeed in 1907, during his time on the Municipal Council, he asked the President of the Board of Municipal Commissioners if any attempt had been made to induce the Hackney Carriage gharry owners to apply rubber tyres to their vehicles.

FIGURE 2.10. (a) Henry and Rosa's Singapore wedding in 1903, taken by G.R. Lambert & Co. (b) Grave of Henry Lake Coghlan and his wife Rosa in Weybridge Cemetery, England.

Until shortly before his death, he was still an active Director of Eastern Industries Limited. His effects on Probate amounted to just over £63,207.

He is buried in Weybridge Cemetery in Surrey, England, along with his wife Rosa, who died in 1926 (see Figure 2.10(b)).

Coghlan had married Rosa (full name Rosa Emily Smith, who was originally from Chelmsford, England) at St. Andrew's Cathedral at 4 pm on 4 April 1903 in front of 250 guests. Their wedding reception was held at "Burnbank" on Paterson Road — this was the home of Mr. & Mrs. Diss, who were friends of the bride. Indeed, it had been Mr. Diss (who ran a naval and military tailor business in Singapore) who gave the bride away. The wedding was described in *The Straits Times* as one of the prettiest weddings solemnised in Singapore. The best man was Lieutenant C.M. Phillips, SVI, and there was a large turnout from the Singapore Volunteer Corps, given that Coghlan was a Lieutenant and a Quartermaster in the Corps. Coghlan even wore his Singapore Volunteer Corps uniform for the wedding. The bride and groom honeymooned in Telok Kurau. Their wedding photograph is shown in Figure 2.10(a).

Henry and Rosa had a son, Kenneth Lake Coghlan (b. 1911, Hampstead; d. 1985), who was a Barrister, and another son, Patrick Boyle Lake Coghlan (1907–93), who took over Eastern Industries Limited when his father died.

Contributions of Alfred Lermit and Henry Lake Coghlan to the Surveying Profession in Singapore

Alfred Lermit and Henry Lake Coghlan had a great influence on the property industry in Singapore and were held in high respect across all business sectors. Lermit's legacy is still visible in the surviving buildings that he designed and built with his partners, and both Coghlan and Lermit were pioneers in the establishment of professional, general practice surveying firms in Singapore. Coghlan's work for the Government as a valuation expert was the first example of the professional expertise of a

Chartered Surveyor being used for public good in Singapore. Lermit was often called upon as an expert witness in cases of boundary or ownership disputes, given his knowledge of land and property in Singapore and his expertise in such matters. Both dedicated most of their lives to the growth of Singapore for the benefit of its people.

It is noteworthy that they both forged strong business links with Eurasians, although it would still be some years before Singapore would witness the pre-eminence of locally born Chartered Surveyors. During their time in Singapore, the public and private sectors attracted many surveying professionals, engineers and architects. Primarily at that time, they were mostly UK expatriates, many of who settled in Singapore or Malaya permanently. However, a good number of them were Eurasians from Malacca and Sri Lanka, as well as a few Straits-born Eurasians who were of mixed Indian, British, Dutch, Portuguese, German and Irish ancestry. The next Chapter will focus on some of these individuals. They laid the foundations for the growth, establishment and professional reputation of Land and General Practice Surveyors in Singapore.

As a footnote, two other Chartered Surveyors were working in Malaya in the 1880s and 1890s: Arthur Arch and John Koch. Neither was based in Singapore, but they operated across the Malay Peninsula during their time in the region. So, their story will be explored further in the R.I.C.S. Malaysia book when it is written. In brief:

John Godfrey Koch (1856–1913) was born in Ceylon (now Sri Lanka) in 1856 to a long-established Eurasian family of mixed German, Dutch, British, Portuguese and Sri Lankan ancestry. He worked in the Survey Department in Ceylon from 1875 to 1880 and then for the Straits Settlement Revenue Survey Department from 1881 to 1883. After that, he was the Surveyor and Settlement Officer in Kuala Kangsar (Perak, Malaysia) from 1894 to 1895 and was promoted to District Surveyor in Matang, Perak by 1896. After some time in Larut (also in Perak), he moved to Lower Perak, and by 1900, he was the Federated Malay States (FMS) District Surveyor for the entire Perak District, where he remained for several years. Koch became a Fellow of the Surveyors' Institution in the 1890s. His wife was a member of the Van Cuylenburg family, another well-known Eurasian family. Koch passed away in 1913 in Sri Lanka.

Arthur Joseph Edwin Arch (1862–91), AMICE, was a Minerals Surveyor who worked for the Rio Tinto Copper Mining Company. He qualified in April 1885 as a member of the Surveyors' Institution, and in August 1885, he joined Rio Tinto as a surveyor. After surveying tasks in Spain and China (Tientsin) in December 1888, he arrived on the Malay Peninsula, where he produced a concessions map of 2,000 square miles of Pahang for the Malay Peninsula Prospecting Company. *The Straits Times* in 1890 mentioned that the map contained much valuable information that the Pahang Government was only too glad to acquire. Arch had connections with prospecting operations in Pahang for some time. However, in December 1890, he returned to England and then to Brazil to work for Messrs. Brunlees and McKerrow on the Brazilian Imperial Railway Extension. He died tragically by drowning in the Rio Paraguassu, near Bahia, on 23 March 1891.

The diversity of skills that Chartered Surveyors carried with them around the world was evident even back in the 1880s, barely 20 years after the founding of the Institution. Tin mining had been a major industry in Malaya for centuries, and the decision by the Colonial Government to regulate the industry through concessions and the FMS Survey Departments resulted in work for Land Surveyors such as Koch and Arch in the late 19th century.

Chapter 3

The Story of Regent Bidwell

At the turn of the century in 1900, there were only two Chartered Surveyors in Singapore (Alfred Lermit and Henry Lake Coghlan), but by the end of World War I in 1918, the number of Chartered Surveyors in the Straits Settlements had reached double figures. However, only a handful remained in Singapore — one of them, Regent Bidwell, was also a renowned architect responsible for some of Singapore's finest colonial-era buildings.

Regent Bidwell

The detailed story of Regent Alfred John Bidwell's work for Swan & Maclaren is mentioned in many different books and publications, but Julian Davison's *Swan & Maclaren: A Story of Singapore Architecture*, published in 2020, captures Bidwell's career, including all the architectural detail of each of his works. I will not attempt to duplicate or compete with Davison's excellent work, which contains numerous photographs and reproductions of Bidwell's original building plans, many of which are held in the National Archives of Singapore. Instead, I will add some different perspectives to the fascinating story of Regent Bidwell, his life and time in Singapore, and his legacy.

Regent Alfred John ("R.A.J.") Bidwell (1869–1918) was born in June in Camberwell, London. He was the son and eldest child of John and Jane Bidwell. Bidwell's family was from an ordinary background — his mother was a laundress and his father was a cork cutter. By 1881, Bidwell was living with his uncle and aunt (Arthur and Eliza Bidwell) in West Ham. Despite his humble surroundings, he was able to secure a position as a trainee architect in London at Lockyer Sons and Cox, then later with another architectural firm, Crickmay & Sons. He then became Chief Assistant to Walter Henry Woodroffe, FRIBA, of London, and finally, he was assistant to the Superintending Architect of London County Council and based at their offices at 21 Whitehall Place in London. He was a member of the Architectural Association of London — indeed, he had been placed on their honours list for design during his time at Lockyer Sons and Cox, where he started his professional training in the early 1880s.

Although his primary training was as an architect, Bidwell was, in 1902, elected as a Fellow of the Surveyors' Institution, hence his inclusion in this book.

He came to Kuala Lumpur from London in 1893, and almost every year from his arrival to his untimely death in Singapore in 1918, he left a legacy to the built environment, either in Kuala Lumpur or Singapore. However, 1897 to 1904 were arguably the most productive and significant years of his career, with the design and construction of a new main central building for the Raffles Hotel being perhaps not just his most famous design but also the one for which Swan & Maclaren is most remembered.

Bidwell joined the Public Works Department (PWD) in Selangor on 16 February 1893 on a three-year agreement as Chief Draughtsman for a salary of $1,500 per annum. Later that year, he was first mentioned in the local newspapers as "playing cricket at Selangor Cricket Club for the 'Under 3s'" (referring to those having been in Selangor for less than 3 years). By 1894, he was Honorary Secretary of the St. Mary the Virgin Church Committee in Kuala Lumpur, and he was present at the laying of the foundation stone. The Church still stands today and is now a Cathedral.

During his short time in Selangor, Bidwell (or "Reg" as he was known to many) was involved in several landmark design and construction projects. The most famous is the Government Office building, now known as the Sultan Abdul Samad Building in Jalan Raja, one of the most well-known landmarks in Kuala Lumpur. Constructed between 1893 and 1896, the first plans had been drawn up by Bidwell under the supervision of the state architect, A.C. Norman. The plans were based on a classical Renaissance style, but the PWD State Engineer (Charles Edwin Spooner) considered the design unsuitable for the Malayan climate, and he instructed Bidwell to rework the building in an Indo-Saracenic or Neo-Mughal style based on Norman's ground plan. Although the building is credited to A.C. Norman (only his name appears on the foundation stone), the appearance of the building is largely the work of Bidwell, with significant contributions from A.B. Hubback, who joined the Selangor PWD upon Bidwell's departure in 1895.

In March 1895, Bidwell resigned as Chief Draughtsman, expressing unhappiness with his pay and additional job scope as Acting Architect. Bidwell cited an offer from a private firm for $275 a month as proof that he was underpaid. Despite the acrimonious nature of his departure, his Certificate of Service upon his departure stated that during his time of service, his work was "thoroughly satisfactory". Indeed, on Bidwell's return to Kuala Lumpur on 3 April 1897 for the Government Office building's opening ceremony, Spooner made a point of saying that "his greatest thanks were due for the valuable cooperation and the help that he had received from Mr. Norman and Mr. Bidwell during the original design of the building". Had it not been for them, Spooner was perfectly certain that they would never have had the building they saw before them. He added that Bidwell put ideas into his head and, likewise, he had put ideas into Bidwell's head, which Bidwell was most happy in carrying out. He added that both Norman and Bidwell seemed to catch his ideas "as if by magic". Newspapers of the time described the building as containing many pleasing features, the *"coup d'oeil from every side being almost equally successful from an architectural point of view"*. Much of the same could be said for most, if not all, of Bidwell's designs, in that he would ensure that every aspect of a building's external features was pleasing to the eye, even those sides hidden from the public view.

Arrival in Singapore

Bidwell joined Swan & Maclaren in Singapore on 10 April 1895, and his position there strengthened in 1899 when he became a partner alongside J.W.B. Maclaren. In these early years in Singapore, he became actively involved in sports (cricket, bowling, tennis and golf) as well as most of the key social activities in Singapore's calendar. By 1897, he was on active jury service in Singapore.

His earliest surviving work in Singapore was the 1895 construction of a godown in South Road (now called Read Street) — this building is now part of the Clarke Quay Festival Village, facing onto Read Bridge. Julian Davison's *Swan & Maclaren* includes the original building drawings by Bidwell.

On 7 August 1897, at St. Andrew's Cathedral, he married Edith Allan, daughter of the late Charles Martin Allan (1839–92), who came from the highly respected Allan family from Geylang. Charles had come to Singapore early on in his life as an assistant to the world-famous naturalist and collector Alfred Russel Wallace; however, Charles decided to settle down in Singapore rather than continue to support Wallace on his global explorations. 1897 was to be a busy year for Bidwell. Besides getting married, he oversaw the construction of what was described as a piece of "Municipal Renaissance" — the new Department Store for the Katz Brothers on the corner of Chulia Street (then called Kling Street) and Bonham Street, opposite Raffles Place. The Bonham Building (see Figure 3.12) was a 3-storey building of the "classic Renaissance style of architecture". The original designs were said to have been done by **Alfred Lermit** (who features in the previous Chapter) in 1893, and construction started in 1894 after the Great Fire of 15 November 1893, which saw some of the buildings onsite destroyed. A further building catastrophe on 28 September 1895, which saw a "considerable loss of life", resulted in what was almost a complete rebuild, in particular, on the Kling Street side.

The newspapers of that time stated that the Bonham Building designs were by Mr. Osborne, formerly of Crane Brothers, so it is likely that Osborne's building designs superseded those of Lermit. For sure, it is known that after the 1895 building catastrophe, work was completed under the joint supervision of Bidwell and Osborne and without further loss of life, which no doubt added to Bidwell's credentials. Sadly, this magnificent building was lost to Singapore in the 1970s to make way for UOB Plaza 2.

Teutonia Club, Scotts Road (Now Known as the Goodwood Park Hotel)

In 1899, Bidwell and his Swan & Maclaren colleagues were successful in winning a competition to design and oversee the construction of a new club building for the Teutonia Club, funded by Club members, who were all members of Singapore's German community. It is stated in Arnold Wright's *Twentieth Century Impressions of British Malaya: Its History, People, Commerce, Industries, and Resources* that the build cost (to replace the older, dilapidated building) was $20,000 and was subscribed by present and past Germans residents in the "colony". Swan & Maclaren had produced the winning design despite stiff competition from Alfred Lermit, and Bidwell's design, European in style, had a strong Germanic feel, in particular, a tower that rose above the building. See Figure 3.1. On 26 September 1900, the Club held an "At Home" celebration to welcome its members back after a 15-month period of construction, during which time Club activities were hosted at the nearby Tanglin Club. Of note in Bidwell's design during this ceremony were the lofty Assembly Room (with the stage at the far end), electric lighting, new Tower, spacious apartment rooms and Drawing Room facilities for women, which the older, dilapidated building did not have.

Two new paintings, portraits of H.I.M. the Emperor of Germany and H.M. the Queen, were also unveiled at the ceremony — toasts to both were raised at the celebration. Sadly, 14 years later, the Teutonia Club was confiscated as enemy property with the outbreak of World War I. Of course, the building has survived, and for over 60 years, it has been the home of the Goodwood Park Hotel.

FIGURE 3.1. (a) Goodwood Park Hotel, Scotts Road in 2018 and (b) in 1908, when it was the Teutonia Club — another of Regent Bidwell's designs, completed in 1900.

Raffles Hotel (see Figure 3.2)

Between 1897 and 1899, Bidwell designed and superintended the construction of the building for which he is perhaps best remembered — the Raffles Hotel. In 1897, Messrs. Sarkies Bros, owners of the existing Hotel, commissioned Swan & Maclaren to produce plans for the replacement of the central block of the Hotel (which was then a Beach House bungalow) with a much larger three-storey building in the "Renaissance" style of architecture. One newspaper expressed the view that when finished, it should be "equal to any hotel in the East, and an ornament to Singapore".

FIGURE 3.2. Views of the Raffles Hotel, Singapore. (a) Regent Bidwell of Swan & Maclaren designed a new main body of the Hotel in 1899. The new main building was a first for Singapore, as it incorporated powered ceiling fans and electric lights. He also designed the Bras Basah Road extension of the Hotel, completed in 1904. (b) The Billiard Room was a later addition to the Hotel, designed by Tomlinson & Lermit and constructed on the Bras Basah side of the Hotel in 1906. Tomlinson & Lermit's design includes a verandah along Bras Basah Road and a Carrara marble floor to the Billiard Room itself.

During the construction works, the Beach House bungalow was used as a dining room for the hotel. Swan & Maclaren took great care to use the best materials, with teak verandahs and hardwood throughout, mostly from Perak, and with "no expense spared in obtaining the best fittings for the buildings". The staircase newels were formed as pedestals to carry bronze statues, which were created to form part of the electric lighting scheme. Bidwell was very much a pioneer in the introduction of electric lighting systems to his building designs, and for the Raffles Hotel building, he introduced a double electric lighting system that could power 800 16-candle-power incandescent lights, in addition to five arc lights of 2,000 candles each.

Completion was originally targeted as July 1898 but did not happen until November 1899, partly delayed by an incident in August 1899, when a glass section of the roof collapsed in a storm and fell onto the new marble floor below. However, it was worth the wait as the Hotel soon became *the* place to stay in Singapore, a position it still holds today. Swan & Maclaren was not only commissioned to design the new Hotel building but also to "superintend" the construction.

The inaugural dinner was held on 18 November 1899, with the *Singapore Free Press and Mercantile Advertiser* reporting that the electric lighting worked without a hitch.

Bidwell followed up this success by adding a new extension to the Hotel on the Bras Basah Road's side, work for which started in 1902 and finished in 1904. This new extension provided an additional 112 rooms and a magnificent new kitchen. The verandahs were different from those in other buildings — they were resplendent with "pretty Minton tiles instead of wood, so as to ensure the maximum of cleanliness and coolness". The provision of shops on the ground floor of the new extension were a fascinating innovation, and to this day, they continue to be a success. Also, the position of the extension allowed the land behind it — enclosed between the extension and the main building — to be turned into a beautiful garden with plants, trees and palms, which is still a main feature today. The provision of en-suite bathrooms and dressing rooms was another feature hitherto unseen in Singapore, and the ability to provide water for shower-baths from a rooftop water tank was another impressive step forward for Singapore.

Methodist Girls School, Short Street

In March 1900, as the building architect, he was congratulated at the opening of the new Methodist Girls School on Short Street at its junction with Selegie Road, on land that the Methodist Church Mission had acquired in 1898 from members of the Johor Royal Family. The cost of the new building was $8,878, with a total project cost, including land purchase, of $13,119. The previous building was a shophouse on Short Street, and this new building was required to enable the expansion of the school to house 150 children, including a kindergarten.

The building was described as "handsome, modest in size and designed in the neo-palladian style" and "excellently adapted for the use for which it was intended". Sadly, the School needs outgrew the building, and it was demolished by 1933 when they moved to another location.

The Memorial Hall and Town Hall

1901 was not a good year for Bidwell, as he was ordered home to England in July due to ill-health. However, he was not away for too long, and whilst 1902 was also a relatively quiet year for him, 1903 was a significant year because he became involved in the design of the new Victoria Memorial Hall (see Figure 3.3). Discussions first took place about where to locate it as far back as H.M. Queen Victoria's Diamond Jubilee in 1897, not long after the Diamond Jubilee Ball in June, which had been hosted by H.E. the Governor at Government House. That event was attended by Bidwell.

Things came to a head in 1901 when Queen Victoria passed away — the earlier discussions had pertained to a Jubilee Hall or Victoria Hall, so the word Memorial appears to have been added at this point. Work on the Memorial Hall began in 1902, by which time the decision had been

FIGURE 3.3. Victoria Theatre and Concert Hall Singapore was designed by Bidwell of Swan & McLaren. The Clock Tower (centre) was also designed by Bidwell.

made that on the completion of this new Hall, the old Town Hall (which was in a poor state on the adjoining site) would be converted (in effect, rebuilt) into a Theatre. By the time the foundations of the Memorial Hall were completed in March 1903, there were already "some disquieting rumours" about the building project, which the PWD sought to dispel. An important meeting was held in the Town Hall on 16 September 1903, presided over by the Governor, to discuss the design and latest building plans. The Governor stated that when the original model was exhibited by its creator (Colonial Engineer Alexander Murray), the working plans had not been produced, and so the (PWD) architect had produced building plans based on his own interpretation of the model, but the Governor and the Building Committee were unhappy with what they saw.

So, Swan & Maclaren was invited rather than "called in" (as *The Straits Times* puts it), and they presented their version of the plans. Bidwell was the architect, and he had made considerable alterations to the original plans because the original versions were considered "not suitable or very ornamental" by the Building Committee, whose main concern was to ensure quality and value for money for the many people and organisations that had contributed funds. The meeting was a success, and the latest designs were strongly recommended — Bidwell's designs clearly found favour with the Committee — and the plans were approved, with a round of applause to conclude the proceedings. At that stage, the total project costs, including construction, foundations, architects' costs and furniture, were estimated at $314,786, which left a subscription balance of $45,214. Bidwell worked alongside the PWD staff to oversee the entire construction project through to completion in 1905.

Once the Memorial Hall was completed, the decision was made to completely gut the old Town Hall and rebuild it with a new façade in the same style as the Memorial Hall, as part of a proposal to convert the Town Hall into what was called the Victoria Theatre. The original discussions around this took place between 1889 and 1890 with Swan & Lermit, who were appointed as architects to draw up the plans, but their proposal did not progress due to the high (and unexpected) level of projected costs. The Town Hall project commenced in 1906 and was completed in 1910 — again, Regent Bidwell was involved from start to finish and was commended for his success in the complete renovation and conversion of an "ancient" Town Hall into a modern Theatre.

The Memorial Hall and Theatre were joined together by the construction of a central clock tower, designed by Bidwell and completed in 1905, along with the Victoria Memorial Hall.

The Memorial Hall and Town Hall (now known collectively as the Victoria Theatre and Concert Hall), including the central clock tower, were gazetted as a National Monument on 14 February 1992.

Chesed-El Synagogue, Oxley Rise

One of Bidwell's lesser-known masterpieces is his design of the Chesed-El Synagogue on Oxley Rise. The synagogue was built in 1905, originally for the private use of Menesseh Meyer and his family. Figure 3.4 shows external and internal views of the synagogue. Jane Beamish and Jane Ferguson's *A History of Singapore Architecture: The Making of a City* tells us that Bidwell's theme was Byzantine (though other sources say Renaissance or Palladian style), both internally and externally, and it also states that from what was very much a tightly constrained and difficult site to develop, Bidwell created a "stimulating and harmonious whole".

Externally, the synagogue is heavily rusticated, with "stones" set in the plasterwork, alternating rough and smooth. Projecting from the front is a covered porch. Internally, two rows of towering white Corinthian columns line either side of the main body of the Synagogue, with each column having three rings of gold at intervals along its length and a gold-painted leaf design at the top, where the columns are joined by arches decorated with delicate gold motifs. The walls and the marble flooring are in brilliant white. To the sides of the covered porch are staircases leading to the second-storey women's gallery, separating the women and children from the men during services.

The building was gazetted as a National Monument on 18 December 1998.

FIGURE 3.4. Chesed-El Synagogue, Oxley Rise — designed by Bidwell for Swan & Maclaren and built in 1905 for Menesseh Meyer.

Singapore Cricket Club, New Pavilion (see Figure 3.5)

Bidwell also designed the Pavilion at the Singapore Cricket Club, which was officially opened for member use on 25 March 1907. The plans for the new Pavilion were approved in June 1905, and it was constructed around the existing much smaller pavilion — indeed, apart from the cast iron verandah columns or brackets, all that remained of the old pavilion were the foundations of three of the walls of the Bar, except for brick sections that were either built into the new foundations or broken up to be included in the new concrete foundations. The impressive new pavilion was built projecting out in two sections towards the pitch, with three flights of granite steps leading down to the lawn.

Bidwell, as well as being the architect, was also on the Club's Building & Finance Committee, along with F.J. Pigott, who is mentioned in Chapter 1. Although Bidwell took part in many sports, cricket was his favourite. Indeed, he was largely responsible for bringing the Australia

FIGURE 3.5. Singapore Cricket Club dating from 1913, showing Bidwell's Pavilion.

national cricket team to Singapore to play in 1909, and during time spent in Melbourne a few years earlier, he was made a "sometimes" honorary member of the East Melbourne Cricket Club, one of the best clubs in Australia at that time. On that trip, he promised Frank Laver, a famous Australian cricketer, an enjoyable time for the Australia team should they come to Singapore.

The success of the Singapore Cricket Club from the turn of the century onwards meant that the membership's needs had outgrown the previous facilities; hence, the need for a much larger new Pavilion. One notable change, aside from the improved facilities for club members, spectators and players, was that the annual meeting of the Club could now be held on club premises for the very first time, instead of being held at the Tanglin Club.

Bidwell's design also included a ladies' room ("apartments"), and despite the efforts of some Committee members to get the plans revised to remove the aforementioned room due to a perception that it would be little used, the ladies' room (including toilets and a dressing room), was indeed included. *The Straits Times*, 14 March 1907, tells us that "... there is no doubt our fair friends will find that their comfort has been carefully studied ...."

Another significant addition was the spacious front Porch that would provide the "Clubman a chance of getting into his gharry or ricksha dry". From the Porch, Bidwell designed a set of stairs to the Entrance Hall paved with tessellated pavement, with the Cricket Club mat in the centre. The spacious new Entrance Hall had doors leading to the Bar, which was 54 feet in length and almost free from columns, except for behind the Bar counter, which had a horseshoe, marble-topped surface finish.

Completed on 10 February 1907 and opened to members shortly after, one new feature of the building design was the inclusion of cast iron ventilating boxes in between each joist of the upper floor to allow foul air to get away. Shortly after opening, the Club held an international cricket tournament — there was no doubt the availability of the new Pavilion helped boost attendance.

Adelphi Hotel

The (second) Adelphi Hotel was located on the corner of Coleman Street / North Bridge Road during the 1880s and is one of Singapore's lost treasures. Through Bidwell's designs, it was extensively altered between 1903 and 1906, with the addition of a three-storey extension on the Coleman Street elevation, with work starting soon after being purchased by Messrs. Sarkies, Johannes & Co. in 1903. The budget set by the new owners was $200,000.

Work on the new extension was carried out in a number of stages so that the hotel itself did not have to close during the works, despite this being almost a complex project involving demolition work, a new building structure and integration with what remained of the original Hotel structure. The project included the installation of an electric generating plant for lights and fans. Julian Davison's *Swan & Maclaren* suggests that Bidwell's plans were very much inspired by the architect Alexander Thomson's Grecian Chambers in Glasgow, an example of Greek Revival architecture.

Sadly, unlike the Grecian Chambers, which still stands today in Glasgow, the Adelphi Hotel was closed in 1973 and demolished in 1979. Today's Adelphi Complex was constructed on its site in 1980. Many of the R.I.C.S. Malaya Branch Annual Surveyors' dinners were held at the

FIGURE 3.6. Adelphi Hotel, Coleman Street, circa 1950s. This Hotel was designed by Swan & Maclaren during Bidwell's time with the practice.

Adelphi Hotel, from the inaugural 1936 R.I.C.S. Malaya Branch dinner through to 1958.

Figure 3.6 shows the Adelphi Hotel during the 1950s when it was a frequent venue for the R.I.C.S. Malaya Branch annual dinners as well as the venue for Branch Committee meetings.

The New Church of St. Joseph, Victoria Street (see Figure 3.7)

On 21 August 1904, Bidwell represented Swan & Maclaren at the laying of the Foundation Stone of the new Church of St. Joseph, a ceremony overseen by Bishop Azevedo of the Portuguese Mission. The new Church was finished in 1912 and is now gazetted as a National Monument.

It is one of Singapore's oldest fully operational Catholic Churches, albeit closed from 2017 until 2022 whilst major renovations took place. Constructed in Romanesque style, the main Church building is laid out in the form of a Latin cross, with the nave arcaded by large masonry pillars from which pointed, richly moulded arches spring.

FIGURE 3.7. The new Church of St. Joseph, Victoria Street.

The architect was David Craik of Swan & Maclaren, though it is likely that Bidwell influenced Craik's design considerations.

Stamford House, at the Junction of Stamford Street and Hill Street

Bidwell was also the architect for Stamford House at the junction of Stamford Street and Hill Street, which as of 2021, is the home of The Capitol Kempinski Hotel (see Figure 3.8) The building was constructed in 1904 and designed as a mixed-use development in a part of Singapore, something that the anchor tenant (the Australian department store chain of Messrs. Whiteaway, Laidlaw & Co.) had not yet ventured into. Whiteaway, Laidlaw & Co. had been operating successfully in Singapore for some years and had pioneered the "cash" system of payment to replace the "chit" method.

The expectation of the building owner (T. Paul of Stephens Paul & Co., a multinational import and export firm that operated in Singapore from the 1890s up until World War II) was that this development — described in newspapers as a "well-ventilated ... lofty white block" — would be a catalyst for similar developments in that part of town, given that rents would be lower.

There is also no doubt Bidwell's magnificent design, both externally and the state-of-the-art facilities inside would attract a wide range of customers.

The development did not just accommodate Whiteaway Laidlaw but also housed Messrs. Weill and Zerner (jewellers), Thompson Thomas and

FIGURE 3.8. Stamford House, Stamford Road / Hill Street Singapore.

Company (an Australian store) and Messrs. Watts and Co. (bootmakers and saddlers), as well as residential accommodation.

Feedback from the new occupiers in 1905 was that trade was better than expected, and the building ventilation design was ideal for goods storage.

Much has been written about this magnificent building, and this book will not seek to repeat what others have already written. However, one feature of the building worthy of mention is Bidwell's desire to ensure that every part of the building's exterior — even those hidden from public view — would be of equal aesthetic quality.

For those with limited knowledge of the building, a trip to Stamford Road is recommended to take a closer look at the ornamental leafy Stuccowork that dominates the exterior of the building — without a doubt, this building is one of Bidwell's finest works.

Bidwell and Singapore's Black and White Houses

Black and White houses and bungalows are a feature of the Singapore built environment that continues to attract interest and fascination. The first Black and White houses and bungalows were built in the late 1890s and continued through to World War II. The original inspiration for them came from British architects who came to Singapore and Malaya in the 1890s, having learned their trade in London in the 1870s and 1880s, where they "copied" a style that was commonplace during the "Tudor" revival period in London and South-East England during that period. Back then, and to this day, these houses are most popular with families despite the huge cost of renting one! The "black" element of the design is a reference to the dark timber that dominates the structure of the houses, and the "white" element refers to the white-painted walls of the houses. Julian Davison's *Black and White: The Singapore House 1898–1941* is a great source of information for those seeking to find out more about these houses and bungalows.

A significant number of Black and White houses were commissioned and built in Singapore between 1898 and 1941, principally through Swan & Maclaren. Within that practice, Bidwell was the initial proponent and architect of these houses and bungalows, and he (and Swan & Maclaren) was a major influence on the innovative design of these houses, thus creating a catalyst for their popularity in Singapore during the early part of the 20th century.

Indeed, Bidwell designed the oldest surviving examples of Black and White houses in Singapore — Atbara, No. 5 Gallop Road, adjacent to the Singapore Botanical Gardens and Glencaird at White House Park.

Atbara, 5 Gallop Road (see Figure 3.9)

Resplendent with a Gothic-style "lych-gate" porch, Atbara is a very unusual L-shaped three-storey structure atop a grass bank on a large building plot. It was designed and built in 1898 for J. Burkinshaw of the legal firm of Aitken, Donaldson & Burkinshaw. No. 5, along with the house at No. 7 Gallop Road known as Inverturret — another Swan & Maclaren masterpiece completed in 1906 — see Figure 3.10) were bought by the Straits Trading Company in 1923 and then leased to the French government for its Embassy and Ambassador's residence between 1939 and 1999. For Atbara, Bidwell took a slightly different approach to Glencaird, as he incorporated a number of different architectural styles, and he included masonry piers that elevated the building in the same fashion as the Anglo-Malay bungalows that had hitherto been the standard design of colonial residences in Singapore. In 2021, a programme of several years of renovation for Atbara and Inverturret was completed and overseen by National Parks to enable both houses to become home to the Singapore Botanic Gardens' collection of rare and endangered trees from around the region. Atbara now serves as the Forest Discovery Centre@OCBC Arboretum, and Inverturret now serves as the Botanical Art Gallery.

Glencaird, White House Park (see Figure 3.11)

FIGURE 3.9. The side and front views of Atbara, 5 Gallop Road. The latter shows the Gothic-style lynch-gate porch.

FIGURE 3.10. Inverturret, 7 Gallop Road.

Another Bidwell masterpiece is Glencaird at White House Park. Designed and built around 1902/03, it is the only one of the original four houses in the White House Park estate still standing today. The house has so far retained its original features despite alterations and restoration programmes carried out throughout the years. One of its unique features is an entrance at the corner rather than the centre of the house. In 1947, the government of Australia bought Glencaird and it became the official residence of the Australian High Commissioner. It was given conservation status in 1991. The Australian High Commission put up the Glencaird site for sale in 1994 and sold it to Wheelock Properties / Marco Polo Developments in 1996, which retained Glencaird but developed 11 bungalows within the overall land plot.

Bidwell made great efforts to recreate the late-Victorian English country house, with adaptations to make it suitable for the Singapore climate, and in doing so, he moved away from the architectural style of residence that had been adopted in Singapore since the early colonial days.

There is little doubt that Bidwell's innovative design for Atbara, Glencaird and Swan & Maclaren's later design of Inverturret paved the way for the trend for Black and White houses, which lasted until the late 1930s. Other architects in the Swan & Maclaren practice took over from Bidwell when he left the practice, and they became the driving force behind the design of post-war Black and White residences. These included the EEA&C Telegraph Company Estate houses in Telok Blangah, the design for which Bidwell had initially led back in 1908.

During his time with Swan & Maclaren, Bidwell and his family lived at Omdurman, Dalvey Road, a substantial five-bedroom property built on the hillside opposite the junction of Dalvey Road and White House Park. Its large grounds (including a tennis court) has long since been redeveloped — the landed properties numbered 54 to 54C Dalvey Road are now situated there. Sadly, no photographs are known to exist of Omdurman, so we do not know anything of its design or even who designed it.

FIGURE 3.11. Glencaird, White House Park, in 2018, following a major restoration programme. This house was designed by Regent Bidwell.

Bidwell's Career From 1907 Onwards

The period from 1907 onwards was a period of uncertainty with regard to his career, although he continued to design and deliver iconic buildings. Having left the partnership of Swan & Maclaren when it was dissolved in 1907, he remained with them as an employee and not as a Partner. In 1910, Bidwell informed *The Straits Times* that he had received a telegram informing him that Maclaren had become gravely ill during his return to Scotland. One can infer from this article that even though Bidwell was no longer a partner but an employee, his relationship with Swan & Maclaren appears to have been at least cordial.

Indeed, from 1909 to 1911, whilst as an employee of Swan & Maclaren, Bidwell still played a major role in five noteworthy Swan & Maclaren projects in Singapore and another in Surabaya, Indonesia.

Singapore Sporting Club

Bidwell was involved in the design and construction of a new Grandstand (plus road and drainage works) for the Singapore Sporting Club on Racecourse Road, an area of land now known as Farrar Park in Little India. Bidwell and V.A. Flower (the other architect in Swan & Maclaren at that time) were on the Building Committee that made the decision to accept a tender from Howarth Erskine & Co. in the sum of $55,421. This tender included the provision of additional roadworks to accommodate changes to the existing approach road brought about by the decision to build the new Grandstand at an angle to the racecourse, which was "very much the angle ... adopted in most race clubs at present". There was some concern concerning the funds for the new Grandstand, and discussions took place over whether $55,000 would suffice. Ultimately, on 15 June 1910, the decision was made to empower the General Committee to raise $55,000 through an initial debenture and complete the project by issuing further debentures to a maximum of $80,000.

Crescent Flats at Meyer Road, Tanjong Katong

These crescent-shaped flats, which faced the sea next to the Seaview Hotel, were constructed between 1909 and 1912 on the instructions of the owners, Mr. and Mrs. Meyer. Bidwell had been the architect. The flats, a total of 36, were constructed to a high standard for those wishing to live away from the city. They had the luxury of ceiling fans powered by electricity, installed by the Siemens Brothers.

They are now recognised as the very first example of beach-side apartments in Singapore and were designed to replicate The Royal Crescent, a residential development in Bath, England, one of the greatest examples of Georgian architecture in the United Kingdom.

Sadly, despite many pleas for the conservation of these flats and the Hotel, they, along with the Hotel, were demolished in the early 1990s to make way for condominium developments.

Modernisation of the Famous John Little Department Store on Raffles Place (see Figure 3.12)

Completed in September 1910, it included the installation of an electrical system for the building, provided by the Siemens Brothers. Incidentally, one of Bidwell's first tasks for Swan & Maclaren in 1895 had been to design the front façade for this building. The Store was eventually demolished in the 1960s, and Singapore Land Tower now lies on its site, although Bidwell's front façade design lives on (in replica form only) as the front façade of Raffles Place MRT Station.

FIGURE 3.12. Raffles Place. The Chartered Bank building is at the centre of the postcard, and to its right is the façade of John Little Department Store, designed by Regent Bidwell. To the immediate left of the Chartered Bank building is a partial view of the Bonham Building.

Medical Hall Building, 265 to 271 South Bridge Road in Chinatown, Singapore (see Figure 3.13)

Constructed in 1910, it was purpose-built to house the very first traditional Chinese medicine Eu Yan Sang outlet in Singapore, which had been set up by wealthy businessman and philanthropist Eu Tong Sen. Described by Julian Davison as an "elegant three-bay, three-storey building executed in the Renaissance manner", this building was of particular importance to Swan & Maclaren, as in later years, Eu Tong Sen would become a valued client. Fortunately, the building survives to this day and is a recognisable landmark for visitors to Chinatown.

Grand Hotel Europe (Hotel L'Europe), Junction of High Street and the Esplanade (Now St. Andrew's Road)

Bidwell was also involved in the ambitious project to rebuild the Hotel L'Europe on behalf of Sam Tomlinson, who had just taken a new 30-year lease of the Hotel after the previous proprietors' lease had expired in

FIGURE 3.13. Eu Yan Sang Building, 265–271 South Bridge Road, undergoing renovations in 2020.

1903. The project was completed in 1907, and the new Hotel building was definitely stiff competition for the Raffles Hotel, given the excellent facilities, including a spectacular roof garden with uninterrupted views.

Renamed Grand Hotel Europe upon its reconstruction, it was sadly demolished in the early 1930s for the construction of the Supreme Court building, which still stands there today. Bidwell is not confirmed as the architect, but he was definitely involved in the construction of the new Hotel building, as well as approving any items of building expenditure.

The Oranje Hotel, Surabaya (Then Known as Sourabaya), Indonesia

Constructed on a 1,000 m^2 plot of land on Jalan Tunjungan in Surabaya acquired in 1910 by Lucas Martin Sarkies, the Oranje Hotel opened on 1 July 1911. Bidwell had been commissioned to deliver a Dutch colonial art nouveau Hotel with a budget of 500,000 Guilders. Given the shape of the plot, his design sought to arrange the hotel rooms around three sides of a courtyard, with the main body of the Hotel in the centre. The Hotel was subject to various additions and changes at different points in its history, and after several name changes, it is now known as the Hotel Majapahit.

The Period From 1912 Until His Death in 1918

Newspaper reports from the time tell us that by 1912, Bidwell had left Swan & Maclaren to operate on his own from No. 3 Finlayson Green. He seems to have retained some favour with his old clients because he continued to work, and so it was that in 1912 he delivered the famous "Butterfly house" at 23 Amber Road, on the East Coast of Singapore, close to Mountbatten Road and Tanjong Katong Road. Amber Road was once on the sea front, and so Bidwell designed the "butterfly wings" — the house's semi-circular verandah — to face the sea in order to allow the sea breeze into the house.

The house had been built for the Cashin family and occupied a huge plot of land. It was therefore a prime candidate for development in later years and the site was sold for $9 million in 2006. After much discussion with the new owners and other stakeholders, the site was redeveloped. Sadly, the butterfly wings were demolished, but the central part of the house remains, fronting Amber Road and was integrated into the design of the new development (The Aristo). Figure 3.14 shows the remaining part of the original 23 Amber Road structure.

Bidwell's last major work was a 1913 Black and White house in Cairnhill for Dr. D.J. Galloway. By this time, he had already marketed Omdurman, his Dalvey Road house, to be let out (he also offered for sale some of its contents) to allow him to move into a much smaller house that he owned in Tanjong Katong — a clear sign that his fortunes were changing.

Indeed, having completed the Black and White house for Dr. Galloway in 1913, the year 1914 saw Bidwell as a defendant in the Supreme Court, following an action by Dalman & Co. to recover from him the sum of $989 —the final outstanding sum for cement delivered for the Galloway house construction. Bidwell had been due to appear in Court on 16 July,

FIGURE 3.14. No. 23 Amber Road — the "butterfly house", designed by Regent Bidwell and built in 1912. The photograph shows what remains of the house (sans the butterfly wings) integrated into The Aristo condominium development.

but he did not appear. It was alleged and backed by witnesses that the supply of materials was made in conjunction with Choo Sing Teck, a partner of Bidwell, although Bidwell denied any such partnership. The Court later made an award in favour of the plaintiffs. Again, Bidwell did not appear in Court.

Against this backdrop, it was in 1915 that Bidwell's financial circumstances came to a head. At a Public Examination in the Bankruptcy Court on 8 January 1915, the full explanation of his financial circumstances became public. Bidwell described himself as a married architect with four children. He explained that he was a partner with Swan & Maclaren until 1907, and thereafter, he was an employee of the firm but left them altogether in 1912. He owed a total of $16,000, of which $10,500 was owed to Mr. Watkins of Swan & Maclaren. His house (no doubt the one in Dalvey Road) was subject to a second mortgage, but the first mortgagee had repossessed the house. He admitted that not only was he now insolvent but had pretty much been insolvent since he arrived in Malaya in 1893. When he arrived in Kuala Lumpur with the Selangor PWD, he had borrowed money from a *chetty* (the Chattiars were individuals who loaned money to those who were unable to secure a loan from a Bank), which led to even more debts. As time passed, the problem worsened due to major losses in his share investment in the Royal Johore Tin Mining Co., which had suffered heavy losses for several years due largely to the price of tin being significantly lower than in previous years. His salary then from Swan & Maclaren had been around $700 to $1,000 per month, which did not cover his cost of living. His Hong Kong and Shanghai Bank account was empty, and he had no possessions.

The Straits Times and other newspapers of the time often covered Mr. and Mrs. Bidwell attending major social events and society weddings, and with his club memberships and sporting activities in Singapore and overseas, his family's frequent charitable donations, his membership of the three Freemason's Lodges in Singapore (St. Michael, St. George and Zetland), and his Surveyors' Institution fees, his personal expenses must indeed have been considerable. Indeed, he told the Court Examiner that in good times, his household expenses were $500 per month, a considerable sum in those days.

By 1915 he was working at 11 Raffles Quay with Peter J.B. Sow. Although the newspapers stated it was a partnership, *The Straits Times* report on the Examiner's questioning confirms otherwise. In fact, Sow was only paying him on an ad hoc basis for services rendered, even though Bidwell hoped to receive around $1,000 per month by year-end, as well as a share of the profits, but this was not confirmed. When asked about how much he was paid to become a partner of Swan & Maclaren upon Swan's departure, Bidwell could not recall the amount but did say that Swan was not paid any cash for goodwill (or any other sum) in return for relinquishing his share in the business to Bidwell. The agreement was that Maclaren should take 60% and Bidwell 40% of Swan & Maclaren when Bidwell stepped in to take over Swan's share of the business.

At the closure of the Hearing, he said that he owed $1,000 in rent and did not think he could pay off any of his debts for at least six months.

One cannot help but wonder why he did not receive help from Swan & Mclaren, given that he was an integral part of their success story during these years, and much of his social activities significantly raised the profile of Swan & Maclaren between 1895 and 1915.

Nothing is known of his activities in 1916 and 1917, apart from Bidwell's appearance at the Bidadari Cemetery in April 1916 for the funeral of C.W. Edwards, who was the husband of his wife's sister Annie. However, newspaper reports from 1916 to 1918 do mention the wartime activities of Bidwell's eldest son, who was in England at that time with the Royal Naval Air Service.

It is known that Bidwell was in poor health. According to newspaper reports, "after a long period of ill-health", Bidwell died at his home in Tanjong Katong in April 1918 at the age of 48. The site of his family home is believed to have been on the opposite corner of Haig Road and Mountbatten Road to where Katong Shopping Centre now lies.

He was described in the Press upon his death as a well-liked man who would be missed by many.

Given that he died at a relatively young age, it is remarkable that he left so many magnificent buildings behind.

He was buried at Bidadari Cemetery (see Figure 3.15) — the remains of those buried there were exhumed in recent years to make way for a major housing development on its site. However, Bidwell's headstone survives

FIGURE 3.15. Gravestone of Regent Bidwell, now lying in the Bidadari Memorial Garden.

along with those of other notable individuals inside the Bidadari Memorial Garden, which is the very last part of the old cemetery that survives. His legacy in Kuala Lumpur (through his work for the Selangor PWD) will be covered in a future book on Chartered Surveyors in Malaysia. His wife Edith died during the Japanese internment in Singapore in 1944.

Other Bidwell Designs

Swan & Maclaren designed other iconic buildings during this time: Bidwell is not stated on the building drawing as the architect — in fact, many of Swan & Maclaren's building plans of this era did not contain the signature or name of the architect involved.

However, Bidwell's influence and success within Swan & Maclaren during these years have led to other building designs being often attributed to him. It is quite possible that the true designer was one of the other very talented architects at Swan & Maclaren during these years, but at the very least, given his presence in the drawing office, Bidwell's influence was perhaps a factor.

The list is long, and Julian Davison's *Swan and Maclaren* provides a fascinating insight into each one individually. However, to mention just two:

Telephone House, Hill Street

Built in 1907 for the Oriental Telephone & Electric Co. Ltd, Bidwell's Moorish-style arches were a key feature of the building, a throwback to some of Bidwell's earlier designs. The building was innovative for its incorporation of all the latest telephone technology.

Telephone House was demolished in 1979 to make way for a new 14-storey Central Telephone Exchange or Telephone House Complex, operated by Singtel — this was also demolished in 2020 to make way for a Hotel development. An interesting feature of the Central Telephone Exchange, constructed in 1979–80, was the design of three Moorish arches at ground level, which sought to replicate the Moorish arches, a prominent feature of Bidwell's 1907 Telephone House. See Figure 3.16.

FIGURE 3.16. The front entrance to Singtel's Hill Street Telephone House complex. Constructed in the 1980s on the site of Telephone House, which was demolished in 1979.

Bidwell is sometimes credited with the design of the Telecoms (TAS) Building on Robinson Road, Singapore, but this is not correct. The architect of the TAS Building was Frank Lundon of Swan & Maclaren, and it was built in 1927, nine years after Bidwell died. Described by Jane Beamish and Jane Ferguson in *A History of Singapore Architecture* as a building of dignified solidarity, the TAS Building was extensively renovated and re-opened in 2016 as Sofitel Hotel (as shown in Figure 5.9(b)). Lundon's design was perhaps influenced by some of Bidwell's work at Swan & Maclaren.

House for Ezekiel Manasseh on Nassim Road, Now Known as Eden Hall

In 1903, Swan & Maclaren produced drawings for the construction of a villa residence at Nassim Road on land owned by Ezekiel Manasseh, which bears all the hallmarks of Bidwell's design, both internally and externally, and for which Bidwell has been credited in recent years. Completed in 1904, this residence passed through the hands of his wife's family following Manasseh's death during the Japanese internment in World War II and was sold to the British Government in 1953. It is now known as Eden Hall and is the official residence of the British High Commissioner in Singapore.

Bidwell's Children

Regent and Edith had four children, all born in Singapore and baptised at St. Andrew's Cathedral:

Regent Svante Charles Allen Bidwell (b. 12 July 1898) was a schoolmaster and lens reel caption and comment writer. He died in Middlesex, England in 1958. He is shown in Figures 3.17 and 3.18.

John Roy Allen Bidwell (b. 18 April 1900) was originally a planter in Johor but later became involved in the Motor Trade in Singapore and was one of the first Sales Managers for the new Ford Motor cars that arrived in Singapore in the late 1920s. He married in Titusville, New Jersey, US, to Alice Stout and returned to Singapore in the mid-1930s. His wife and son left Singapore prior to the Japanese Occupation, but Roy stayed and was interned by the Japanese from 1942 to 1945. They resettled in New Jersey after the war ended, where Roy died in 1957.

FIGURE 3.17. Regent Svante Charles Allen Bidwell (foreground, right) in Singapore (circa late 1920s).

FIGURE 3.18. (a) and (b) Regent Svante Charles Allen Bidwell with his wife and son, and (c) in Singapore, circa late 1920s (Regent is in the centre of the group photograph). The group photograph was taken at the Singapore Races.

Gwendoline ("Gwennie") Edith Allen Bidwell (b. 2 October 1904) married a Singapore-based engineer (Thomas Munn Kinnear). Like Roy, they were also interned by the Japanese (at Changi and later at Sime Road Prison) from 1942 to 1945. Gwennie and her husband left Singapore after World War II. **Harold Martin Allen Bidwell** (b. 17 October 1907) worked for the Malacca Municipality. It is believed that he moved to England in the mid-1930s, possibly to be closer to his eldest brother, Regent.

Photographs Showing Regent Bidwell

Sadly, no photographs that show Regent Bidwell are known to exist. It is thought that family photographs were in the possession of his daughter Gwennie and her husband, Thomas. Unfortunately, during their internment in Changi Prison from 1942 to 1945, it is believed that all their possessions were either destroyed or lost after their home was taken over by the Japanese Army. Furthermore, Swan & Maclaren do not have any photographs of Bidwell in their archives.

There is, however, a photograph of the 1907 opening of the new Singapore Cricket Club Pavilion in page 582 of Arnold Wright's *Twentieth*

Century Impressions of British Malaya, which shows a large group of members and dignitaries in the foreground of the photograph. Given that Bidwell was the architect of the above structure, he is sure to be in this photograph, but the quality of the photograph does not allow him to be identified.

Fortunately, photographs do exist of Regent Svante Charles Allen Bidwell, Bidwell's eldest child.

Bidwell's Legacy to Singapore

Regent Bidwell was an individual of great ability, drive and vision.

His legacy to Singapore as a professional architect and Chartered Surveyor goes way beyond the iconic Singapore buildings that he is known to have designed – he was also a major influence within the practice of Swan & McLaren and inspired other architects and Chartered Surveyors working for the firm to create even more iconic Singapore buildings right through the 1930s.

Chapter

4

The First Eurasian Chartered Surveyors in Singapore

In 1904, whilst Regent Bidwell was already well recognised for his work, the important role of surveyors was starting to be recognised in Singapore, and the numbers employed in public service grew. Many of them are in the 1904 list of Singapore Jurors. The increasing number of Eurasians is particularly noticeable, especially those involved in land and building surveying as well as engineering and architectural work. A number of them are mentioned in Chapter 1 in the context of the Singapore Survey School, where most, if not all, studied at some point.

Whilst some of the Eurasian Survey School students listed in the 1904 list of Singapore Jurors had already decided to take up other professions instead of surveying, many remained in the profession. Three Singapore-based Eurasian surveyors went on to become Chartered Surveyors — their stories are told here.

John Pennefather

John Pyne Pennefather (1858–1940) (see Figure 4.1) was a Colonial Fellow of the Surveyors' Institution from 1914 to the 1930s. He was born in 1858 in Singapore and spent most of his surveying career in public service in Penang and Malacca.

He and his wife Anne de Souza (they married in Singapore in 1881) had 12 children; many of his descendants are still living in Singapore. Some

FIGURE 4.1. John Pennefather.

of his children and grandchildren became famous sporting figures in Singapore.

Arnold Wright's book —*Twentieth Century Impressions of British Malaya: Its History, People, Commerce, Industries, and Resources* — informs us that in 1906, Pennefather was the Senior Surveyor in the Survey Department at Malacca and was the holder of one of the Straits Settlements Government Scholarships for the years 1871–73. He entered Government service in 1875 as an apprentice in the Survey Department in Singapore and became a second-class surveyor in 1878, an Overseer and Surveyor in 1881, and a District Surveyor by 1882.

By 1888 he had become Chief Surveyor, Singapore, and Senior Surveyor by 1891. By 1893 he was back in Malacca as Senior Surveyor, and then by 1894/7, he was Superintendent of Works and Surveys. F.J. Pigott elaborated on the Survey Department's work in the Straits Settlements and individuals like Pennefather in Chapter 1 of this book.

In 1912, Pennefather set up his own practice, having retired as a senior Government surveyor at the age of 54. He was still practising in 1924, at which time he held an appraiser's licence in Singapore until at least 1932. After his retirement from public service, he returned to Singapore. He died in Singapore on 25 February 1940 at his home in Pasir Panjang and was buried at Bidadari Cemetery (following a service at the Roman Catholic Church of St. Teresa, Kampong Bahru) on 27 February. Many from the Eurasian community in Singapore attended his funeral, and hundreds of wreaths were sent, including wreaths from the Survey Department.

His obituary in *The Straits Times* was under the heading "Death of Well-Known Eurasian". He was a member, at one time a Quarter Master, of the Singapore Volunteer Force (SVF). His sons also served in the SVF — several saw action during World War I.

John Lincoln

John Albert Lincoln (1874–1929) (see Figure 4.2) became a Colonial Fellow of the Surveyors' Institution around 1910. He was Chief Surveyor at the Municipality and a qualified Singapore Juror in 1904. In 1908, he is listed as being in Singapore and working in the Municipal Engineer's Department in a team of 14 surveyors and draughtsmen. As an aside, most of the team, like him, were Eurasians of mixed British, Irish, Dutch, Portuguese, German and Indian ancestry.

At that time, according to Wright's book, the work was split into the management of Government buildings (roads), works within the Municipality, works outside of the Municipality, and the Drawing Office. One cannot be certain, with the passage of time, what projects Lincoln was involved in, but examples of his work done at this time are shown below. Given his engineer-surveyor role, it is highly likely he was involved in most of these projects from start to finish, with surveying, levelling, setting out works, as well as the planning and supervision of construction works.

FIGURE 4.2. John Lincoln pictured on the day of (a) his first wedding in 1907 and (b) his second wedding in 1917.

- **Johnston's Pier Reclamation**: A $248,000 project carried out between 1901 and 1905. As well as land reclamation, a road was built to connect the Pier with Collyer Quay with a new bridge over the Singapore River.
- The (new) **Tan Tock Seng Hospital**: Constructed in 1908/09, providing accommodation for 432 patients and staff. Built on a new site of 27 acres to replace the old Hospital, the total cost of construction and land acquisition was $484,485. The Hospital had concrete surface drains with two outfalls, treated in septic tanks and passed over continuous filter, with the discharge running into nearby watercourses.
- **Quarantine Station at St. John's Island**: Works costing $322,160 were carried out at the Station, including a new pier. The Station then housed 2,000 potential new migrants.
- **Fort Canning Lighthouse** built in 1903 at a cost of $64,420.
- **Singapore General Hospital** for Europeans and natives: The sixth version of the General Hospital was built in 1880–83, but it was extended in 1906 at a cost of $102,310 to provide beds for 270 patients.

By 1911 he had set up on his own account, as he was advertising himself in *The Straits Times* as the "late Senior Surveyor of the Municipality" and was open to "undertaking survey works on moderate charges". His office was at 189 Queen Street and he was living at that time at 10 Niven Road.

Lincoln was born in Bangalore, Karnakata, India, on 31 August 1874 into a Eurasian family. His parents were Elisabeth and John Henry Lincoln. His many descendants live in Singapore and around the world today. John was married twice. His first marriage was to Whilemena Grace Pestana, with whom he had five children: Albert Eric Lincoln (b. 25 March 1909), who later worked at Singapore City Council Sewerage Department as a draughtsman, Edward Terry Lincoln (b. 5 July 1910), Sybil Grace Lincoln (b. 26 May 1912), Cyril Frederick Lincoln (b. 9 June 1913) and David Rayman Lincoln (b. 16 January 1915). David had been born in Sarawak when Lincoln was there to "open up the land" for future development using his land surveying expertise. Lincoln had been posted to the Sarawak

Government Service in early 2014, taking over from James Zehnder, who had taken ill with Bright's disease and died shortly afterwards, aged 42, from complications arising from the disease. Zehnder, from Lundu, Sarawak, was from another Eurasian family with links to Singapore, and he had been the Sarawak Government surveyor since 1899.

Sadly, Whilemena passed away, aged 24, in Kuching, Sarawak, on 26 December 1915. Lincoln's second marriage was to Catherine Petronelia de Souza on 26 December 1917, and they had two girls (Gwen and Ursula) and two boys (Clyde Maurice and Henry).

Lincoln died on 16 June 1929 at the age of 54 at his residence — 468 Serangoon Road, Singapore. He was buried at Bidadari Cemetery the next day. In 2011, his remains were exhumed and placed in a columbarium niche at the Catholic Church of St. Mary of the Angel.

Both he and Wilfred Nicholas (see later in this Chapter) were both Land Surveyors and were the first Asian-born and locally trained Chartered Surveyors.

Wilfred Nicholas

Wilfred Dunstan Nicholas (1875–1956) was a Colonial Fellow of the Surveyors' Institution from 1910 onwards. He was born in Pulau Tikus (a suburb of George Town), Penang, on 28 November 1875 and died on 30 September 1956 at his home at 44 Wilkinson Road, Singapore. The bulk of his career was with the Singapore Harbour Board, based in Tanjong Pagar. In 1939, he was awarded the MBE for his services to the Board. He described his award as "one of the biggest things that happened in my life". This followed the work he did in 1938 for the Empire Exhibition in Glasgow, for which he built a model of the entire Singapore Harbour area.

He came from a well-respected Eurasian family, and his descendants live on in Singapore, United Kingdom, Australia, United States and other countries around the world. His parents were Augustin Francis ("Bokoko") Nicholas (b. 1840, Penang; d. 1935, Malacca) and Catherine Madeline Valentine Rangel (1859–1879), who was from another Eurasian family.

"Bokoko" was a wealthy plantation owner in Malacca and had 11 children with four wives. All four wives died relatively young in life — the first two most likely having succumbed to puerperal fever.

His grandfather (Fernandez Nicholas) came to Penang with the East India Company. It is thought but not confirmed that his surname was anglicised to "Nicholas" from the Portuguese "Nicolas" or "Nikolas".

Nicholas' education was at St. Xavier's Institution in George Town, Penang, and then at the Government Survey School, after which he joined the Government service in 1894, working as a surveyor in Penang, Malacca, Singapore and Lumut for the next ten years at a salary of $66 per month. In 1904, he became Chief Secretary and Draughtsman with Coode Son & Matthews and was involved in the construction of Swettenham Pier in Penang. He first came to Singapore to work in 1905, when the Tanjong Pagar Dock Company was taken over by the Straits Settlements government to create the Tanjong Pagar Dock Board, which evolved into the Singapore Harbour Board in 1913. It is now known as the Port of Singapore Authority (PSA).

Nicholas married Eliza Emily Peterson (b. 1880) in Penang on 8 November 1898, and they had six children. Photographs of Wilfred and Eliza are in Figure 4.3. Sadly, Eliza passed away in Singapore on 19 October 1927. Nicholas then married Amelia ("Tot") Ethelred Minjoot on 3 August 1929 — they had no children. Both his wives came from prominent Eurasian families with strong links to Penang and Malacca.

FIGURE 4.3. Wilfred Dunstan Nicholas and Eliza Emily Peterson.

Nicholas and Eliza's six children were Wilfred Aloysius Nicholas (1889–1960), Bernard Edward Nicholas (1902–80), Violet Alexandra Machado/Farrer née Nicholas (1904–81), Balfour Elias Nicholas (1909–54), Patricia/Priscilla Starline Summerell née Nicholas (1913–65) and Doris Aurora Nicholas (1917).

Nicholas had eight grandchildren, who all affectionately called him "Pop-Pop". A number of his children (and grandchildren) shared the large family home at 44 Wilkinson Road in Katong that Wilfred and Eliza had built, while others lived in the Katong area, where many of Singapore's Eurasian families had settled at that time. This house, set in large gardens, has since been demolished, and the site has been redeveloped. It had four separate apartments, each with its own bathroom, and a large central dining room that could comfortably seat 12. Figure 4.4 shows photographs of Nicholas with his children, their spouses and his grandchildren.

The dining room was dominated by two portraits, one of Eliza and the other of the Sacred Heart (the Nicholas family were Roman Catholics). Like

FIGURE 4.4. (a) Photo of Wilfred Nicholas with his grandchildren, taken at 44 Wilkinson Rd, Katong, Singapore, circa 1940. Standing: Wilfred Anthony Nicholas. Seated: Richard Nicholas, Wilfred (holding Caroline Nicholas), Angela Machado, and Eleanore Nicholas. Wilfred Anthony, Richard and Eleanor are the three grandchildren from Wilfred's oldest son, Wilfred Aloysius Nicholas. Angela is the daughter of Wilfred's oldest daughter Violet Alexandra (Machado/Farrer) Nicholas. Caroline is the eldest daughter of Wilfred's fourth child, Balfour Elias Nicholas. (b) Wilfred Nicholas and all his grandchildren circa 1948/49. Back Row: Richard, Angela, Eleanor and Wilfred Anthony. Front Row: Caroline, Peter Nicholas, Pamela Nicholas and John Farrer.

most Eurasians in Singapore, the family attended the Church of the Holy Family in Katong, a focal point for the Eurasian community. Most Eurasian children in the post-World War II years would have attended Katong Convent (girls) or St. Patrick's School (boys) nearby. After World War II, this home also housed other family members. After Priscilla (Nicholas' second-youngest child) married Gerald Summerell and relocated to Bristol, England, the fourth apartment was rented out to, usually to single British soldiers.

Nicholas suffered a stroke in 1948/49 and moved around in a wheelchair, although he still held his walking stick with him in the chair. On 8 August 1950, in an interview with *The Straits Times*, he recounted a story of encountering a group of Japanese officers during the Japanese Occupation. This group had beckoned him to come down the Harbour steps to a waiting boat, where he was pushed into the water by three of the officers. Given the fast, swirling waters, he was sure he would drown as the current carried him away, but he managed to drag himself out at Hammer & Company's pier. He expressed no anger at the Japanese and told *The Straits Times*: "I leave their punishment to God".

During his 34-year career with the Singapore Harbour Board, Nicholas had been involved in many projects, such as the construction of new harbours, docks and wharves during that period. His work involved surveying, levelling, setting out works and sounding as well as planning for many of the Board's construction projects. He was still working for them in 1950 at the age of 75 — when asked why, he said it was because the Board found him a "useful old hand" and showed no inclination to get rid of him. He believed that there was no job better than the one he did for the Harbour Board, as it was "creative and lasting".

Taking a step back to 1905, this was the year that Nicholas came to Singapore from Penang when he became involved in the Tanjong Pagar Dock arbitration survey and the engineering survey of Mount Wallich and Mount Erskine.

In 1908, he joined the Forest Office in Taiping but returned to Singapore to join Coode Son & Matthews, consulting engineers, where he was Chief Surveyor and Draughtsman on the development of the Tanjong Pagar Docks and wharves.

In 1912, he received his Diploma in Civil Engineering, at which time he was working for Coode Son & Matthews on the King's Dock project.

In 1916, his work on the King's Dock project led to him being offered the role with the Singapore Harbour Board, where he stayed for 34 years until his retirement in 1950, after which he continued to work for the Board on a consultancy basis until his death. At the time of its construction, the King's Dock was the world's largest graving dock in the world apart from the Gladstone Dock in Liverpool. Figure 4.5(a) shows the King's Dock as it was in the 1930s and Figure 4.5(b) shows the site of the King's Dock today.

FIGURE 4.5. (a) An early 1930s photograph of King's Dock (looking back towards Telok Blangah). (b) Photograph of the former King's Dock in 2020 (looking out towards Sentosa island) and now situated next to the condominium, Corals at Keppel Bay.

The King's Dock was opened on 26 August 1913 by then-Governor and Commander-in-Chief, H.E. Sir Arthur Young, KCMG. It was announced that the Dock cost £400,000 to build, was 876 feet (length) by 128 feet (breadth) and had an entrance width of 100 feet, which could be divided by a central caisson into an inner and outer portion that could dock ships of 502 feet and 346 feet in length, respectively; it had a depth of 34 feet at high spring tide (24 feet at the lowest tide), and was constructed of concrete, with the exception of two entrances of Cornish granite ashlar.

Three rows of granite keel block courses also ran throughout the dock under a corresponding number of heavy timber keel blocks, with the rest of the floor being mass concrete. Whilst no longer an operational Dock, it is being preserved and is now a home for marine life. For those familiar with Keppel Bay, the condominium residences known as "Corals at Keppel Bay" sit immediately next to the former King's Dock.

Nicholas was also heavily involved in Telok Ayer Reclamation programmes of this period and also the construction of the Detached Mole, i.e., the substantial Breakwater built to create the inner and outer shipping lanes. The Detached Mole was "lost" during the land reclamation projects of the 1990s, and its remains were finally removed for the construction of Marina Bay Sands and Marina Bay Financial Centre.

After his retirement, Nicholas continued to work for the Singapore Harbour Board as a consultant. His grandchildren recall visits from the Board's staff, bringing plans and documents for his review, comments and approval. Nicholas was now able to fully enjoy his main hobby — horse racing. On his death in 1956, he was interred in Bidadari Cemetery. With the recent redevelopment of the Cemetery, his remains were exhumed and placed in the Choa Chu Kang Columbarium.

Henry Neubronner

Henry ("Harry") Alfred Neubronner (1871–1919) (see Figure 4.6) was born in Malacca on 19 April 1871 to a well-known Eurasian family of mixed Dutch, English and Indian descent. Whilst he never practiced as a surveyor in Singapore, he is mentioned in this book because of his Singapore links — indeed, he spent his final years in Singapore.

Neubronner had spent all his working life in Penang, so there will be much more about him in the forthcoming book about Chartered Surveyors in Malaysia. Neubronner was a renowned architect and surveyor, having qualified with both the Surveyors' Institution and the Royal Institute of Architects whilst in London, where he worked for Chambers & Roberts and later for H.M. Office of Works. His architectural and civil engineering firm in Penang, Wilson & Neubronner (co-founded with Alan Wilson), designed and built many public buildings in Penang.

Neubronner also excelled at sports. When in London, he played football for Clapham Rovers and the Civil Service Association, obtained

FIGURE 4.6. Henry Neubronner.

his Middlesex County FA badge in 1897, and apparently also played for England against France in Paris in 1898. This match is not recorded and might have been a very early amateur representative team before the official formation of the England amateur team in 1906. So, the question remains — was Harry Neubronner the first (and only) Chartered Surveyor to play football for England?

Neubronner's mention in this book will be brief, as he was only living in Singapore towards the end of his life, passing away in Singapore on 23 September 1919, aged 48.

Part II

The Work of Surveyors in the Early 20th Century

Chapter

5

1900–29: The Story of Other Chartered Surveyors of This Period

As mentioned at the start of Chapter 4, a small but growing number of Chartered Surveyors worked in the Straits Settlements from 1900 to 1929, which at this time included parts of Borneo as well as Singapore and the Malay Peninsula. Most of them were based in Singapore.

The previous Chapter tells the story of the Eurasian Chartered Surveyors of this period. This Chapter is dedicated to the story of the British (and one Australian) Chartered Surveyors in Singapore during this period. Some of these individuals stayed for only a year or so, but others stayed longer and left their mark alongside Alfred Lermit, Regent Bidwell and Henry Lake Coghlan. This Chapter will focus only on those who were in Singapore for at least two years as Chartered Surveyors at some point between 1900 and 1929.

Charles Barkshire and John Rea

Charles Robert Barkshire and **John Todd Rea (1868–1948)** were both Fellows of the Institution who worked in Singapore in the early 1900s. Barkshire was in Singapore from 1901 to 1904 as Assistant Surveyor for the Civil Surveyor and Engineer services in the Royal Engineers, a position that Rea inherited from him in 1904 when Barkshire left for a position in Hong Kong. During his time in Singapore, Barkshire was a prominent member of the Merchants and Officers Cricket team. Both of them would have been involved in a wide variety of duties ranging from building construction and maintenance at Tanglin Barracks to engineering works associated with the British military presence in Malaya.

Rea, born in 1868, was from Belfast, and like other surveyors of this period, was a qualified architect and surveyor and worked for the Royal Engineers. He became an elected RIAI member in December 1899, having been proposed by G.P. Sheridan and seconded by R.M. Butler. He ceased to be an RIAI member after 1904. He served a four-year pupillage in the 1880s with Young & Mackenzie in Belfast, followed by a further year in the office as an assistant. At this time, he lived at Glenside, Holywood, Co. Down. He spent a year as a draughtsman in the Royal Engineers Office in Belfast. After passing the qualifying examination in 1891, he joined the permanent staff of the Royal Engineers' Department as a surveyor. In the 1890s, he worked in the West Indies, where he wrote two books: *Building in St. Lucia* (Inverness: Association of Surveyors, 1898) and *Glimpses of the West Indies* (Dublin, 1901). In 1899, he was appointed as assistant surveyor of the Curragh District, County Kildare, where he remained until the end of 1903 when he was promoted to a Royal Engineers' post in Singapore. He arrived on the *Dilwara* via Colombo on 8 January 1904.

In addition to the two books mentioned above, several items by Rea appeared in the *Irish Builder* during his time with the Curragh District. In 1888, when he was living in Holywood, Co. Down, he published drawings of the ruined church at Holywood and the Priory of St. Columba, Newtownards. His final contribution — "Memoranda for road construction" — to the *Irish Builder* appeared in 1910. Perhaps his finest work was *How to Estimate: Being the Analysis of Builders' Prices, Giving Full Details of Estimating for Builders, and Containing Thousands of Prices, and Much Useful Memoranda*, which was first published in 1902 and is one of the earliest books of its kind by a Chartered Surveyor. Indeed, it predates by 20 years the first edition of *Standard Method of Measurement of Building Works*, published by the R.I.C.S. in collaboration with the Quantity Surveyors Association and the building construction industry. Rea's work contains 18 Chapters that explain how to estimate construction costs. Chapter 2 is packed with information about construction costs for the whole range of different buildings in 1902, with the remaining Chapters breaking down the process of estimating costs for the different elements of building construction, from preliminaries through to carpentry and plumbing. The original book is now extremely difficult to get hold of, although those interested in this pioneering book can read a copy online at the Internet Archive (https://archive.org).

John Todd Rea died in Bournemouth, England, in December 1948, at the age of 80.

Edgar Walton

A surveyor with Swan & Maclaren, **Edgar Thomas Walton** was a Professional Associate of the Surveyors' Institution based in Singapore from 1910 to 1914.

John Taylor

John Taylor (1884–1917) was born on 8 May in Holme, Westmorland, England. He initially worked for David Craik in Singapore from 1912 as an assistant architect and surveyor, and by 1913 he had moved with Craik to Swan & Maclaren as Craik's assistant architect. He was a Professional Associate of the Surveyors' Institution from 1910. All this time, he was also in the Singapore Volunteer Corps (SVC) (Royal Engineers) as a sapper. The SVC had been mobilised at the start of World War I, but his role was limited to guarding the Electrical Works, so in August 1915, he returned to Kendal, Westmorland, and enlisted with the 233rd Field Company, Royal Engineers as a sapper. By 1917, he had reached the rank of Second Lieutenant. He was awarded the Croix de Guerre avec Palms (France) for his gallantry. Sadly, he was badly wounded on 11 May 1917 by a gunshot wound to his abdomen, and he died as a result of his injuries. He is buried at the Lijssenthoek Military Cemetery near Ypres.

He left behind a wife (Constance L. Moorhouse), whom he married in Kendal in 1916 when he was on home leave. His death was remembered in *The Straits Times*, and his name is inscribed on the Singapore Cenotaph (see Figure 5.1) as well as on a remembrance plaque inside the Singapore Cricket Club. His Army records tell us that he was 5 feet 4 ½ inches tall, and he was an excellent sportsman; indeed, he was Singapore's hundred-yard sprint champion at the time of his departure to England in 1915.

FIGURE 5.1. (a) John Taylor is remembered on the Singapore Cenotaph, Esplanade Park, Connaught Drive. (b) The Cenotaph was designed in 1922 by Denis Santry of Swan & Maclaren, for whom Taylor worked for prior to World War I.

Knightley Maule-ffinch

Knightley Heneage Maule-ffinch (1871–1936) (see Figure 5.2) spent most of his career in Asia. Born in Ceylon (now Sri Lanka), he was the son of Harry Maule-ffinch, an engineer in the Public Works Department (PWD) in Ceylon and the grandson of Colonel A.B. Fryers, RE Surveyor-General of Ceylon. He joined the PWD Survey Department in Ceylon and married Evaleen Massey O'Dowd in 1895, and they had a large family together. One of their children, Harry (full name Knightley Harry O'Dowd Maule-ffinch), was in later life a highly respected BBC News Editor who started his career as a journalist at the *Malay Mail*, where he reported on the 1915 Singapore Mutiny. Another of their children, Eric Herbert Justus Maule-ffinch, was killed in action in World War I, aged just 18 years. He joined the Malay State Volunteer Rifles at 17 years of age, and his name is remembered on the Kuala Lumpur Cenotaph. Maule-ffinch moved from Ceylon to Singapore in 1901, where he had been offered a position in the Straits Settlements Survey Department. Working in Government service, he was very much involved in land survey work that led to Government land deals for rubber plantations.

FIGURE 5.2. Knightley Maule-ffinch.

He left to set up his own account in 1903, but by 1904, he was working for Tomlinson & Tian Fook, 1 Raffles Place, as a property agent. In 1905, he was operating from Winchester House, 16 Collyer Quay as a Licensed Surveyor and Leveller, Estate Agent and Broker, calling himself a Fellow of the Surveyors' Institute and "guaranteed accuracy, dispatch and neatness". He added that he could also offer "terms lower than any other Engineering Firm in Singapore". In 1906, he launched Mauleffinch & Co. (Note: without the hyphen.) The launch of Mauleffinch & Co. on 9 July 1906 at the Hotel de

L'Europe, with forty guests in attendance, was covered in great detail in *The Straits Times*. Aside from Maule-ffinch, the other partners of the new firm included J.E. Westerhout (an architect related to "Birch" Westerhout, who is mentioned in the previous Chapter) and Vicomte de Bondy, the former French Consul-General for Singapore. The dinner guests were entertained by the Austrian Orchestral Band. *The Straits Times* even listed the music played that night and all 12 dinner courses, along with the champagne and fine wine.

After the fifth course, Maule-ffinch asked the guests to light their first cigars and raise a toast to the King. A toast was also raised to the new Company by Mr. Lowell, who said, "May a long and prosperous life attend them." The dinner proceeded with references to shareholder funds being distributed, but whilst the report on that is ambiguous, Maule-ffinch did state quite proudly that his new firm would supply Singapore with "anything and everything", and that as Singapore continued to prosper, so would his new firm. The press was honoured, as was the French Consul-General, with a rendition of "La Marseillaise", which led the Consul-General to comment favourably on the "*entente cordiale*". Part of the grand finale was a "Strongman" display by J.T. Matthew, in the guise of "Sandow", a world-famous Strongman who spent time in Singapore in 1906 during his world tour.

Sadly, he found himself at a Bankruptcy Court in 1910, seeking a discharge. In fact, the *Singapore Free Press and Mercantile Advertiser*, on 14 July 1910, reported that a receiving order was made on a creditors' petition. The Official Assignee (someone appointed in Singapore by the High Court to administer a bankrupt's affairs in bankruptcy) told the Court that Maule-ffinch had been doing business in Ipoh and elsewhere as a broker and had made significant amounts of fee money from the sale of landed estates. He informed the Court this was correct and offered to settle all his debts in full as a condition of being discharged from bankruptcy. Indeed, a sum of money had been deposited with the Official Assignee that was expected to cover all his proven debts, plus interest and costs. Justice Fisher granted Maule-ffinch's application without conditions.

At this point, Maule-ffinch rejoined public service with the Federated Malay States (FMS) and had a very successful career in Kota Tinggi, Johor. He was also a teacher at the Kuala Lumpur Survey School (see Chapter 1) under Colonel Jackson, RE, Surveyor-General of the FMS, where he was responsible for training the surveyors of the future.

He was also one of the first Scoutmasters appointed in Asia, having founded one of the first Scout Troops during this time.

He and his wife lived in a PWD bungalow in Kota Tinggi, where in 1920, they celebrated their Silver Wedding anniversary. Due to ill health, he left Johor circa 1921 and went to live in Hendon, England, where he joined the surveying staff of Hendon Borough Council.

Maule-ffinch died in Hammersmith, London, in 1936, aged 63. *The Straits Times* reported that with his death, Singapore's last link with Hendon was broken — Sir Stamford Raffles' last days were also in Hendon after his time in Singapore. *The Straits Times* also mentioned that Maule-ffinch would walk past the Raffles Statue every day on his way to work in Singapore.

Harold Carpmael

Harold Carpmael (1871–1927) was born on 13 February in Deptford, Kent, England, the son of Frederick Carpmael (1839–1916), a mechanical engineer from Streatham, London, and Mary Carpmael née Anderson (1844–1934). He attended Dulwich College along with his brothers. A photograph of him during his Dulwich days is at Figure 5.3.

He first arrived in Singapore as Managing Assistant at H.L. Coghlan & Co. in 1906 to work for Henry Lake Coghlan (who is featured in Chapter 2), and he was a Colonial Fellow of the Surveyors' Institution in Singapore by 1913/14. The origin of the title "Colonial Fellow" is unknown, but it appears that it applied to those who were awarded Fellowship of the Surveyors' Institution whilst based overseas.

His arrival via the P&O was heralded by the *Singapore Free Press and Mercantile Advertiser* on 4 January 1906. It confirmed that he arrived to join

FIGURE 5.3. Harold Carpmael (top left) photographed in 1886, in the Shooting VIII at Dulwich College.

Coghlan's practice. It also said that he had many years of experience with eminent firms of London surveyors and valuers, and of late, he had been engaged in the transfer of the London Water Companies to the recently constituted Water Board.

Around 1910 or 1911, he became an accountant, agent, and Singapore/ Malaya representative for a number of London rubber and tin companies, but in 1912, he joined the Singapore Municipality as a Municipal Assessor. *The Straits Times*, on the event of his death, reported that he held this position with "conspicuous success, his thorough knowledge of local property values and his courteous and likeable personality earning him the strong confidence and respect of the public".

In 1922, as the Municipal Assessor, he was involved in a high-profile Supreme Court case. The Supreme Court heard an appeal by Abraham Frankel (who was at that time one of Singapore's wealthiest individuals) against the assessment of the Rural Board in respect of a portion of his Siglap estate. For those familiar with Singapore, two of the main roads through Siglap are Siglap Road and Frankel Avenue. At that time, Frankel's estate extended to all the land and property on either side of those roads.

The appeal focused on one key question: would buildings (even pigstys) built on land, which would ordinarily be regarded as building land, operate to prevent the land from being assessed as building land?

The appeal was heard by Chief Justice Sir Walter Shaw, with the Rural Board represented by Solicitor General Major Langley and the appellant represented by R. Page. This was to be the first appeal against an assessment by the Rural Board following the 1919 ordinance delegating the Municipal Board's powers (including rating assessment/power to levy rates) to the Rural Board. The appeal centred on land that was used as a coconut/rubber plantation, with over a hundred squatters' huts and three substantial stone buildings.

The appellant's representative had to sub-poena Carpmael to appear. The Court heard that he had been the Municipal Assessor since 1916 and was extremely knowledgeable of land and property values in Singapore.

The basic argument that the appellant put forward was that whilst the land was indeed building land, it was not the most economical use of the land due to a number of factors, evidenced by falling rents for houses in the vicinity and the only (limited) demand for houses in this area being from Europeans for "joy houses" where they went for the weekends from central Singapore. So, the appellant argued that the assessment should reflect the economic and market factors against the appellant selling the land as building land and thus against it being assessed as such.

Harold Carpmael's time in the witness box does not appear to have been a pleasant one, as he was questioned and challenged at length about his view that the Frankel land could be developed as houses. He admitted that the authorities were "feeling their way" as they had only started land rating in 1919, and his assessment was based on a "rough estimate", which may have resulted in the big jump in the assessment from its 1912 and 1914 levels. The 1912 assessment for the original 443 acres was $200, then it was $440 in 1914, but by 1921 it had jumped to $3,500 for the (building) land, excluding the coconut plantation.

The result of the appeal was not captured in *The Straits Times* or other publications, and it does appear that the Board's position was weakened by a lack of proper land use planning by the Rural Board as well as a lack of reliable transactional evidence. Also, the assumption that a property owner's wealth was a determining factor in whether his/her land was building land or rural land would not have gone in the Rural Board's favour. The issues around property tax were still clearly a problem in Singapore, and it was not until Singapore's independence that these issues were fully closed out (see Chapter 19).

In April 1924, Carpmael, in his capacity as Municipal Assessor, presented his annual report, which stated that the rateable value of the city within the Municipal boundaries was $23,310,720, showing only a slight increase on the previous year's figure despite a number of new properties being assessed, such as the Ocean Building on Collyer Quay and the YWCA building — this net drop was due to severe depression in trade in Singapore, resulting in falling rents, with sale prices thus having an impact on rateable values. He also reported on a large number of land and property acquisitions for municipal purposes, thus illustrating the wider role of his Department, which was also responsible for administering the Rent Restrictions Ordinance (repealed at that time) to ensure that landlords and tenants complied with the system of rent restrictions. The report noted fines for non-compliance. At the time of his report, the falling price of tin and rubber was having an impact on Singapore, but things got worse in the early 1930s due to the global economic depression, leading to a sharp reduction in Singapore's public works in the early 1930s.

Carpmael's death was reported in *The Straits Times* on 12 December 1927. He had died two days earlier after an operation at the Singapore General Hospital. The newspaper article calls him a "Valued Municipal Official". He was single and 56 years old at the time of his death. His funeral was attended by 150 mourners (despite the December rain), including Charles Frederick ("Fred") Snell (more on him in Chapter 7) and the staff of the Municipal Assessment Department. He was a leading figure in the masonic circles, where he was described as an enthusiastic worker. Also in attendance at his funeral were representatives from the Lodges of Freemasonry in the "Eastern Archipelago". His mother, along with his many brothers and sisters (residing in Dulwich in Kent, England), sent wreaths. He was buried at Bidadari Cemetery in Singapore. Although his grave is long gone due to the redevelopment of Bidadari, his parents' gravestone in West Norwood Cemetery in London refers to his death in Singapore.

Lorong 202, off Joo Chiat Road, was renamed Carpmael Road in his memory. Carpmael Park is also named after him.

George Grey

Born in Chelsea, **George Willis Grey (1878–1954)** was a Fellow of the Institution based in Singapore in the early 1920s.

By 1926, he had become an architect for the Hong Kong Realty and Trust Co. Ltd. (now part of the Wheelock Properties organisation).

He was a practising architect in Hong Kong from 1927/28 onwards and a RIBA member from 1924 to 1929 (and again from 1932 to 1941).

He was also a member of the Hong Kong Institute of Architects.

He authored *Building Supervision* (E & F.N. Spon, 1913), although it is now extremely difficult to locate copies of the book, even though further editions were published in 1922 and 1929. He was also a member of the Royal Sanitary Institute.

During his time in Singapore, it is documented that he was involved in the design and construction of the Post Office, and although his exact role is unknown, it is presumed that he worked for the Singapore Municipality in a supporting role to Major Keys and F. Dowdeswell, who are both credited with its design. In 1927, Keys and Dowdeswell left the public service to form Keys & Dowdeswell, where they continued to work on the Post Office project, and then later on, the Singapore General Hospital building project, by which time Grey had left for Hong Kong.

Grey served in World War I as a Lieutenant. In 1941, he was a member of the Hong Kong Volunteer Defence Corps (HKVDC) at the time of the Japanese invasion and was interned at Shamshuipo prisoner-of-war camp for the remainder of the war. After World War II, he returned to the UK, where he passed away in 1954.

Bernard Dyke

Bernard Frank Dyke was a Fellow of the Institution, based in Singapore from 1924 to 1933. He worked at H.M. Naval Base Singapore. The *London Gazette* in 1911 mentions him as an Assistant Surveyor in the Admiralty Works Department, right at the start of his career (Note: Admiralty was the British Government department with responsibility for the command of the Royal Navy). His appointment was mentioned in *The Straits Times* in October 1923, along with a large number of other appointments to the Admiralty "in connection with the establishment of a Naval Base in Singapore".

For a number of reasons, the British Government had made the decision to invest £11,000,000 in the establishment of a Naval Base, which also included the construction of King George VI Dock at Sembawang, for which work started in 1928 and was completed in 1938. The surveying skills of Dyke would have been essential in planning the construction of the infrastructure to Sembawang that was essential for the Dock to be fully operational once constructed. He was also have been closely involved in the work of the main contractor for the Naval Base construction works, Sir John Jackson (Singapore) Ltd.

He was appointed as the Acting (Chief) Surveyor in 1929 whilst in Singapore. He returned to England in 1933, and by 1941, he was Chief Surveyor of the Admiralty. Born in Thanet, Kent, England, in 1889, Bernard died in Worthing, England, in 1979, aged 90.

Chandos Bone

Chandos Eric Bone was a Colonial Fellow in Singapore and Malaya from 1919 onwards. Although he spent most of his career in Malaya, he is included in this book because he was based permanently in Singapore in the early 1930s whilst working at the Government Survey Office. He arrived in Malaya from Australia circa 1915. Indeed, it is known that he was in the 1915 "List C" of new Army recruits from the Straits Settlements, enrolled for enlistment in England. He was then working for the Survey Department in the Federated Malay States and called E.C. Bone rather than C.E. Bone.

By 1940, he was Assistant Surveyor-General for the Federated Malay States and the Straits Settlements at the Survey Office, Malacca. After World War II, he returned to his Survey Office role, and was based in Kuala Lumpur. Prior to that, in 1937, he was the Superintendent of Surveys in the PWD Survey Department and was based in Malacca, and he was managing a team of 50 people. His annual salary was $10,800 (a considerable sum at that time), plus a car allowance of $720 per year.

The life of Bone is worthy of a book in itself. There is much to tell that has yet to be told outside his immediate family, and events in his life seem to be continually intertwined between Singapore, Melbourne and England. What is known is that he was one of the very first (if not the first) Australian-born Chartered Surveyors, having been born in Richmond near Melbourne, Victoria, on 12 August 1889 to Ava (née Merton) and George John Bones. He was the oldest of their three children. He attended Wesley College, where he won prizes and was involved in rowing and athletics.

He registered as a surveyor with the Victoria Lands Department in 1912 and changed his name from Bones to Bone in 1916, by which time he was already in the Federated Malaya States working as a surveyor with the Government.

The Military Cross (MC) was awarded to him in 1919, when as Temp. Captain Chandos Eric Bone of the Royal Field Artillery, he was recognised for "valuable services rendered in connection with Military Operations in Salonika".

After World War I, Bone returned to Australia, where he married Blanche Speed in 1920 before returning to the Malay States to a role as the Acting Superintendent of Surveys. He then worked in Malacca through to the late 1930s. After time with the Malay States Volunteer Rifles, he joined the (British Army) 39th Royal Field Artillery. In 1924, he was appointed Major Chandos Eric Bone, MC.

He managed to escape Malacca prior to the Japanese Occupation, and then enlisted as a Major with the Australian Army during the Second World War.

He was also a Licensed Surveyor in Melbourne, Victoria, Australia, from at least 1925 (if not earlier) onwards, and indeed, he was still registered as a Licensed Surveyor up until his death in Malvern East in May 1955, aged 64.

Sidney Edwards

A Professional Associate of the Institution in Singapore during the 1920s, **Sidney James Edwards (1887–1953)** was a built environment professional who was a qualified architect (RIBA) and Chartered Surveyor. He qualified as a Chartered Surveyor in 1921 and became a Fellow of the Institution in 1926.

He was born in England in 1887, the son of Professor Joseph Edwards (1854–1931) of Harrow-on-the-Hill, London. Professor Edwards was a renowned author and Professor of Mathematics at Queen's College London. Sidney studied architectural design at the Royal Academy in London under Sir Aston Webb, during which time he was an articled pupil of London City Architects' firm Gordon and Gunton.

In taking the world-famous Mathematical Tripos at the University of Cambridge, he was able to make a special study of concrete design. This study led to a move to Singapore in 1914 to work on the construction of the Chartered Bank's new reinforced concrete building facing Raffles Place (see Figure 3.12). This new building for the Chartered Bank (now Standard Chartered Bank) was needed because the Bank had outgrown its existing building, which had been designed and built on Battery Road between 1893 and 1895 by Swan & Maclaren. The new building was to be constructed slightly further along Battery Road and was to be designed by David Craik of Swan & Maclaren. However, Craik was unable to see the project through to fruition as he left for active service in France in World War I. It fell on to Edwards to finish the task, and at the opening ceremony on 6 May 1916, Governor Sir Arthur Young expressed his indebtedness to Edwards for the "careful manner in which he with Swan & Maclaren supervised all the work" to deliver what the Governor described as a "stately structure".

Edwards left Singapore soon after this to return to Europe, where he too served in World War I on active service in Mesopotamia, a region that is now part of Iraq, Kuwait, south-east Turkey and parts of the Turkish–Syrian and Iran–Iraq borders. He also joined the Indian Army Reserve of Officers, where he was a garrison engineer in Kolkata (then known as Calcutta). Edwards returned to Singapore in 1921. Indeed, on 11 March 1921, we see an article in the *Malaya Tribune* informing its readers that Edwards had just returned from England to Singapore to join Ralph Booty & Co. In November 1921, Edwards presented plans to the Singapore Cricket Club to extend Bidwell's Pavilion. His extension plans were approved, and the work was completed in 1922.

In September 1923, the *Malaya Tribune* (and *The Straits Times*) carried a notice under the heading "Ralph Booty & Co.", confirming that the partnership business henceforth would operate as Booty and Edwards, the firm of architects, surveyors and valuers of Singapore, Colombo and Penang. Edwards came from a "highbrow" academic family — he was himself a Cambridge scholar and associate member of the RIBA. Julian Davidson's book tells us that this apparently led Edwards to feel a sense of entitlement that was not reflected by the commissions given to him by the partners of Swan & Maclaren.

Stepping back to 1922, a decision was taken to build a new Town Hall on the North-Eastern corner of Viharamahadevi Park between Colpetty and Cinnamon Gardens in Colombo, Ceylon (now Sri Lanka). However, the architects of the PWD were fully stretched on the design and construction of a new Parliament and Government Secretariat on the Galle Face, and as a result the design of the Town Hall was put out to competition. A total of 25 submissions were received from architects based in various parts of the British Empire (there were no local entries). Austin Woodeson, the Chief Architect of the PWD acted as the judge, and the eventual winner was Booty and Edwards.

The Straits Times of 25 November 1922 reported that the estimated construction cost was Rp 1,000,000. It was at this time that Edwards first added the letters "PASI" to his signature — before that, only the letters "ARIBA" followed his name. Anyone who has visited Colombo will know that this building is one of the most magnificent buildings in the city, if not the country. Edwards' design draws heavily from the design of the Capitol Building in Washington, DC, and despite alterations to the building since its construction, it is still a truly remarkable structure.

Sidney Edwards was, along with George Grey and Walter Brown (both of whom feature in this Chapter), one of the 11 founding members of the Singapore Society of Architects, formed in 1923. The other eight founder members were Messrs. Keys and Dowdeswell (architects of the Singapore Post Office [now the Fullerton Hotel] as well as the Singapore General Hospital construction that is now the home of the SGH Museum), J.S. Ruchwaldy, Messrs. Meadows and Knight from the Singapore Municipality, Frank Dorrington Ward (Government architect), D.W.V. May and R.M. Love. The Singapore Society of Architects became the Institute of Architects, Malaya, in 1931, and in turn, this is now Pertubuhan Akitek Malaysia (PAM) — the Institute of Architects Malaysia.

The Colombo Town Hall competition rules required Edwards to open an office in Colombo, and for a time, he practised under the name "Booty and Edwards", renting an office in Chatham Street, Colombo.

However, Edwards severed his ties with Booty, and in 1927, he formed a partnership with H.H. Reid and R.G. Booth under the name "Edwards, Reid and Booth". They soon established themselves as the most successful and prolific architects in Colombo, attracting work from the Government, the commercial sector and private clients. In 1932, they moved to a new and larger premise on Prince Street. Their success continued even after Booth's resignation in 1936 to return to England, with Edwards and Reid going on to design and build other commercial buildings within the Fort Business District, including the Times Building and the Chartered Bank. A 1933 photograph of Edwards, Reid and Booth is shown at Figure 5.4.

The partners enjoyed prominent positions in Colombo society and lived in large bungalows in Colombo's southern suburbs. They worked a three-year tour of duty, followed by six months of fully paid leave in the UK. Indeed, it was during one such period that Edwards married Barbara J.V. Wing of Clifton, Bristol, in January 1929.

Edwards was active as a Chartered Surveyor in Ceylon. Indeed, he became a Committee member in the R.I.C.S. Ceylon Branch in 1932.

FIGURE 5.4. Sidney Edwards (centre), H.H. Reid (right) and R.G. Booth (left), circa 1933.

Edwards, Reid & Booth became Edwards Reid & Begg in 1936, with the opening of an office in Madras (which was to be operated by a British architect named Kenneth Andrew Begg) alongside the Colombo office. Edwards retired in 1937, although he retained a financial stake in the practice.

Reid continued alone in Colombo. As a hang-over from

his connection with Booty, Edwards continued to run the branch office in Madras but in 1937 Begg took over its management. Begg spent only two years in Madras, but his name continued to be used for a further 50 years.

Reid continued to run the practice as Edwards Reid & Begg until his death in 1952, at which point the practice was taken over by Jimmy Nilgiria, and then after 1958, by the renowned architect Geoffrey Bawa. Bawa practised under that title until 1990.

At the same time, the practice of continued under the name "Booty, Edwards & Partners" despite the loss of Edwards and became a major force in the development of South-East Asia's architectural structures in the 20th century. Following its acquisition in 1930 by Arthur Oakley Coltman, the practice prospered, and over the years, it has been responsible for some of the major governmental and commercial buildings in Sri Lanka, Malaysia, and Singapore through its offices in Colombo, Penang, Kuala Lumpur and Brunei.

In 1969, the firm was renamed BEP Akitek Sdn Bhd, the name it still carries today.

In 2021, BEP celebrated its 100th anniversary. A monograph was issued by BEP to celebrate this anniversary, and this monograph contains further information about Edwards' time with Booty, Edwards & Partners.

Sadly, Edwards' wife, Barbara, died in Colombo on 19 July 1931. He retired and returned to London in 1937 and never returned to Singapore or Ceylon to practice.

After his retirement, Edwards continued to practice from an office in Lambeth Bridge House in London and continued to do so after World War II. It is believed that he passed away in England in 1953.

Edwards will be remembered not just for his talent as an architect and surveyor but also as a successful business leader who upheld the highest technical and professional standards. He and Reid were both very influential in the establishment of the architectural and surveying professions in Ceylon.

Charles Frederick Snell and Geoffrey Polglase

Charles Frederick ("Fred") Snell was a Professional Associate from 1913 (and a Colonial Fellow from 1922) who was based in Singapore from 1922 to 1946 before returning to England permanently. He was based at the Municipal Offices. Alongside him in the Municipal Offices from 1931 to 1932 was **Geoffrey Polglase**, a Professional Associate of the Institution. Geoffrey moved to Hong Kong in 1933 to perform a similar role there, and at the start of World War II, Geoffrey became an RAF pilot. Tragically, Geoffrey was killed on active service in May 1940 (just prior to the start of the "Battle of Britain") and is buried at Blindley Heath (St. John) Churchyard, Godstone, Surrey, England. There will be more about Snell in Chapter 7.

Walter Brown

Walter James Brown was a Colonial Fellow of the Surveyor's Institution in Singapore in the early to mid-1920s. However, he is best known as an architect and was active in England, then Singapore and later in Ireland from 1919 until his retirement in 1961.

Born in 1894, he served in World War I. He then came to Singapore as an Assistant Municipal Architect, where his work on the architectural drawings of the **Municipal Building (later known as City Hall and now part of the National Gallery)** — see Figure 5.5 — secured him his RIBA qualification. He was a RIBA associate from 1919 and a Fellow from 1927. He was an RIAI member from 1925 to 1934.

The Municipal Building was planned and designed by the municipal architect S.D. Meadows, and then by Alexander Gordon, who took over Meadows' position in 1925. It is now a National Monument and, therefore, a building of great significance in Singapore. Whilst Meadows' and Gordon's names are the ones that are recorded in history as those who planned and designed it, Walter Brown should also get a mention, as his work on the

FIGURE 5.5. Municipal Building, later City Hall, and now the National Gallery, Singapore.

architectural drawings was important in the creation of this magnificent building.

Sadly, Brown's contract with the Municipality was terminated whilst he was in England on annual leave in 1924. Whilst there, he was married in Old Chesterton, Cambridge, in April 1925, to Mary Kathleen Adams. Brown later moved to Ireland, where from the late 1920s, he worked for the Irish Sailors' & Soldiers' Land Trust in Dublin. He designed cottages for the Trust at Ballyshannon, Athlone, Galway and Cabra, Co. Dublin.

He left Ireland in 1933/34 to live in Edgeware, Middlesex, London, and served in Whitehall, London, during World War II, at which time (amongst other things) he was ordered by Winston Churchill to find more small ships for the evacuation of Dunkirk. He later worked for the Ministry of Housing and Local Government in England and drew up plans for council estates such as Falcon Lodge, Birmingham. He was given the freedom of the city of Lichfield, and as the Ministry's Principal Regional Architect for the South-Western Region, he was awarded the OBE in 1956.

John Caseley

John Reginald Caseley (1898–1972) was born in England. He qualified as a Chartered Surveyor in 1920, arrived in Singapore in the mid-1920s and was based at 550 Havelock Road, where he worked for Fogden,

Brisbane & Co. Ltd., a well-known Construction Company in Malaya and Singapore during the inter-war years. However, his first job upon arrival in Singapore was in the PWD as an Assistant Engineer, starting on 4 January 1924, for which his annual salary was $4,800 plus $1,560 per year as a car allowance.

As a senior employee of Fogden, Brisbane & Co. Ltd. and later a Partner, he would have been involved in many of its construction projects from the late 1920s until the late 1930s, such as the Elgin Bridge project (see Figure 5.6). Indeed, an article in the *Singapore Free Press and Mercantile Advertiser*, 19 October 1927, reported on an unusual ceremony that was held 80 feet below ground at the bottom of a caisson.* The ceremony was to celebrate the sealing of the first of six caissons of the Elgin Bridge. Caseley, as the Resident Engineer, was there.

After the ceremony, they started the process of filling the caisson with concrete. The article also described the many difficulties faced in the project that were not encountered at the boring stage.

However, the article confidently informed its readers that the Bridge would be completed in 12 months' time (i.e., October 1928). Sadly, by April

FIGURE 5.6. Elgin Bridge, a construction project that John Caseley was heavily involved in.

*A caisson is an engineering term for a watertight retaining structure often used for working on the foundations of a bridge pier and is constructed in a way that the water can be pumped out, keeping the working area dry.

1928, the *Malaya Tribune* carried a Municipal report recommending that the contract between the Municipality and Fogden, Brisbane & Co. Ltd. be terminated by mutual agreement, with the firm being paid a fair and reasonable profit for work done.

The Bridge was actually completed on time and opened for traffic a few months later. The Municipal Bridge Engineer, T.C. Hood, must have been relieved! There had been difficulties with construction. The concrete Caissons rested on very hard clay, but above that, there were 5 feet of sand topped by soft and treacherous blue clay. The weight of buildings on either side of the river caused the blue clay to play tricks on the engineers, making it necessary to carry out much more excavation than the results of the borings had told them to.

Another significant construction project involving Fogden Brisbane & Co. Ltd. was the BMBC Broadcasting Station at Caldecott Hill, which was the first of its kind in Singapore and around which the Mediacorp complex now exists. The original building, designed by Swan & Maclaren and opened officially by the Governor on 1 March 1937, has since been replaced. Caseley himself was present at the official opening. As an aside, the Caseley family lived in a house on the estate at Caldecott Hill. This estate, comprising Caldecott Close, Andrew Road, Olive Road and Joan Road, was also constructed by Fogden Brisbane & Co. Ltd. in the late 1930s and largely still exists today. Around this time, Caseley was Vice President of the Engineering Association of Malaya.

Caseley was a renowned organist and pianist, and when in Singapore, he often played at the YMCA and wedding ceremonies at St. Andrew's Cathedral. In June 1930, he married Dr. Katherine Spence, MD, of the Singapore Child Welfare Department at the Cathedral. In February 1942, Caseley and Katherine escaped Singapore just before it fell to the Japanese. Their story is in the BBC World War II archives, and as recalled by their daughter Jane:

> Less than 48 hours before the surrender of Singapore to the Japanese, they packed their one permitted suitcase, each with what little belongings

> they could gather, and managed to get a place on board *The Gorgon*. The decision to leave behind some of their treasured possessions included their wedding presents, his organ music, the family grand piano, and, sadly, their cat.
>
> After a traumatic period when the engine failed, and they drifted in the Java Sea (by then controlled by the Japanese), they managed to reach Australia. They stayed in Perth (using a banker's note from their Singapore Hong Kong & Shanghai Bank to survive), and whilst they contemplated staying in Perth, they decided to return to the UK, where all their family still resided. They left Australia and headed for Auckland, after which they sailed east to Cape Horn, up the west coast of South America, through the Panama Canal into the Caribbean. The *Stratheden*, a P&O ship that had been requisitioned for War service, was fast enough to sail alone (not in convoy) and was able to avoid U-Boats that were known to be in the vicinity.
>
> They finally reached Clyde in Scotland, where the dockyard workers cheered the ship home — this was three months after they left Singapore.

The BBC archives recalls the story that when in Glasgow, the Customs Officers confiscated the Caseleys' camera film, which had captured the devastating scene of Singapore's impending fall to the Japanese — how valuable would that film be to historians today!

The Caseleys bought a smallholding near Edenbridge in Kent, got furniture from second-hand or antique shops, and collected their daughters from the Spence family in Scotland, who had looked after them since they had returned from Singapore in January 1938 to attend school. The Caseleys never returned to Singapore — the memory of what happened was no doubt too painful.

By 1949, Caseley was the UK's Deputy Director of Open Cast Coal Production, Ministry of Fuel and Power, and in the same year, he was awarded a CBE in King George VI's New Year's Honours List. He died in England in 1972, and his wife Katherine (who became a GP locum for Doctors in Edenbridge, Kent) passed away in 1996.

Their daughter, Dr. Jane Rosamond Caseley (who passed away in 2019), married Dr. Kurt Hellman in 1961. Dr. Hellmann was one of the most eminent pioneers in metastasis research and treatment. He was perhaps best known for his discovery of the biologic activity of two important drugs: Razoxane and Dexrazoxane.

Herbert Atkin-Berry

Born in Richmond, Surrey, England, **Herbert Courtenay Atkin-Berry (1888–1960)**, the son of renowned architect William Henry Atkin-Berry, attended Dulwich College from 1898 to 1904. He was born with the surname Berry, with Atkin as his middle name, as was the case with his father and other family members. His family name had been changed by Deed Poll on 18 February 1910 from Berry to Atkin-Berry. Photographs of Atkin-Berry are shown in Figure 5.7.

FIGURE 5.7. Herbert Courtenay Atkin-Berry.

He arrived in Kuala Lumpur in 1912 to work for Philip Russell (who was Swan & Maclaren's Kuala Lumpur agent, a practice thenceforth known as Russell & Berry), and in April 1915, it was announced that he was in List A of "gentlemen to whom the Governor of the Straits Settlements has, on the authority of the Army Council conveyed on him by the Secretary of State for the Colonies, offered temporary Commissions for the duration of the War in the new service battalions now being raised". He joined the Royal Fusiliers and saw front-line action in France. In a letter published by the *Singapore Free Press and Mercantile Advertiser* on 28 April 1916, he gave a harrowing account of his awful experiences "somewhere in France". He was awarded the DSO (Distinguished Service Order), and in 1917, he was also awarded the MC for gallantry.

He married Edith Mary ("Duffie") South in Chelsea, England, in October 1917, and after World War I ended, he returned to Malaya to resume his work for Russell and became Swan & Maclaren's "architect-in-charge" and manager of their Kuala Lumpur office. Aside from his architectural skills, he was also a Chartered Surveyor — he was a Colonial Fellow of the Surveyors' Institution from 1929 onwards. He also held an appraiser's licence in Singapore from the 1920s onwards, and he was an expert (valuation) witness in land compensation claims in Singapore. During his time in Singapore, he lived with his family at a residence called "Kashmir" at Dalvey Road.

In 1924, when Harry Robinson retired from the Swan & Maclaren Singapore office, Atkin-Berry moved to Singapore to become Managing Partner of the Swan & Maclaren practice from their office at Hong Kong Bank Chambers on Collyer Quay. Thereafter, he split his time between Kuala Lumpur and Singapore, a position he held until his retirement in 1939.

His role managing the entire Swan & Maclaren business meant that he was not directly involved in the design of many notable buildings by the company during this period. This he left in the very capable hands of David Craik, Frank Brewer and Frank Lundon, as well as the very capable architects who came into the practice later, including D.S. Petrovitch, FRIBA, Denis Santry and C.Y. Koh (Koh Cheng Yam), who joined in 1938, and was Swan & Maclaren's first Singaporean architectural assistant. The latter led the firm after World War II.

Atkin-Berry became a Colonial Fellow of the Institution in 1929, but in addition to being a Chartered Surveyor, Atkin-Berry was also an architect and an expert witness in land (value) compensation claims in Singapore. Indeed, he held an appraisers' licence from the 1920s to the 1930s, as did Frank Lundon and Messrs. Keys and Dowdeswell, who are mentioned elsewhere in this book.

During the 15 years with Atkin-Berry at the helm, Swan and Maclaren's success story continued, albeit they had very strong competition from the likes of Keys & Dowdeswell. Amongst Swan & Maclaren's notable achievements during Atkin-Berry's time at the helm, and one in which Atkin-Berry was directly involved in himself, was the design of the former Tanjong Pagar Railway Station, which is now a National Monument (see Figure 5.8). Atkin-Berry and Denis Santry both supported the work of

FIGURE 5.8. Former Tanjong Pagar Railway Station. (a) Main front entrance to the Station. (b) The Station Hall Ticket Office. (c) Close-up view of one of the Panels in the Station Hall, depicting the economic activities of Malaysia and Singapore.

D.S. Petrovitch, the Swan & Maclaren architect who was responsible for the design of the railway station or "Terminus", which was completed in 1932, and with the deviation of the line to the new Terminus, the result was the closure of the previous Terminus at Tank Road.

The trio sought inspiration for the new building from existing railways stations in England; Atkin-Berry even visited several English stations in 1927 for this purpose. The result was an impressive structure resplendent with Art-Deco features that possessed all the facilities needed to ensure the comfort and convenience of not only the passengers using the new Terminus but also meeting the needs of its staff. These ensured its smooth and effective operation.

At the opening ceremony on 2 May 1932, the speech of Sir Cecil Clementi — Governor and Commander-in-Chief of the Straits Settlements — visualised a world-wide importance for this new Railway Station. His vision was to see the Station as the Terminus of two lines, one reaching across Asia and Europe to the English Channel and the other reaching through what is now Thailand, Vietnam, Cambodia, Laos and China, connecting to the Trans-Siberian Railway in north-east China.

With the global resurgence and rebirth of rail transport in recent years, Clementi's vision could well be achieved by 2032 (one hundred years after he shared his vision), with Singapore as a global Terminus, albeit at Woodlands, not Tanjong Pagar. Its status as a National Monument will ensure its preservation and eventual re-use when linked to the new Cantonment MRT station upon its completion in 2025.

Atkin-Berry as an Appraiser

In addition to his role as Managing Partner of the business, he was also a licensed appraiser, which led to him being called upon over the years to act as an expert witness on land values in Singapore, during matters of dispute between landowners and the Government when land was being requisitioned for public purpose.

One such case was a 1930 Supreme Court matter between a local landowner and the Malayan Planters' Provident Fund Board in relation to an alleged breach of the terms of a mortgage relating to land off Serangoon Road to be developed as the Woodsville Estate, for which construction (40 houses) had already started. Atkin-Berry's involvement was three-fold: he had valued the land in 1927 for the Hong Kong Bank, Frank Lundon of Swan & Mclaren had also valued the land in 1930, and furthermore, Swan & Maclaren was project managing the construction. (Valuers reading this book may be interested to know that the fee for the 1927 valuation was ½ per cent on the first $50,000 and ¼ per cent on the remainder.) Problems started in 1929 when property values fell sharply, and work stopped on site, at which point Swan & Maclaren was owed $11,000, with little hope of being paid. The dispute centred on a supposed 1928 Swan & Maclaren valuation, against which building funds has been loaned. Atkin-Berry proved to the Court's satisfaction that no such valuation had been provided by Swan & Maclaren to either the landowner or any Bank or other lending body in 1928, and the strong inference was that the 1927 valuation report had been "altered" without Swan & Maclaren's permission. Indeed, Justice Palgrave Simpson used the words "criminal roguery" in summing up his decision in favour of the Provident Fund Board. The market crash of 1929 meant that the fraudulent 1928 valuation was in the spotlight. Atkin-Berry was fully exonerated, and Justice Simpson made a point of stating his belief in what Atkin-Berry had said on the matter. The question of Swan & Maclaren's potential conflict of interest was mentioned in passing but was not deemed a matter for the Court.

Other Noteworthy Buildings From Swan & Maclaren During Atkin-Berry's Time at the Helm

It has already been mentioned that Atkin-Berry was the Managing Partner of Swan & Maclaren from 1924 to 1939 and was not, therefore, directly involved in the everyday activities of the architects and surveyors working for Swan & Maclaren. However, during his time as Managing Partner, despite increasing competition, Swan & Maclaren was responsible for the design and construction of innumerable commercial and residential buildings in Singapore as well as the full length and breadth of the Malay Peninsula; in particular, Ipoh, Penang, Klang/Kuala Lumpur and Johor.

At this time, one could argue that Singapore's innovative and creative architecture rivalled any city or town in Asia, as Singapore's architects strove to emulate the design of new buildings in the UK, Europe and even in cities such as New York. Swan & Maclaren played a key role in this and created a legacy for Singapore, a good proportion of which has survived to this day.

Aside from Tanjong Pagar Railway Station, Atkin-Berry is known to have been directly involved in the design of only one or two buildings, which is logical given his role in managing the business and huge talent at his disposal within Swan & Maclaren during the inter-war years.

Part Three of Julian Davison's book goes into great detail about the work of Swan & Maclaren from the end of World War I in 1918 to the start of World War II in 1939, and so this book will not seek to repeat or plagiarise any of this great work. Indeed, there is a long list of buildings that Swan & Maclaren designed during Atkin-Berry's period, alongside a long list on buildings not designed by them but where Swan & Maclaren managed the construction project.

Below is a selection of some of the noteworthy construction projects:

- **St. Patrick's Boys' School, East Coast** — constructed in 1932. Designed by Swan & Maclaren.
- **Majestic Theatre and Southern Hotel, Eu Tong Sen Street** — completed in 1929, designed by Swan & Maclaren. The coloured tiles on the front elevation depicting Chinese scenes were made in Belgium, copied from Chinese artistic work supplied by Swan & Maclaren, and then imported to Singapore and fitted *in situ* by artisans from Shanghai.
- **Nunes and Medeiros Buildings at the junction of Malacca Street and Cecil Street** — office buildings designed by Petrovitch and constructed

in the 1930s, introducing the Continental Cubist architectural design to Singapore. A special feature of the Nunes and Medeiros Buildings was their external panels (designed by Cavalieri Rudolfo Nolli), symbolic of commerce, agriculture, shipping, building, electricity and other industries. One of these panels survives to this day and is held in the National Museum of Singapore.

- **Union Insurance of Canton Building, Collyer Quay** (see Figure 5.9(a)) — Denis Santry won a competition for its design in 1921, and Swan & Maclaren managed the construction project. United Engineers was the Principal Contractor for the construction. Comprising seven floors of office space, the Union Insurance of Canton Building was the largest commercial office building at the time of its completion in 1924, and alongside the new Hong Kong (& Shanghai) Bank Building (adjacent to it), it attracted most, if not all of the most sought-after tenants in Singapore. The Tower at the top of the new building could be seen many miles out at sea.
- **The new Hong Kong (& Shanghai) Bank Building, Collyer Quay**. On completion in 1924, Swan & Maclaren became one of its first occupiers.
- **Telegraph House, Robinson Road** (also see the reference to this building in Chapter 3) — the architect was Frank Lundon. See Figure 5.9(b).
- Swan & Maclaren designed many **Black & White houses** built in the 1920s and 1930s. They also designed and built small housing "estates" for senior staff of major corporations, including those at Bukit Tunggal and Holland Road, plus individual houses for a diverse range of clients.
- **Amber Mansions, Orchard Road** (where Dhoby Ghaut MRT is now situated).
- **David Elias Building, Middle Road.**
- **Shophouses** at Tanjong Pagar Road, Murray Street and Lavender Street.
- **Meyer Chambers, Raffles Place/Malacca Street** — constructed between 1929 and 1930; the architect was Petrovitch.

FIGURE 5.9. (a) Union Building, Collyer Quay, Singapore, circa 1950s. (b) Telegraph House (now the Sofitel). (c) Chinese (Institutional) Methodist Church, Telok Ayer Street/Cecil Street. (d) Presbyterian Church, Prinsep Street.

- **Presbyterian Church, Prinsep Street** (see Figure 5.9(d)) — Swan & Maclaren did not design this Church, but they supervised its construction, and Santry laid the Foundation Stone in 1930.
- **Chinese (Institutional) Methodist Church, Telok Ayer Street/Cecil Street** (see Figure 5.9(c)) — designed by Swan & Maclaren. Santry was closely involved in its construction, and he was present at the laying of the marble foundation stone on 9 January 1924.

Aside from his role as Managing Partner (later Senior Partner) of Swan & Maclaren, Akin-Berry was on the Board of various rubber and tin mining Companies. He was also the inaugural Vice President of the European Association of Malaya in 1937.

Akin-Berry and his family returned to England for good in October 1939, and he was given a big farewell celebration at the Royal Singapore Golf Club on 15 October 1939 — he was an active golfer. They settled down at Storrington, Sussex. One of his sons (Harold Leslie Atkin-Berry) was an RAF Pilot Officer who died tragically aged 20 on 10 July 1940 whilst on active war service with the 107th Squadron RAF 41360.

Atkin-Berry's brother (Henry Gordon Atkin-Berry, MC, ARIBA) came to Singapore in 1935 to join Swan & Maclaren. At that time, he was with Bird & Atkin-Berry architects in London and was also a Lecturer at the Architectural Association. Tragically, Henry did not manage to escape from Singapore prior to the Japanese Occupation. He tragically died of dysentery on 20 April 1942 in the Palembang Men's Camp whilst he was a Sumatra Japanese internee. Henry had designed the Seamen's Chapel on Anson Road, opened in 1937 behind the Seaman's Institute building.

Herbert Courtenay Atkin-Berry died in Worthing, Sussex, England, in 1960, aged 72.

There will be more on Atkin-Berry later in Chapter 7 — he was the inaugural Chair of the Institution of Surveyors' Malaya Branch and thus played a vital role in the establishment of R.I.C.S. in the region.

Alexander Cullen

A Professional Associate of the Institution since 1927 and in Singapore from 1929, **Alexander Cullen** worked as a surveyor for the Singapore Municipality and was married in the Singapore Presbyterian Church on 18 June 1931 to Ada Dorothy Band (see Figure 5.10). A Scotsman, he was the Singapore Scouts' District Commissioner ("Running Hare") and wore a kilt as part of his scout uniform. In 1930, he vigorously urged all Scoutmasters to move away from the classroom atmosphere that was prevalent in Malaya and Singapore Scouts at that time and encouraged Scoutmasters to "get out to camp… get out onto the trail of buried treasure and gather round the Council Fire at night, tired and happy with the doing of things".

FIGURE 5.10. Alexander and Dorothy Cullen (centre, front table) following their wedding in June 1931 at the Presbyterian Church, Orchard Road. Also shown in the photograph is Revd. Stephen Band, the bride's father, who was Minister at the Wedding.

He believed strongly that the pursuit of outdoor activities was what scouting should be, and not by learning about it in the classroom. In the National Archives of Singapore, there is a photograph of Cullen in his scout uniform (minus kilt) at Purdy Scout Camp, Changi, in 1929. In her book, *Out in the Midday Sun: The British in Malaya 1880–1960*, Margaret Shennan covers the life of Alexander and his wife Dorothy in some detail, including their marriage in Singapore, her move to Australia just prior to the Japanese invasion and the tragic end to Cullen's life (see Chapter 9 to read more about this).

Philip Pank

A Professional Associate of the Institution, **Philip Roy Pank (1903–1980)** was born in Norwich, England.

Little is known of his surveying work in Singapore from 1926 to 1930, but a clue to his occupation is that as a prominent member of the Singapore

Cricket Club, he was involved in organising cricket competitions and matches. In doing so, he was in the "Government and Municipal" cricket team, so it is likely that he worked for the Municipality of Singapore as a Chartered Quantity Surveyor.

As mentioned, he was a very active member of the Singapore Cricket Club. He also played doubles in Lawn Tennis tournaments at the Singapore Cricket Club and was a well-known hockey player at the Club too. Indeed, the extent of his sporting activities makes one wonder if he had any time left for his day job as a surveyor? However, by early 1930, having recovered from a period of illness, he made the decision to leave Singapore; an illness had possibly forced the decision on him, or perhaps it was his employers. On 24 June 1930, a farewell dinner was held at the Raffles Hotel hosted by the Riff-Raff Club, of which he was the Secretary. The Club was set up in the 1920s in Singapore; its original purpose is unknown, although the name would suggest that the main focus of the Club was dinners at the Raffles Hotel. The Club Chair presented Pank with a silver cigar case and paid tribute to Pank's unequalled keenness in the hockey and cricket sides, his fine sporting qualities and Pank's claim to be a "prince of good fellows". Modesty was clearly not one of his qualities! He set sail onboard the *President Johnson* that same day. This cruise ship was originally the *Manchuria* and was operated by the Dollar Steamship Company (now APL).

By 1931 he was back in Norwich, where he set up his own practice, Philip Pank & Partners, which is now one of the oldest quantity surveying practices in East Anglia, becoming a partnership in 1951. He died in Norwich in 1980.

Frederick Forsaith

A Professional Associate of the Institution in 1929, **Frederick** ("Fred" or "Leslie" to his family) **Leslie Forsaith (1893–1980)** was born in Croydon, England, on 13 January. In 1911, he is listed in the Croydon Census as a clerk in a firm of architects and surveyors. He enlisted in the

Royal Army Medical Corps for the duration of World War I (at which point his occupation was stated as being an "architect's clerk") and was demobilised in early 1919 after service with the RAMC in North Africa and Gallipoli. He married Phyllis Agnes Pennock (1898–1967) in Edmonton, England, in 1922.

FIGURE 5.11. Frederick Forsaith in 1935.

He joined the Air Ministry in 1928 and then came to Singapore, where his role was as an Assistant Surveyor with the Air Ministry Directorate, W&B Department, RAF Singapore.

Nothing is known of the surveying role he played in Singapore for the Air Ministry, but it is known that he was an accomplished pianist and organist, and aside from performing as a pianist at Victoria Hall concerts, he was also the organist at the Orchard Road Presbyterian Church.

Their son Frederick Ian Forsaith was born in Singapore on 11 August 1929. The family all returned to the UK in 1932, and by 1939, Forsaith was working in Cambridgeshire, England, as a surveyor for the Air Ministry. A 1935 photograph of him is shown at Figure 5.11.

He was on the RAF lists as PASI (acting) during World War II. He died on 1 March 1980 in Bedford, England, aged 87.

His grandson David Forsaith is now also a Chartered Surveyor based in Cambridgeshire, England.

Jabez Turner

A Fellow of the Institution by the time he was in Singapore, **Jabez Tennyson Turner** first worked in England in the Prison Service in 1903 as an Assistant Surveyor and the Admiralty as a surveyor. He arrived in Singapore in 1924 but returned to England only four years later. By

1939, he had progressed to Chief Surveyor, Civil Engineer-in-Chief's Department, Admiralty, where he was made a Companion of the Imperial Service Order (ISO).

His time in Singapore with the Admiralty saw the early stages of the development of the Naval Base at Sembawang, where he would undoubtedly have been involved in overseeing the early stages of the construction of the Naval Base infrastructure.

Outside of work, he was heavily involved in theatrical and musical performances, in particular, the Vaudeville Nights at the Victoria Theatre. His wife, Maggie Taylor, was a singer of some note, and she was also heavily involved in the performances of the Singapore Amateur Wireless Society from 1924 onwards. The Society were the pioneers of radio broadcasting in Singapore and was founded in 1924 by amateur enthusiasts from the expatriate community supported by commercial enterprises dealing in radio equipment. The Society's founders worked for companies with a vested interest in wireless technology, such as General Electric, Marconi's Wireless Telegraph Company, and Standard Telephone, with whose support the Society secured its temporary transmitting licence.

Jabez Tennyson Turner was born in Bethnal Green, London, in 1880, married Maggie in London in 1904, and died in Portsmouth in 1958, aged 78.

Chapter 6

Langdon Williams and his Work with the Singapore Improvement Trust

Despite the efforts of many individuals to provide Singapore with good infrastructure and improved standards of living, there had continually been poor and inadequate housing for the low income-earners of Singapore. Instead of getting better, the problem worsened year on year with the population growth. By the early 1920s, the problem had reached a point of crisis, but the creation of Singapore Improvement Trust (SIT) in 1927 offered some hope. It was created with the passing of the Singapore Improvement Ordinance. The Ordinance provided the Trust with some of the executive powers it needed to carry out improvement schemes and to prepare the general town plan, but it did not address the issue of housing for the general population, apart from those displaced by the improvement schemes — and so it continued until the mid-1930s, by which time the SIT had only rehoused 4,500 people via the earlier mentioned scheme.

Initially, the SIT had no mandate to construct public housing, as the municipality served as an agent for the SIT in building accommodation to rehouse people affected by improvement schemes. However, due to the serious shortage of housing, the SIT began building houses and flats for lower-income families. An amendment to the Singapore Improvement Ordinance in 1930 allowed the SIT, with the approval of the Governor, to erect buildings "as the Board may think fit".

Notwithstanding this, the SIT had only scratched the surface of the problems of slum dwellings and people living in squalor, and no other part of the Government was willing or able to step in. The problem was being exacerbated by continued population growth. The idea of private sector intervention failed, even though the SIT was empowered to improve land outside the town and sell it to private enterprises.

A major part of the problem was that the 1927 Ordinance did not convey effective powers upon the SIT to compulsorily purchase slum dwellings that were unfit for human habitation, which (as was the case in the UK), would have entitled the slum owners to land value only and not the value of the land and buildings. This came to a head in what is known as the "Bugis Street case". In 1933, the SIT declared 94 buildings in the Bugis Street area (246–282 Victoria Street, 193–217 and 194–216 Rochor Road, 519–559 North Bridge Road, 15–31 and 16–32 Bugis Street, as well as 56–72 and 57–71 Malabar Street) unfit for human habitation, but three building owners contested this.

The building evaluation that the SIT utilised to assess buildings for fitness for human habitation was based upon a set of criteria around the adequacy of the internal light and air (or ventilation) and on satisfactory water supply, sewage, drainage, refuse and nightsoil removal arrangements. Advice and a report was sought and obtained from Dr. Dawson, the acting Municipal Health Officer. Dr. Dawson's report also made it clear that the reasons for unfitness could not be remedied.

The Chief Justice granted writs of prohibition against the SIT on the grounds that the owners were not informed of the basis on which their properties were deemed insanitary and that the SIT had made the declarations based on extraneous factors and as such, the owners (through their legal adviser) claimed that the SIT Board, in making its decision, was acting *ultra vires* and not within the powers bestowed upon it under the 1927 Ordinance. They also claimed that the SIT Board had acted in a quasi-judicial capacity, which it was not allowed to do. The owners' legal team argued that it was hard to believe that (apart from dilapidations or some kindred cause) a building that had satisfied the Municipality (in terms of building construction approvals) 50 years ago could now be unfit for human habitation. Among the contentions was that the houses all had an internal drain (with a cover) running through the house, taking the kitchen and bathroom sewage away, which in the opinion of the owners, was commonplace in Singapore, and also was something that could not physically be changed in any event.

The decision was appealed but was upheld by the Court of Appeal. The case eventually went all the way to the UK Privy Council, which ruled in 1937 in favour of the three slum-dwelling owners on the grounds that the UK Ministry of Health's standard for assessing unfitness for human habitation, which had been adopted by the SIT in the absence of a set of criteria in the 1927 Ordinance, was not applicable to Singapore. As a result, the SIT did not have the legal right to declare the buildings unsanitary.

This was a major blow to the SIT, and the issues were debated at length in the Legislative Council.

However, in 1935, even prior to the "Bugis Street case", a decision had been made by the SIT to change the strategy of demolishing unfit slum dwellings to it providing new housing to alleviate the problem.

Although the first houses built by the SIT were at Lorong Limau in 1932 (the SIT had built 558 artisan quarters by 1940), the next major scheme, at Tiong Bahru, is said to be Singapore's first public housing estate. A total of 784 flats, 54 tenements and 33 shops were built there from 1936 to 1941. The SIT also constructed "artisan quarters and tenements" in Balestier and other congested parts of the city. By the time of the Japanese Occupation during World War II, the trust had built 2,049 houses and 54 shops.

FIGURE 6.1. Langdon Williams. This photograph was reproduced from the *Singapore Free Press and Mercantile Advertiser*, 27 October 1941.

The person who was responsible for delivering these successful SIT housing projects during the late 1930s was **Leonard Langdon Williams (1893–1941)** (see Figure 6.1), a Chartered Surveyor from Wales. Due to increasing public and political pressure to do something to alleviate what was becoming a major issue for Singapore, Williams was despatched to the UK at the end of 1935 and stayed until February 1936, arriving back in Singapore on the P&O *Corfu* after a period of home leave.

During this time, he visited Cardiff, Leeds, Liverpool, Birmingham, Manchester and Edinburgh to gather information on the best practices in town planning, slum clearance and housing for those on low or no income, and he also attended the International Town Planning Congress. His time in the UK resulted in his comprehensive report to the SIT, which was very timely, given that in his absence, the Singapore Ratepayers Association was again putting pressure on the SIT to do something to resolve this major problem.

Williams' report argued that affordable housing for working-class people could be built only by the authorities with state subsidies and not by private enterprises.

Background on Williams

Williams was born in the Welsh town of Newport (then in the county of Monmouthshire) in 1893. He was the son of William Henry Williams — a commercial traveller — and Flora Augusta Williams. He studied at Newport High School and lived with his parents and siblings (Dorothy Langdon Williams and Henry Clifford Williams) at The Bryn, No. 10 Gold Tops, Newport, a substantial house that was later converted to commercial offices. It is still standing today.

He served as a Captain in the 2nd Monmouthshire Regiment in World War I. Nothing is known of his professional training but he had reached Singapore by 1922 because there are several newspaper articles from that year confirming him being the Municipal building surveyor, as well as stories of him partaking in golf competitions in Singapore. He was also a member of the Singapore Society of Architects and a Council member (alongside Denis Santry of Swan & Maclaren) by 1924. Indeed, the *Journal of the Singapore Society of Architects* (April 1924) included a cartoon figure of him, which attests to his self-deprecating sense of humour.

In July 1923, Williams married Marion Galloway Black at the Singapore Presbyterian Church. The best man was Frank Brewer, FRIBA, a renowned architect of Swan & Maclaren who later designed the Cathay as well as many other buildings in Singapore. Langdon and Marion Williams were very well-known in Singapore — their golfing and hockey exploits were covered in the Singapore press throughout the 1930s right up until his untimely death. Even his holidays were a subject of media interest; for example, when he and his wife went to Europe on a "flying holiday" in August 1938, a photograph of him at the airport leaving for his holiday appeared in the *Morning Tribune*! The *Singapore Free Press and Mercantile Advertiser* reported that if they had gone by sea, they would have missed the British summer, but they chose to fly back instead, against tradition, and this would given them an extra 3 to 4 weeks of extra summertime. Their devotion to golf was such that they lived in a purpose-built "spacious and airy" house off Singapore Golf Course Road.

In 1934, he was elected as a Colonial Fellow of the Institution of Surveyors.

During his time as Municipal Building Surveyor, his 1928 report to the Municipality mentioned that the increased building activities and development in Singapore were creating problems; for example, building owners were increasingly employing architects to prepare plans, but when it came to the actual development, they relied on the Municipality to do the inspection to save on the architect's fee. The problem with that was that the Building Inspector would visit each job once every four days on average, which was an inadequate amount of time to properly check the brick sizes and proper use of timber, sand and cement. Williams mentioned a case of where 20-feet *bakau* (timber from mangrove trees) piles were needed, and whilst the Building Inspector saw 20-feet piles being used on his visits, cracks appeared later. What this meant was that these 20-feet piles were used "only here and there", i.e., shorter piles were used when the Building Inspector was not around to intervene. He made the point that it was penny wise and pound foolish to do this, as the building would have to be demolished if the problem was severe. His main concern was that this was a matter affecting public safety. Indeed, he made the same point in 1929 when he used the occasion of the Municipal Commissioners' annual report to plead with building owners to use skilled supervision during the erection of reinforced concrete buildings. During that year, there were 751 prosecutions for various safety related offences.

In August 1926, as the Municipal Building Surveyor, he introduced new Building Bye-Laws that increased the minimum width of a house from 12 to 16 feet, with the provision of more kitchen, toilet and bathroom facilities on a sliding scale, depending on the number of rooms. The Bye-Laws also provided separate access to toilet and bathroom facilities so that night soil, when being removed, was not taken through the living areas. His 1936 annual report stated that though this was not popular with property owners, this resulted in healthier housing conditions.

In January 1931, he was appointed to the role of Manager of the SIT — this was on the back of a decision to terminate the contract of then-Manager Walter H. Collyer whilst he was on home leave in Australia.

The 1936 Williams Report

Williams' landmark 1936 report was in line with public opinion, as evidenced in a number of newspaper publications from that time. Furthermore, the Straits Governor, Shenton Thomas, declared that the SIT should put up a 10-year housing programme that would take "concrete shape in much larger schemes that have been attempted hitherto". The presentation of this report to the SIT was captured by *The Straits Times* on 29 April 1936. It was a turning point for the country, although sadly, Williams was unable to fully implement the report to the extent that he wanted to. His untimely death put paid to any chance of driving it through after World War II ended.

The key points of his report (see below) were eagerly awaited in the light of the Bugis Street litigation, and the Singapore media said that Williams' report, if accepted, was likely to prove of great importance in influencing future policy in Singapore:

- Owners of slum property should not be compensated when their houses are pulled down due to their insanitary condition. Williams stated: "The butcher is not compensated if his meat is condemned. Bad food is dangerous to the health of the public. So is an insanitary house. So the law says: 'no compensation.'"
- Under Clearance Orders "at Home", where conditions are so bad that the only possible remedy is wholesale clearance, the property owners will get compensation for only the land, and not the property.
- It urged the importance of rehousing schemes, similar to those for working class people under Government subsidy in Great Britain, and declared that Tiong Bahru was an ideal location for that policy.
- Rather than express his own views in the report, which might be regarded as being biased, he decided to give the city administrators in Great Britain during his visit a questionnaire, which would inform his report on what worked effectively in those cities.
- He compared the Singapore position to the one "at Home", with reference to a British Ministry of Health circular that describes slums in exactly the same way as in Singapore: "The slum is in the main a relic of the time when general law and bye-laws did not prevent the establishment of housing conditions, which are intolerable according to present standards".
- Based on the experience "at Home", the problem of providing new houses for the working classes can only be tackled by the authorities — the number of houses required could not be found by private building.
- He added that he did not believe the problem should be tackled by building high tenements. He instead proposed that three-storeyed houses should be built (starting in Tiong Bahru) close to the Clearance Areas, and that these houses should be built before the slums were demolished.

- The plans for Open Spaces in Tiong Bahru (and those already in Balestier) were comparable to those in Manchester, Leeds, etc., and were to be encouraged.
- On the "absurd" cry from private property owners that the SIT would be competing with the private sector, he stated that there was ample room for both, and like "at Home", the public sector (affordable) housing developments were not taking "bread and butter" from the private sector developers. No doubt with the thought of the Bugis Street houses at the forefront of his mind, he stated that private property owners were mostly absentee landlords who cared nothing about sanitation or affordability.
- The big concern was affordability. Williams said: "The casual labourer who cannot afford to pay an economic rent. This results in his crowding into insanitary illegal cubicles. No one worries about him". He added that it was a great problem that had to be resolved if they were to remove this blot on Singapore's fair name.
- Finally, he stated that English Law principles should apply, but the local Ordinance needed to be changed on the same lines as in Great Britain to ensure that these principles could be enforced in Singapore too.

Williams' report largely became policy in 1938 through the Weisberg Building Policy Committee, which was chaired by Hyman Weisberg (1890–1976), who was the Colonial Treasurer and Financial Secretary to the Straits Settlements Government. The Committee was established to look at the problem faced by the large number of 99-year ground leases in Singapore coming to an end. The Committee's belief was that Singapore's housing problem was not "primarily a problem of destroying bad houses, but rather a problem of building good ones". The Committee recognised that housing for Singapore's poorest people had been let to the private sector, and if the situation were to continue, there would never be an acceptable solution, i.e., the poor would continue to live in overcrowded slums. The problem was not so much the ability of the family to pay rent but their need for sufficient, habitable space. Weisberg had been born and raised in the New Road, Whitechapel area of London, of Jewish parents who migrated to London from Odessa. In his early life, he must have lived in severely unhygienic vermin-infested conditions, with a dire lack of clean drinking water. This, undoubtedly, must have been on his mind when his Committee lent its support to Williams' proposals.

Going back to the Tiong Bahru housing (see Figure 6.2), Williams, who was the architect, described them as providing "modern comfort" because each unit had a balcony, there was plenty of light provided by light wells in the middle of the blocks, and there was a playground for children. Williams also developed (and perhaps perfected) the Singapore shophouse design. However, at a rental fee of $25 per month, only those earning a clerical income could afford them.

FIGURE 6.2. (a) The famous "Horse-Shoe Block" creation of Langdon Williams — Block 78 Guan Chuan Street/Moh Guan Terrace, Tiong Bahru. It was built between 1939 and 1940. (b) Block 55 at the junction of Tiong Bahru Road and Tiong Poh Road. The oldest SIT Block in Tiong Bahru. It was built in 1936. (c) Side view of 79 Chay Yan Street — situated at the opposite end of the "'Horse-Shoe Block".

Living in Singapore

Langdon Williams built a family holiday house on Pulau Ubin in the English Tudor style. They had two motor launches and a jetty at the end of the garden. That house is still there today (see Figure 6.3), managed by the National Parks Board (NParks) as a visitor centre, and is called **House No. 1, Chek Jawa**, with magnificent views of Singapore and its surrounding islands. One fascinating feature of Williams' house is its fireplace and chimney, a rare sight in Singapore. Even more unusual — when it was built, the fireplace and chimney were in full working order. Indeed, in April 1935, the *Sunday Tribune* shared with its readers a photograph of the working fireplace inside the main living room.

FIGURE 6.3. House No. 1, Chek Jawa, Pulau Ubin. Holiday retreat of Langdon Williams and his family in the 1930s.

Back on the mainland, Langdon and Marion lived in a large residence off the Singapore Golf Course Road (now Sime Road), which was handy for them, given their predilection for golfing. Their residence was described in *The Straits Times* in 1938 as one of the new trend of "modern flat-roofed reinforced concrete Singapore homes" that were slowly but surely starting to replace the classical or Black and White houses that had been favoured over the previous 25–30 years by those in Singapore with the financial ability to build or purchase their own home. The Williams' house was "spacious and airy" as a result of the roof projections that allowed cool air into the lower part of the house, and as with other such new houses, the ability to air-condition them was made easier with less "open space" than in the classical houses, albeit the affordability of air-conditioning was still not quite there yet.

By 1940, news had spread of Williams' work with the SIT, resulting in the Sarawak Government paying the SIT to bring Williams to Sarawak to help in the replanning of Kuching. At the same time, he had also submitted plans for the replanning and development of Johor Bahru.

Sadly, at the height of his career, tragedy struck. On 25 October 1941, he was found dead in a stream off 17th mile Kota Tinggi Road in Johor, some 40 km north-east of his home in Singapore. There was a Coroner's Inquest, and his death was believed to have been caused by drowning. He was found face down and fully clothed. Newspaper reports at the time of his death stated he was looked upon as a man who gave many better living conditions, and that Singapore families had benefitted considerably from the policies he enacted, such as slum clearance, the Tiong Bahru housing, and other housing schemes in Chinatown, which at the time were considered ground-breaking.

His funeral was on 26 October 1941, and he was interred in Bidadari Cemetery. Among the mourners were Frank Brewer (his best man at his wedding in 1923) and Charles Frederick Snell. As with Regent Bidwell, his grave can no longer be visited, as Bidadari exhumations were completed by the end of 2006 for the redevelopment of the land, and the cremated remains from the exhumed were placed to rest at the Choa Chu Kang Columbarium. A large number of funeral wreaths were sent to his funeral, including the ones from Sir David and Lady Galloway of Johor (they were Marion's uncle and aunt) and the Sultan of Johor.

Jim Fraser (see Figure 14.11) took over from Williams after his sudden, unexpected death. His first task was to review all the work done by the SIT over the previous 20 years. This task was unfortunately interrupted by World War II and the Japanese occupation of Singapore, when many SIT staff were interned by the Japanese, with others employed in administrative tasks. Fraser successfully took up the task after World War II and continued in the role until his retirement in the late 1950s.

Further tragedy ensued with the Japanese invasion. Marion Williams was captured by the Japanese and taken to the Palembang Women's camp in Muntok, Sumatra, where she died on 7 February 1945.

Conclusion on Williams

It is clear from his landmark report that Langdon Williams was a catalyst for change, and in pressing ahead with the Tiong Bahru, Chinatown and Balestier affordable housing schemes, he has left a lasting, positive mark on the Singapore built environment. His "public interest" philosophy was also very true to the Institution of Surveyors' Charter, and it was the driving force behind everything he did.

Sadly, he never lived up to his full potential due to his untimely death, and in his lifetime, he was also shackled (and very frustrated) by the Colonial Government's Ordinance and the frustrating outcome of the Bugis case that left the SIT with limited powers. This was despite the Weisberg Committee supporting Williams' report and advocating a more active government approach to Singapore's housing problems. The response from the Colonial Government was disappointingly slow, albeit there was some success in Tiong Bahru.

Another part of the problem was the funding required to fulfil these aspirations, given that it was funded by the levying of an annual "improvement rate" of 2% on the value of all land and buildings within the Singapore Municipality. Additional British Government funding could be sought to deliver on the recommendations in the Williams report, but securing this was not easy as burdensome financial conditions were attached.

It would only be from the late 1950s onwards that the SIT and, later, the Housing & Development Board (HDB) would press on and get things done as Singapore gradually broke free from the shackles of the Colonial Government's administration.

The National Archives of Singapore's Oral History interviews include a series of interviews with Soh Wah Seng (a draughtsman in the SIT survey office) and Winston Arnold Reginald Matthews, a Eurasian assessor at the SIT). Both of them worked under Williams, albeit they did not have much direct involvement with him. Their interviews include assertions about Williams. Whilst only hearsay and unproven, each one is explored below:

Assertion	Author's Response
"When Walter H. Collyer retired in 1931, Jim Fraser was overlooked for the role in favour of Williams — this was because Williams' uncle was Sir David Galloway and strings were pulled to get him the job."	Walter Collyer was close to Jim Fraser (he gave away Alice Ross when she married Fraser in Singapore in 1929), so it is likely that he would have looked favourably upon Fraser as his successor. However, newspaper reports of the time state that Collyer did not retire. Instead, his contract had been terminated while he was on leave in Australia, and he was replaced by Williams. So it is quite likely that Collyer had no say in the appointment of his successor. However, it is possible that Sir David Galloway might have pulled a few strings for his niece's husband (he was not Williams' uncle). This, in turn, would have created some friction between him and Fraser, especially as Fraser was apparently better qualified at that time — and in those days, seniority (age and years' service) and qualifications were key factors in promotion decisions.
"Collyer was popular, down to earth and openly engaged with all his employees, but Williams was aloof and autocratic and thought himself better than his staff — particularly locals — because of his long list of qualifications (FRIBA, FSI, MTPI, MI Struct. E.)."	It is likely that Williams adopted the autocratic and distant style of senior expatriate managers in British Government Departments of that era, both in the UK and in Colonial Government offices. If he did so, then he would have been like this with everyone of lower rank and seniority to him (not just locals but towards expatriates too). This style of management prevailed in UK Government Departments and Colonial Government offices of this era.
"Williams committed suicide because he seemed to succumb to an association with Chinese contractors and involved himself in some sort of bribery or corruption, which was about to be investigated."	The newspapers of the time state that it was death by drowning. He was found face down, fully clothed, some 40 km from home, but because the results of the Coroner's Inquest are not known, we are unable to verify if drowning was deemed to be the cause of death or if the Coroner suspected suicide, or indeed foul play. Furthermore, there is no public record of an investigation subsequent to his death into the affairs of the SIT to indicate any form of corruption or collusion involving Williams.

Williams and Empress Place

One final thought on Langdon Williams: If his ideas for the Government buildings on Empress Place had been accepted and come to fruition, the Singapore river would now look very different. On 8 November 1930, *The Straits Times* reported that Williams (representing the SIT) was part of a Committee along with the Colonial Secretary (Chair), the Colonial Engineer, the President of the Municipal Commissioners, a Singapore Chamber of Commerce representative, the Chair of the Straits Settlement (Singapore) Association, and the Municipal Architect Mr. Gordon. Their plans were to redevelop the entire Empress Place area as well as the Singapore Cricket Club, comprising what is today the Asian Civilisations Museum, the Victoria Concert Hall and Museum, along with the Arts House and Parliament House Block C. The whole project was subject to the approval of the Governor and was essentially a "slump measure", i.e., Government-led construction work to generate employment, but it was stated that even if the economy improved (rubber selling at two shillings a pound), it was still advisable to "scrap the lot" and build something really worthy of the town.

Looking back, it seems bizarre and incredulous that Williams' idea of demolishing these iconic buildings and replacing them with a Civic Centre would have even been given a second thought, but because of the economic slump such ideas were actually contemplated.

It was noted by the Committee that an American Judge visiting Singapore had noted that the island would become the biggest and most important centre of commerce in the East, bigger than Shanghai and Calcutta, and that something on a similar scale to the Municipal Buildings, which had been described in *The Straits Times* as a "fine example of dignified construction that would be a credit to any town", would be appropriate, as there was "plenty of ground to display them" and make the town conspicuously beautiful.

The newspaper described the Victoria Theatre and Memorial Hall (now the Victoria Concert Hall and Museum) as "highly expensive god-children" and suggested that the days of the Theatre had gone, with the advent of "talking picture palaces" and a decline in interest in the amateur productions that were prevalent at the Theatre at that time. Also, usage of the Memorial Hall had declined too.

The Committee's proposal was described as a boon with plenty of money to support it. However, by 1935, it was revealed that this "Civic Centre" idea (comprising a Supreme Court, Town Hall and other Government offices) did not get the support that it needed in 1930–31 when it was first proposed. Instead, the Supreme Court project went ahead on the nearby site of the Hotel L'Europe. *The Straits Times* explained that despite what the Committee had recommended, the feeling was that the Victoria Theatre and Memorial Hall buildings were extremely sturdy, and the cost of buying land and rebuilding them elsewhere in Singapore was too prohibitive. The newspaper hinted that the project might be revisited, but World War II brought such thoughts to an end.

Chapter 7

1930–39: Quantity Surveyors Emerge in Singapore

Aside from the influence and work of Langdon Williams, the number of Chartered Surveyors, particularly Quantity Surveyors, in Singapore increased from the late 1920s through to the outbreak of World War II. Prior to that time, the only Chartered Surveyors in Singapore were Land Surveyors, General Practice surveyors (property agents) or valuers (known as assessors). Quantity Surveyors had yet to make their mark.

The Institution's merger with the (UK) Quantity Surveyors Association in 1921 was the catalyst for Chartered Quantity Surveyors to flourish and deliver their expertise around the world.

Furthermore, in the 1930s, a Branch of the Surveyor's Institution was set up for Malaya (including Singapore), but more importantly, through the efforts of David Waters, local surveyors were trained up to become Chartered Surveyors.

Against a backdrop of the global depression during the 1920s up to around 1931 (and the shadow of World War II looming over the end of the decade), Singapore started to grow again as during the 1930s and was, during those years, regarded as the "Clapham Junction of the Eastern Seas".

Furthermore, Singapore's European population of 7,600 in 1935 included 4,500 connected to the Army, Navy and Air Force — this and the "Clapham Junction" factor were the drivers for the increasing number of Chartered Surveyors in Singapore.

- **The Naval Base at Sembawang:** The 1930s saw a significant amount of work at the Naval Base, although work had already started during the 1920s after land was acquired and provided for the new Base. Construction of the naval base was eventually completed in 1938, but because of scope and market changes, the final construction cost was seven times the original estimate. In 1938, the Governor Sir Shenton Thomas opened what was then the largest dry dock in the world at the Naval Base (King George VI Graving Dock). This, and the RAF, Army and other military projects, led to Singapore being called "the key to the Pacific" and "the Gibraltar of the Orient". The Naval Base did not host any permanent naval fleet as the strategy was to dispatch a fleet to the base when a (Japanese) threat was imminent, but surveyors were very important to the success of all the major construction projects at Sembawang Naval Base and the Royal Naval Air Station there. The Base was used for repairing and refitting the Royal Navy Fleet Air Arm's aircraft carriers. Surveyors were also needed for the construction of the RAF station at Sembawang during the same period.
- **RAF Air Base, Seletar, and RAF Tengah:** Another project linked to a potential (Japanese) invasion started in the 1920s under the designer and planner C.E. Wood. It became operational in 1928 under Group-Captain H.M. Cave-Brown-Cave, and in 1930, it officially became an RAF station. It was an important landing site as it was strategically located along the England-Australia air route. Commercial traffic also utilised its facilities until Kallang was built. The various construction projects there in the 1920s and 1930s included many Black and White residential bungalows for RAF personnel and their families. Most of those bungalows still exist today. RAF Tengah followed in the mid to late 1930s.
- **Military installations at Fort Canning (Military HQ), Changi and Tanglin Barracks:** These locations were already in military use but were developed extensively in the 1930s and enhanced with the construction of **Gillman Barracks** in 1936.

- **Government building projects:** Aside from the SIT projects mentioned in the previous Chapter, there were a number of other major projects that were completed, such as the land reclamation, the Tidal Basin and seawall construction works at Telok Ayer (Basin), Beach Road and Kallang (Basin). New bridge works were also carried out in these locations. The Supreme Court building project also happened.
- **Consumer goods and leisure activity for the wealth:** The 1930s saw a great deal of Hotel development as well as leisure-related development, such as the Cathay Building (including the Cinema), to satisfy the demands of the wealthy locals and expatriates.
- **Protectionism of colonial wealth:** C.M. Turnbull's *A History of Singapore 1819–1988* describes the late 1930s as a time of peace, increasing comfort and leisure for the middle and upper stratas of Singapore society. It also described the levels of complacency that had started to creep into British colonial rule and that the Colony was, by and large, run by and for those who had won through power and wealth. The ensuing construction of commercial properties for colonial employers and luxury residential properties for the wealthy colonial (and, to some extent, wealthy locals) created much work for Chartered Surveyors. Unfortunately, the poor of Singapore, apart from the benefit of the public works of the 1930s and some of Langdon William's projects, saw no improvement to their lives. Indeed, for many, their quality of life had deteriorated whilst the wealthy Colonials (and wealthy locals) continued to prosper.

Against the backdrop of these key factors impacting upon the built environment in Singapore during these years, the need for surveyors grew, and this led to the arrival of two people whose positive legacy to the profession in Singapore would last for years to come. Their names were David Waters and Eric Watson. Their practice (Waters & Watson) was the first Quantity Surveying practice in South-East Asia. Later Chapters will explain how this practice evolved over the years until the merger with Arcadis NV in 2012.

Waters and Watson both arrived in Singapore in 1931. Waters came here to work for Brossard Mopin & Co., a contracting company based in Raffles Place — he was then still a Probationer of the Institution. Watson qualified as a Chartered Surveyor in 1931, at which point he came to Singapore to work for Fogden, Brisbane & Co. Ltd. They met up, and within a short space of time, Waters asked Watson if he would be interested in joining him in his own practice. So it was that on 20 February 1933, Waters and Watson set up their own practice in Singapore, based initially at No. 98 Robinson Road and later at Battery Road. The firm was offering Quantity Surveying services as well as insurance loss assessors, appraisers and arbitrators. Their influence, even to today, is without question, especially given the number of prominent surveyors who joined the practice (and its successors) through the next few decades.

Their families were friends too. Indeed, in September 1936, when Ian Robin Douglas Watson, son of Eric Watson, was christened at St. Andrew's Cathedral, David Waters was present.

David Waters was born in Reigate, England — he was the son of Howard George Waters and Minnie Louise Waters and the husband of Marion Winifred Waters. Eric Watson was born in Sunninghill (and raised in nearby Windsor Forest) in England in 1905 and was the son of Robert and Lillian Alice Watson. He married Kathleen Douglas Fox Clinton at St. Andrew's Cathedral in May 1935.

Horace W. Langdon & Every acquired Waters & Watson in 1936, and after a short break back home in the UK, David Waters went to Hong Kong in 1937 to help set it up. David Waters was a strong presence in Singapore; for example, on 11 November 1937, he spoke at the Singapore Rotary Club about the Quantity Surveying profession in Singapore. In a talk entitled "The Quantity Surveyor and his Work", he made a specific point of saying that Quantity Surveying was the one profession where there were more jobs than men to fill them, and that there were no women, much to his relief. Indeed, he stated that this was one of the few professions that had not been "invaded by women".

However, despite his misogynistic views, he was undoubtedly a passionate advocate of the profession and took time to describe in detail the job of a Quantity Surveyor ("the technical accountant to the building industry" and "expert in the measurement and valuation of building work") to the Club members. He also talked of how old a profession it was, given that from ancient Egypt, there is an illustration of the "Chief Measurer of the Royal Works", and Waters proudly proclaimed that one could safely assume that Quantity Surveyors played their part in the construction of the Pyramids!

He described his own job as one of preparing Bills of Quantities and the settlement of extras and omissions — a job that eliminated a lot of waste in the construction industry through the use of science. He also explained that in Malaya (and Singapore), the most prevalent form of contract form (as in Great Britain) used to be the Lump Sum contract based on drawings and specification, but this had largely been replaced by the Lump Sum contract based on a Bill of Quantities. It was also prevalent then in the Armed Forces and the Colonial Governments around the world. He also described how important a Quantity Surveyor was at all stages of construction: its early stages (building economist), the tender stage (contract draughting), the construction stage (custodian of the money — showing how it is spent), and the end of the contract (quasi-arbitrator between the building owner and the architect). He also told the Rotary Club members that to become a Quantity Surveyor involved passing a number of difficult professional examinations through the Chartered Surveyors Institution, and the profession itself had a strict code of professional conduct.

In his talk in November 1937, Waters confirmed though there were 20 Quantity Surveyors in Malaya, only 12 were Chartered Quantity Surveyors. Including the support staff, there were a total of 60 people working in the Quantity Surveying profession in Malaya at that time.

There will be more about David Waters and Eric Watson later in this Chapter, as well as in Chapters 8 and 9.

During the 1930s, there were no more than 12–14 Chartered Surveyors working in Singapore at any given point in time and of that number, some were in Singapore for only a year or two. Some worked for the Government as assessors (valuers) or Quantity Surveyors, some worked as Quantity Surveyors in private practice, and others came to work here as surveyors for the RAF, Royal Engineers or at H.M. Naval Base. Many only stayed for one or two years, either due to their inability to settle in Singapore, ill health or the increasing need for their services "back home" or other places where British military bases were being expanded.

However, some Chartered Surveyors stayed in Singapore for a longer period during the 1930s — they were:

Chartered Surveyors at the (Air Ministry) Directorate of Works and Buildings, RAF Far East HQ, Seletar

Edgar Hunt

Edgar Victor Hunt (see Figure 7.1) was a member of the Institution's inaugural Malaya Branch Committee in 1936 and a Professional Associate since 1925. He lived at the Grove, 103 Meyer Road, Tanjong Katong, and later in Holland Road. Hunt, as the RAF Command Surveyor, was heavily involved in the construction of RAF Tengah from 1935 onwards. However, on 11 May 1937, his wife Gladys E. Hunt died, and her funeral at Bidadari Cemetery was attended by many from RAF Seletar as well as by representatives of the Surveyors' Institution Malaya Branch. This tragic event no doubt precipitated his return to England, given that he had three young children, and so it was that he left Singapore before RAF Tengah was opened in 1939.

In August 1935, Hunt was present at the visit to RAF Tengah by the Malayan Association of the Civil Engineers and of the Engineering Association of Malaya. The visit was hosted by RAF officials, including Hunt, in the absence of C.E.O. Wood, Chief Engineer for the RAF Far East, who had overall responsibility for all construction projects across the entire region. (Side note: Jalan Kayu in Seletar was named in Wood's honour

CIVIL ENGINEERS VISIT TENGAH AERODROME.

Excavation Almost Entirely By Hand Labour.

(By Our Services Correspondent.)

Members of the Malayan Association of the Institution of Civil Engineers and of the Engineering Association of Malaya last week inspected a new aerodrome being formed largely by hand. This is the island's second Service aerodrome and is off Choa Chu Kang and Lim Chu Kang roads. Started in November 1933, it is now three-parts finished.

Unlike the R.A.F. Station, Seletar, the R.A.F. Station, Tengah, as it is now called, is to be a base for land-planes only, there being no access to the sea. The coast is four miles away.

The area which is being conditioned as a landing ground approximates to 192 acres, of which 85 acres comprise the initial "T" shaped landing area.

Shallow Cuts.

The levels over the site varied considerably when work was started, the difference between the highest and the lowest being 55 feet. Features of the contours were three hills, three valleys and two long ridges.

Even so the total excavation possible on the site was insufficient so [illegible] areas were constituted.

subsoil drainage, return of the top-soil, turfing, etc., is all the direct concern of the labour force, which today numbers 1,900 men excluding supervisors.

Clearing the site of rubber and coconut trees, pineapples and blukah presented no difficulties.

Thus, strange as it seems in this mechanised age and in direct contrast to other Service construction work on the island, the new aerodrome is being fashioned, up to the stage where building begins, almost entirely by hand. The landing ground on the R.A.F. Station, Seletar, was also formed by hand, using trucks and rails.

Modern Engineering Problem.

As the section officer, Mr. C. Chambers, explained to the party, aerodrome making is an engineering problem of comparatively recent times and although there is nothing spectacular to be seen on his site there is much of absorbing interest.

Tengah Station, although its creation has not involved reclaiming large tracts of mangrove swamp common to Service construction work nearer the coast, is an engineering feat for which those responsible may well be proud.

With the exception of clearing, all

The visitors, in company with various R.A.F. works officials. Mr. C. Chambers, Section Officer, is in the front row, second from the left.
—*Straits Times Photograph.*

(Altogether some 400 acres were acquired for the new air station).

The mechanical excavating plant on the site consists of two 3/8 cubic yard Ruston Bucyrus excavators. All other excavating work is carried out by chankol. The use of mechanical plant is limited because of the shallow nature of most of the cuts.

Tractors No Longer Used.

Haulage of the excavated material to the areas required to be filled was initially effected by tractors but trucks and rails are now being used in their

[illegible]ing and levelling, as well as the operations mentioned by Mr. Chambers in his preliminary talk to the party can be seen in progress.

Grass Surface.

The nature of the strata to be excavated varies considerably and blasting is resorted to at times. A number of large pig-ponds, some of them as much as six feet in depth, have had to be filled.

An important part of the preliminary work was the formation of turf nurseries.

The carpet grass (axonopus

(Continued in next column)

SINGAPORE'S NEW AERODROME.

A view of the work at the civil aerodrome, Singapore, showing, in the middle distance, some of the areas already turfed. The excavator in the foreground is one of the two engaged on the job.
—*Straits Times Photograph.*

Some of the 'scrapers' used, in the initial stages of the works programme, in conjunction with the tractors now engaged on earth-work consolidation.
—*Straits Times Photograph.*

PUBLIC PARK IN MUAR.

To Mark Anniversary Of Sultan's Reign.

(From A Special Correspondent.)
Muar, Aug. 19.

The Sultan of Johore has consented to the proposal of his third son, Tunku Ahmad, State Commissioner, Muar, for the establishment of a public park in Muar in order to commemorate the 40th anniversary of the Sultan's reign.

His Highness has requested that the funds to provide the park be borne by public subscription of an entirely voluntary nature.

He has granted to Muar 20 acres of land for the setting of the park.

The State Commissioner, Muar, is calling a public meeting at Muar on Thursday to launch the scheme.

MRS. F. A. FULLER.

Large Attendance At Funeral.

The funeral of Mrs. Fanny Annette Fuller, wife of a senior European warder in the Singapore Prison Department, who died on Sunday from smallpox, took place at Bidadari yesterday.

The funeral was attended by the husband and many friends, including friends from the Prison Department and many others.

The Rev. A. V. Wardle conducted the service.

Wreaths were received from the following:—Her loving Husband, Her loving children Pearl and Freddy, Captain and Mrs. O. L. Hancock, the staff, Singapore Criminal Prison, the Malacca Prison staff, the Johore Prison staff, Mr. and Mrs. W. E. Jenkins, Mr. and Mrs. W. Lindsay, Mr. and Mrs. H. Mansfield, Muriel and Pat, Mr. and

(a)

FIGURE 7.1. (a) An article in *The Straits Times* on 20 August 1935 covering the Malayan Association of the Civil Engineers and of the Engineering Association of Malaya's visit to Tengah Aerodrome, with a photograph featuring Edgar Hunt (centre).

on his departure from Singapore in 1937 — kayu being the Malay word for "wood".) See Figure 7.1.

FIGURE 7.1. (b) RAF Tengah, photographed in the immediate post-war years when RAF 81 Squadron was based there. It is now known as RSAF Tengah Air Base.

The Association's visit was to specifically see the new Aerodrome, for which construction had started in November 1933 and was three parts finished. However, the visitors were also able to see progress with the construction of the landing area. Tengah was created for land planes only due to there being no direct access to the sea, which was four miles away. The landing ground at Tengah comprised 192 acres, of which 85 acres formed the T-shaped landing area. A total of 1,900 labourers (excluding supervisors) were working on the project, which was very labour-intensive, requiring the use of hand tools, perhaps because the team only had the use of two Ruston Bucyrus excavators. The construction of an aerodrome was not too complex from an engineering perspective, but the construction and design of aerodromes was a relatively new challenge, so the visitors were no doubt fascinated to learn more.

After site levelling works for the landing area were completed, carpet grass *(axonopus compressus)* would be used across the topsoil, then consolidated by heavy tractors. The visitors were told that there would be an appreciable period before the filled areas would reach the right formation level.

Chartered Surveyors working for the RAF Directorate of Works and Buildings during this period would have been involved in a wide variety of Quantity Surveying tasks that were key to the construction of all the new buildings and infrastructure associated with the development of RAF Seletar and RAF Tengah. In the case of RAF Seletar alone, the lack of accommodation in 1937 to support additional Squadrons meant that £1,500,000 was allocated over the next three years to build hangars, barracks and workshops. Indeed, new arrivals would have to sleep under canvas sheets until the construction was completed.

The British Government's budget for the military expansion in Singapore was substantial but not generous, so the role of the Quantity Surveyor would have been critical in ensuring construction costs were managed effectively.

Other Chartered Surveyors here in the mid-1930s with the RAF Works and Buildings Department were **Norman Hopkins** (see later in this Chapter), **Herbert Johnson, Frederick Honey** and **Ronald Inglis.**

- **Herbert Thomas Johnson** was a Fellow of the Institution, having first qualified in 1910. He served in World War I when he was awarded the Military Cross. He was in Singapore with the RAF Works and Buildings Department in the early 1930s but little else is known of his career.
- **Ronald Alexander Inglis (1908–96):** He was only in Singapore for a year or two but is mentioned here because he was an expert in railway surveying and engineering. Indeed, later on in his career, he wrote several books relating to railway surveying and engineering, including *An Introduction to Railway Engineering: A Short Textbook of the Theory and Practice of Railway Surveying, Construction and Signalling for the Use of Students and Civil Engineers*, which was published in 1953.

Frederick Honey

Frederick Charles Honey (1913–93) was a newly qualified Professional Associate (PASI) of the Institution when he arrived in Singapore in 1935 to work for the RAF Works and Buildings Department at RAF Seletar. He returned to Runcorn, England, with the Works and Buildings Department in 1939. At Seletar, he would no doubt have been involved in Quantity

Surveying tasks pertaining to its expansion, with the construction of Black and White houses for senior RAF personnel as well additional road infrastructure.

On 6 April 1936, Honey was involved in a sailing incident several miles offshore at Tanjong Rhu that must have remained in his memory for the rest of his life. Honey was with W.J. Hennessey and Dargue Robinson from RAF Seletar while they sailing out from Tanjong Rhu when a sudden squall sprang up as they were sailing through the Outer Roads (that lay beyond the Breakwater) four miles out from Singapore General Post Office [now the Fullerton Hotel). Their intention was to reach Seletar that day, but the seas were extremely rough, and worse still, something struck their yacht, and it capsized. Whilst Hennessey supported Robinson (who could not swim) and his dog Ginger, Honey swam to get help. Several boats and other vessels passed by but did not hear their cries for help. Given the rough seas, it was impossible for Honey to swim ashore, but luckily, he reached the *SS Hong Siang*, where a launch (*Juno*) was used to help him and crew members to search for the Hennessey, Robinson and Ginger. After a long search in the rough seas, they spotted them just in time as they were all exhausted, and it would not have been long before they would have perished. They were brought to Clifford Pier, and the men were taken to hospital, where they fully recovered. The overturned yacht was recovered. Apart from a broken mast, it was undamaged.

Honey was later awarded the Bronze Medal and Certificate of the Royal Humane Society, presented to him a year later by Air Commodore A.W. Tedder.

Chartered Surveyors at the Royal Engineers

Several Chartered Surveyors were in Singapore with the Royal Engineers for short periods of one or two years in the 1930s. One who stayed for a longer period in the 1930s was **Leslie John Barrington Tovey (1894–1947),**

MBE. He was a member of the Institution's inaugural Malaya Branch Committee in 1936. A Professional Associate of the Institution, he was a Senior Surveyor of Works in the Royal Engineers: he was a Captain by 1936 and later attained the rank of Lieutenant-Colonel. Tovey was based at Military HQ, Fort Canning, from 1935 to 1938. Prior to Singapore, he had been a Surveyor of Works, Royal Engineers, based in Cairo, Egypt, from April 1924 onwards.

In October 1935, in Singapore, H.E. Governor Sir Shenton Thomas presented Tovey with his MBE and stated that "*by his unremitting hard work in the capacity of Surveyor of Works, [Tovey] has contributed largely to the smooth and continuous progress of the work in connection with*

FIGURE 7.2. (a) Block 9 of the former Gillman Barracks and now home to the Shanghart Gallery and Vavuz Gallery. (b) Block 6 at the former Gillman Barracks, now the home of NTU CCA Singapore, Office & Research Centre and Mucciaccia Gallery.

the increase of the garrison in Singapore. It is also due to his energy and foresight that the large number of contracts involved have been carried out without a hitch. His exceptionally *valuable services in connection with this work have earned for him the mark of His Majesty's favour*". The reference to the Singapore "garrison" is confirmation that amongst Tovey's tasks were to complete the Gillman Barracks construction project and support and enhance the military facilities at Tanglin and Changi. The barracks complex (see Figure 7.2) included living accommodation as well as sporting facilities. Sir Thomas expressed hope that Tovey would long be spared to enjoy the honour conveyed upon him. Sadly, Tovey died in Cheltenham, England, in 1947 and thus only enjoyed his honour for 12 years.

Chartered Surveyors at H.M. Naval Base, Civil Engineer in Chief's Department (and Superintendent Civil Engineer's Department)

It is known that eight Chartered Surveyors worked at H.M. Naval Base at Sembawang during the 1930s.

Work on the construction of the Naval Base had started in 1928 (see Chapter 5), but the decision was made to expedite the work, which led to an increased number of H.M. Naval engineers and surveyors arriving from the UK to work in the Civil Engineer's Department to help finish the project. The Civil Engineer in Chief's Department was formed at the Admiralty in 1919 when it replaced the earlier Architectural and Engineering Works Department. It was later renamed the Navy Works Department.

The Naval Base was completed in early 1938; as a result, half of the Chartered Surveyors only stayed in Singapore for a couple of years, but the following individuals stayed for a longer period to help finish the task:

- **Edward Victor Pollington (1890–1951):** He was a member of the Institution's inaugural Malaya Branch Committee in 1936 and was a longstanding member of the Institution. He was in Singapore at H.M. Naval Base from 1935 to 1939 before returning to England.

- **John Courtenay Lang (1909–94):** He was here in the late 1930s, right up until the Japanese Occupation. He is mentioned again in Chapter 9.
- **Charles Eric Side (1906–87):** He was in Singapore in the late 1930s and returned after World War II to work at the Naval Base from 1946 to 1948.
- **Eric Cuthbertson Hill (1901–84):** More on him later in this Chapter.

Chartered Surveyors in the Education Sector

Charles David Quarmby was born in 1880, and he was a Professional Associate since 1904. He came to Singapore in 1921, and his role in Singapore was the second Commercial Master at Raffles Institution, where he started work on 12 November that year. His starting salary was $4,800 per year plus living accommodation. His job was to teach basic commercial skills to the students, to form the basis of their future technical pursuits. He later moved to Penang, where he spent most of his career, including a senior position in the Education Department as well as Principal of the Government Commercial Day School. He moved back from Penang to Singapore in the 1930s to become a School Principal. He was also an elder of Presbyterian Church.

In 1942, he was evacuated to Durban, South Africa, but did not return to Singapore. Indeed, he moved to Capetown and then eventually retired to Durban. In September 1970, aged 93, he wrote to *The Straits Times* to ask if they could invite his ex-pupils to enter into correspondence with him. They willingly published his residence address but wrongly stated that Durban was in South Australia, so one wonders how many letters he got! More information will be provided in the future book on the RICS and Chartered Surveyors in Malaysia.

Formation of Malaya Branch of the Institution in 1935

Stepping back to 1935, the growing number of Chartered Surveyors in Malaya meant that as a group, they could see that other professions (notably Architects and Engineers) were forming local Branch Committees, and so a number of these individuals decided that the time was right for them to create a local Branch of the Surveyors' Institution. The *Singapore Free Press and Mercantile Advertiser* reported on 25 January 1935 that members of the Institution residing or practising in Singapore had met in the Adelphi Hotel on 14 January 1935 and announced that they had sought sanction for the formation of a Malaya Branch of the Institution, and they were looking for members based in the Federated Malay States and Johor to cooperate too.

The Straits Times then announced on 12 February 1936 (see Figure 7.3) that the Chartered Surveyors' Institution had been exempted from registration in Singapore under the Societies Ordinance. This meant that the Institution was free to arrange formal meetings and hold events as a Branch.

Later reports state that the Branch was formed in September 1935, and the slight delay in announcing it must have been due to the delay in securing its Societies Ordinance exemption. Their inaugural dinner was due to be held that month but was postponed due to the death of King George V. The newly formed Malaya Branch of the Institution eventually held its inaugural dinner on Tuesday, 9 June 1936, at the

The Chartered Surveyors' Institution (Malaya Branch) has been exempted from registration in Singapore under the Societies Ordinance.

FIGURE 7.3. The notification in *The Straits Times*, 12 February 1936, of the granting of exemption to the Institution of Chartered Surveyors from registration under the Societies Act.

Adelphi Hotel, Singapore. The Branch Chair was Herbert Courtenay Atkin-Berry, and the Vice-Chair was Charles Frederick ("Fred") Snell. Atkin-Berry presided over the dinner and dance on the Hotel Rooftop "with the assistance of the Hotel Band and Cabaret". The 29 guests included prominent architects and engineers such as Mr. Gooch (Deputy Director of Public Works, Straits Settlements), D.C. Rae (Municipal Architect and Building Surveyor) and Frank Lundon (Swan & Maclaren), along with their wives. The Branch's inaugural Honorary Secretary was David Waters.

Prior to that, on 21 April 1936, Waters had given a lecture at a meeting of the Institution's Malaya Branch on the subject of the History and Function of Quantities. Some of what he said was repeated in his Rotary Club lecture in 1937 (this was captured earlier in this Chapter), but he went much further on this occasion, and as a result, the lecture was the subject of much debate on the day of the event and in the local Press afterwards. This lecture was the first-ever given at a meeting of the Institution in Singapore — a landmark event.

Waters talked about how the objective to be attained by a Quantity Surveyor was to define more accurately and exactly (than can be done by the general drawings and specifications) the exact amount and nature of work to be executed. As such, in his view, the Bills of Quantities should form part of the Contract, given that they represent in an exact form the representations of the architect in his general drawings and specifications. He expressed a view that Singapore was a hundred years behind other colonies, for example, Kenya, where Quantity Surveyors were subject to statutory registration. However, he did say that things were changing fast, so Singapore should catch up and, indeed, the reliance on lump sum contracts was already starting to wane, with his firm alone seeing a 300% increase in the preparation of Bills of Quantities work from 1934 to 1935.

Waters explained the functions of a Bill of Quantities as:

- Providing a uniform basis of tender to avoid the past problems of huge variance in quantities assumed by tenderers;

- Providing the building owner with a close forecast of cost before calling a tender. The estimate should be within 1% of the average tender. In his view, the practice of calling for tenders to find out whether the work could be done within a certain figure was a "waste of time"; and
- Defining clearly the obligations of the contract. In his opinion, the contract should not simply include plans, specifications and an agreement alone. He quoted RIBA as a body that supported this view.

He went on to argue that a schedule of rates supplied in connection with a lump sum contract was a farce, as the probability of additions is far greater than omissions, the lump sum system was primitive and unbusinesslike, and that the only reason why the colony had not seen more disputes was because the number of competent contractors was small and they would prefer to cover themselves in the tenders than see an arbitration or court case happen. Contractors who try to run their business on scientific lines prefer quantities.

The lecture was attended by Frank Dorrington Ward (the renowned Straits Settlements Government Architect), who stated that, in his opinion, if Quantity Surveyors were available, then it made sense to use them to have quantities prepared for all work irrespective of value and type. He also clarified what David Waters had said: a contract with quantities was still technically a lump sum contract, the difference being that the work had been gone into more closely and the general drawings and specifications specifically interpreted; if the contract was carried out without any alterations being ordered, the total shown by the addition of the money columns of the Bill of Quantities, which had been carried to the tender form and was the sum for which the contractor agreed to execute the work, would remain unaltered.

Waters asked those present what the minimum value was for which quantities should be supplied, quoting the example of England, where it was then £500. Of the responses to this, Alexander Niven (more on him later) stated that in his opinion, each job should be considered on its own merits — for example, a simple godown costing £40,000 to build might not need them due to the simple nature of the structure, but a £500,000 job would need quantities. The chair of the lecture, Fred Snell, said that one impressive point in all this was that economics was often overlooked with contractors using rule-of-thumb methods.

One final point of interest (or at least to Quantity Surveyors reading this) is that a building owner would pay a Quantity Surveyor 1.5 to 2%, but the rewards in savings were around 5%, which in itself is good enough reason to use a Quantity Surveyor.

The 1937 Second Annual Dinner of the Malaya Branch was again held in Singapore, this time at the Sea View Hotel (Meyer Road). Fred Snell presided over the event. The highlight was the celebration of the success of locally born and trained surveyors — of the five candidates who passed the Institution's examinations, two were locally born Chinese, and one of them passed the Part 1 Intermediate Examination. Guests included Colonel Cave-Brown (Chief Engineer, Malaya Command), Mr. Hallas (Chief Engineer), D.C. Rae (Municipal Architect) and Frank Lundon of Swan & Maclaren.

At that time, there was a total of 25 chartered surveyors in Malaya. In this context, Malaya comprised the Straits Settlements (Singapore, Malacca and Penang) as well as the Federated Malay States (where Kuala Lumpur was situated) and Unfederated States such as Kedah and Johor.

Snell set out the Branch's objectives:

- Advancement and facilitate the acquisition of knowledge
- Promote the interests of the profession for the public advantage
- Bring members in close touch with each other (and with the Institution in London) and
- Provide machinery for the holding of the Institution's examinations locally.

June 1938 saw the Third Annual Dinner being held, again at the Sea

View Hotel. It was presided over by Francis ("Frank") Angell (more on him later) in the absence of the Branch Chairman, Major George Percy, who was in hospital. He announced that of the 15 candidates who sat in Singapore for the Institution's examinations, 3 were Chinese, and of the 6 that passed, 2 were Chinese. The guest speaker was the Solicitor-General (later Sir) Arthur Worley (1892–1976), who expressed the view that these professions were not popular with the laity despite the growing importance of technical efficiency, professional honour and fidelity. He said that as the profession was only 50 years old and still youthful, the first cynical gibe about a Chartered Surveyor had yet to be invented. He added, "I tried hard to invent one for this occasion but without success."

In the context of the popularity of professions, Worley quoted from Winston Churchill's foreword to the 1932 "Chartered Surveyor" booklet, words that Worley said were undoubtedly true: *"In a complicated and expanding civilisation such as ours, the status and character of the learned or scientific professions become matters of great consequence. The importance of high standards of technical efficiency is obvious; but even more important is that sense of professional honour and fidelity which springs from prolonged vocational training and a lively sense of corporate responsibility."*

By 1939, Snell had become the Branch Chair, and at the Annual Dinner in June 1939 (again at the Sea View Hotel), he proposed a toast to "the surveyor" who had no stately edifice of brick or stone to perpetuate his *(sic)* name but rendered valuable service in connection with all such works. He also informed those present that of the 13 people (lower than the 1938 number) who had sat for their professional examinations, 6 were from the Services, i.e., the RAF, the Admiralty and the Royal Engineers, and 7 were local men, of whom 3 qualified as probationers of the Institution by passing both parts of the Intermediate Examination.

The guest of honour was Harold Cheeseman (1889–1961), who spoke about the profession in an extremely positive light. Cheeseman was an Englishman who arrived in Penang in 1907 at the age of 18 without formal teaching qualifications, but through his own determination and ability, rose to become the Deputy Director of Education for Malaya by 1938. After internment at Changi during the Japanese Occupation, he later became the Director of Education after World War II. He was also responsible for the 1946 Cheeseman Report, with the goal of restructuring the curriculum for the vernacular schools.

Cheeseman told the guests that his first knowledge of surveyors was when the Naval Base advised him of their need for Quantity Surveyor clerks, and upon enquiries, a local chartered surveyor educated him on what a Quantity Surveyor was, which impressed him so much that he apprenticed a "Chinese youth" (for whom he was a guardian) to this friend's firm. The youth in question was Seah Mong Hee (more on him later in subsequent Chapters), and he was apprenticed to Waters & Watson.

Seah's father had gained employment with Cheeseman as a butler. Cheeseman added that Seah was now one of those bidding to become the first "Asiatic" chartered surveyor. Cheeseman also pointed out that the examinations were of the same standard as in Great Britain, which was very important, as he felt that local boys (*sic)* would not thank the Institution for letting them in through the back door. He said that they "wanted equality of qualification and to show their worth. To lower the standard would be unfair to Malaya and what they wanted was opportunities, not privilege".

The Evolution of Waters & Watson

The formation of Waters & Watson in 1933 (with an office on the First Floor of 98 Robinson Road, then at Battery Road by the late 1930s) saw the establishment of the very first Quantity Surveying firm in Singapore, albeit individual Quantity Surveyors were already working in Singapore by 1933 for building contractors and for the Government. Later in this book, we will see how the firm evolved into Langdon & Every (Far East) by 1954, Langdon Every & Seah by 1969, Davis Langdon & Seah by 1990, and then Langdon Seah (and ARCADIS company) by 2012/3.

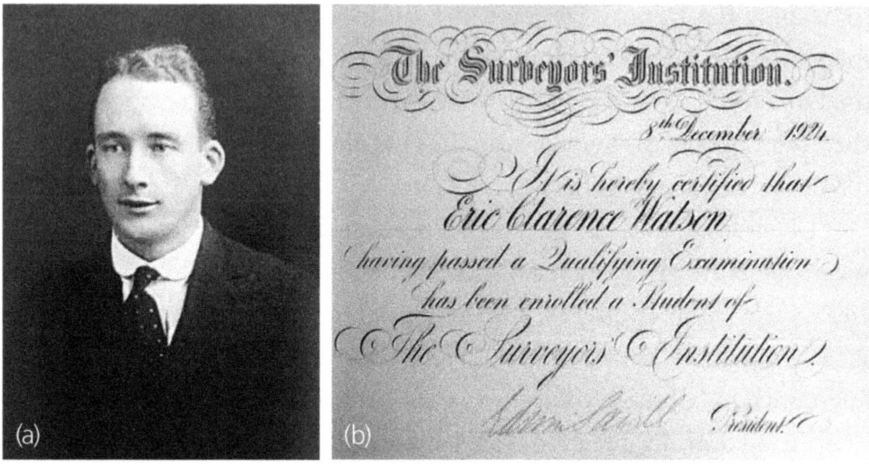

FIGURE 7.4. (a) A young Eric Watson. (b) A copy of his Certificate of Enrollment as a Student of the Surveyors' Institution.

FIGURE 7.5. David Waters (third from right) and Eric Watson (extreme right) in a photograph, most likely taken during a social gathering during the mid-1930s in Singapore.

David Ross Waters (1907–43) and **Eric Clarence Watson** (see Figures 7.4 and 7.5) were true pioneers and deserve a special mention. Waters had arrived in Singapore in 1931 to work for Brossard, Mapin & Co., a contracting company involved in a number of major construction projects in Malaya and Singapore. Watson (born in Sunningdale and raised in Windsor Forest, Berkshire, England) had worked in London in the 1920s for Horace W. Langdon & Every alongside **Francis ("Frank") Angell** (see

Figures 7.6 and 10.1.) where he qualified in 1929 as a Chartered Surveyor. Watson also arrived in Singapore in 1931, where he took up an offer to work for Fogden, Brisbane & Co. Ltd., also a contracting company. Watson became a Colonial Fellow of the R.I.C.S. in 1934.

They had a difficult start — not surprising given that architects, engineers and local contractors in the construction industry in Singapore (and indeed the Government) did not use Bills of Quantities in the tender process for construction projects. They preferred the lump sum contracts, where contracts were mainly agreed upon using a lump sum specification and drawings basis, with the task of exercising control of the contract being undertaken by either the architect or the engineer depending on the nature and complexity of the project. Most resistance came from Singapore-based architects who saw the use of Bills of Quantities as a threat to their contract administration role, particularly in regard to variations and progress payments.

However, eventually, with much perseverance and efforts to promote the advantages of using the skills of Quantity Surveyors (for example, David Waters' talks to the Rotary Club), their new business thrived, and they were involved in a number of major projects in Singapore and Malaya prior to the Japanese invasion in 1942. Their breakthrough came with their Quantity Surveyor role in the construction of Hill Street Police Station between 1934 and 1935 and the European Swimming Club (now the Singapore Swimming Club), which opened in 1936.

Their use of Bills of Quantities demonstrated value and the highest standards in Quantity Surveying, and so word spread, resulting in demand for their services, principally from the Air Ministry, Army and Admiralty. This led to them recruiting more Quantity Surveying staff, including Frank Angell (who had arrived in Singapore in 1937 to work for the Air Ministry) and two local graduates: Seah Mong Hee and Joseph Maschoieff. See Figure 7.6 to see Angell, Seah and Maschoieff in their Singapore office circa 1937–38.

However, the advent of war and the threat of a Japanese invasion led to the families of Angell, Watson and Waters being evacuated from Singapore.

FIGURE 7.6. Joseph Maschoieff, Mrs. J.G. Cox, Francis ("Frank") Angell and Seah Mong Hee in the pre-war Waters & Watson office.

Frank Angell joined the Royal Engineers in 1939 to construct Singapore's defences against the Japanese invasion, and with them, he left Singapore and reached Jakarta. From there, he somehow managed to escape the impending Japanese invasion by boat to Sri Lanka, an extremely traumatic experience. Unfortunately, both Waters and Watson were captured by the Japanese in 1942 and forced to work on the Burma "Death" Railway in Thailand. The fate of both men will be discussed in Chapter 9.

Francis Angell

Francis ("Frank") Leslie Angell (1905–73) was born in Hendon, Middlesex. He was articled as a Quantity Surveyor to Horace W. Langdon of Horace W. Langdon & Every in 1922 and stayed with them until 1936 when he came to Singapore to work for the Air Ministry. He then joined Waters & Watson as a partner on 1 August 1938. Having returned to England after World War II had ended, he returned to Singapore to help restart the Waters & Watson practice.

By 1956, when Watson died, he became Senior Partner of the practice but returned to England later that year to join the Southampton office of Horace W. Langdon & Every. He retired in 1970 and passed away in Hampshire in 1973.

Charles Frederick Snell

The Institution's 1939 Malaya Chapter Chair (and 1936–37 Vice-Chair) was **Charles Frederick ("Fred") Snell (1892–1974)**, a Surveyor with the Singapore Municipality.

Snell (see Figure 7.7(a)) was born in Hampstead, London, in 1892 and was the son of Emma and Edward Snell. Edward was a fishmonger, and Snell lived above his father's shop on Hampstead High Street. Snell qualified as a PASI in 1913, and prior to coming to the region, he was in the RASC from 1914 to 1918 (World War I service). He became a Colonial Fellow in 1922, shortly after his arrival in Singapore. He was a valuer and Government assessor and also served on the Mohammedan and Hindu Endowment Boards.

Fred's wife was Margaret "Peggy" Glenton-Kerr — she was a journalist from Hertfordshire, England, and both were well-known in the expatriate community in Singapore. A photograph of them both is at Figure 7.7(b). They lived in Adam Park in a Government Black and White house and were married in March 1931 in St. Andrew's Cathedral,

FIGURE 7.7. (a) Fred Snell, long-time resident of Singapore and 1939 Chair of the R.I.C.S. Malaya Branch. (b) Fred and Peggy Snell in 1933 at a garden party as Government House being introduced to Sir Cecil and Lady Clementi.

Singapore. They initially honeymooned locally in Tanah Merah and then later a bit further north at Fraser's Hill. Snell was a keen sportsman and played tennis, golf, rugby and football, as well as other sports — his football team was the Loyals, and he is known to have played at the Anson Road Stadium. This long-gone Stadium was once dubbed the "Wembley of Malaya"; the site of the Stadium is now a stretch of dual carriageway (Straits Boulevard) by the interchange with Shenton Way and Maxwell Road, between Singapore Conference Hall and Marina Bay Neighbourhood Police Centre & Fire Station.

Snell's role as a Government assessor was a very challenging one, because volatility in the land and property markets was having a significant impact on the assessment for rates. This had already seen the judgement of his predecessor **Harold Carpmael** being challenged in the Singapore Courts. Indeed, **Snell's 1930 Report** as the Government assessor confirmed that the town's total assessment value had dropped from $30.48 million to $29.96 million after several years of increases, and by 1931 it had dropped even further, thus impacting Municipal revenues. Vacant properties were also having an impact on revenues, with refunds for 1,466 vacant houses in 1929, which increased to 2,694 by 1930. By 1931, it had risen further, mostly due to the slump, but also to new buildings not being occupied (although the number of new buildings had dropped too).

An additional challenge was ratepayer dissatisfaction. So, by 1937, Snell was doing a lot of work to educate the ratepayers of Singapore as to what the revenue was spent on, amid continued dissatisfaction with the burden of rates. When issuing his annual report, he was at pains to point out that it was not just the SIT carrying out slum clearance and town improvements, but the Municipality was also active in this too. Indeed, in 1936, a total of 361 premises were demolished, mainly *attap* huts for anti-malarial measures. Another 57 properties belonging to the Commissioners and the SIT were demolished. The Municipality was also active in acquiring property for public purposes, spending almost $671,291 on compensating owners in the process. At the same time, the Municipal boundaries were extended to include properties in Geylang Serai, off Paya Lebar Road, in the 1937 assessment list.

Snell was also a member of the Rent Assessment Board, and in the late 1930s, up until the Japanese Occupation, the Board was kept very busy with requests from commercial and residential landlords to increase rents and from tenants to either reduce or keep rents at their existing levels. The Rent Restriction Ordinance had been introduced some years earlier and updated to manage the problem of significant rent increases in the post-World War I era due to the demand for space far exceeding the supply. There will be more on this in Chapter 20 in relation to rent control and the views of John Aspinall, who was Snell's successor as the Municipal Assessor after the end of World War II.

Snell joined the Singapore Volunteer Corps in 1924, and by 1939, was still a reservist. Fred and Peggy's only child, Gordon, was born in Singapore in 1932 and went to boarding school in the Cameron Highlands. Gordon is now a famous children's author and was married to Maeve Binchy (1939–2012), also a famous author whose novels sold over 40 million copies worldwide — they both lived in Dalkey, just outside Dublin, where Gordon continues to reside since his wife's passing in 2012. In 1941, with the Japanese invasion imminent, Gordon and his mother, Peggy, managed to get out of Malaya and ended up in Geelong, Australia. Snell was not so lucky and became a Japanese prisoner-of-war. When he reached Australia in 1946, his family was greeted by an emaciated figure, the results of hardship from years of internment. The Snell family returned to England to settle in 1946, where Snell continued to work in public service as a surveyor. They settled in Watford, England, where Peggy passed away in 1969 and Fred in 1974.

Last Malaya Branch Committee Prior to the Japanese Occupation of Singapore

From 1939 to 1940, the last Malaya Branch Committee of the Institution prior to the Japanese Occupation comprised the following members:

- **Alexander Niven** of the Public Works Department, Singapore
- **David Ross Waters (Branch Honorary Secretary)** of Waters & Watson, 8 Battery Road, Singapore
- **Francis ("Frank") Leslie Angell (Branch Vice Chair)** of Waters & Watson, 8 Battery Road, Singapore
- **Herbert Courtenay Atkin-Berry** of Swan & Maclaren, Hong Kong Bank Chambers, Singapore (1938 to 1939 only)
- **George Reginald Percy,** a State Engineer at the Public Works Department, Singapore
- **Charles Frederick ("Fred") Snell (Branch Chair)** of the Municipal Offices, Singapore
- **Frederick Jones-Davies** of Jones-Davies & Smith, Nunes Building, Malacca Street, Singapore
- **Eric Cuthbertson Hill** of H.M. Naval Base, Singapore
- **Norman Frederick Hopkins** of the Air Ministry, Far East RAF Station, Singapore

Elsewhere in this book are stories about the contribution of **Niven, Waters, Angell, Atkin-Berry** and **Snell**, but nothing is known of the careers of **Hill** and **Hopkins** in Singapore, other than that they worked for H.M. Navy and the RAF, respectively, in Quantity Surveying roles.

Frederick Jones-Davies

Frederick ("Freddie") Jones-Davies (1907–68) was born in Nottingham in 1907 and died in Kensington, London, in 1968. He was a partner of Jones-Davies & Smith along with Arthur Kenneth Smith. Both of them qualified as Chartered Quantity Surveyors in 1934, and in 1937, they set up a Quantity Surveying firm in Singapore bearing their names. The firm was based in Nunes Building, Malacca Street, Singapore, a building which survived until the 1980s — the site of Nunes Building is now the rear car park entrance to Republic Plaza. The firm was involved in preparing Bills of Quantities for development projects in Singapore and Johor, including the new Johor Government offices in 1939, for which Palmer & Turner were the architects.

Little is known about Smith. As for Jones-Davies, in September 1939, he moved temporarily to Johor to support the Government office project, and on his arrival, he hosted a grand party at the Royal International Club to make his arrival known. Prior to that, he lived at 33 Paterson Road — this grand property is no longer around. In recent years, the site of this house was developed to build private condominiums (Paterson Suites). Jones-Davies was the cousin of Dr. Donald Honeywell (1903–77), who was the principal of King's School in Singapore in the 1930s up until the Japanese Occupation. Jones-Davies and Wilson (both bachelors) together hosted many social events in Singapore during the late 1930s. From 1940 up until the time of the Japanese invasion, they were both heavily involved in war efforts, raising funds for the likes of the Salvation Army and funds to support the War Funds effort. Jones-Davies also supported Honeywell's teaching activities; for example, he taught Quantity Surveying as part of the "Back to School" effort for the increasing number of servicemen in Singapore at that time.

George Percy

The final 1939–40 Branch Committee member, **Major George Reginald Percy (1890–1947)** qualified in 1913 and became a Colonial Fellow in 1922. He was commissioned in 1937 into the Straits Settlements Volunteer Force as a Major (from 1941, as a Volunteer Reserve). He was the PWD Acting Settlement Engineer. As well as being a Chartered Surveyor, he was also a qualified engineer. In 1938, his salary was $10,200 per annum, with a motor car allowance of $1,260 and an additional "acting" salary of $1,200 per annum. He, his wife, Edith, and his children were in Singapore, Malaya and elsewhere in South-East Asia for many years.

Edith and their children were evacuated prior to the Japanese Occupation and reached Liverpool on the *Duchess of Bedford* on 4 April

1942. He was evacuated on the *MV Redang* on 13 February 1942. This passenger ship was shelled and sunk 50 nautical miles off Berhala Island by Japanese naval vessels, with the loss of 58 of the 89 people aboard. He was rescued by *HMS Tapah* but was captured by the Japanese at Muntok on 17 February 1942.

Major Percy survived internment to become Acting Director of Public Works in 1946, but sadly, he died a year later. However, his daughter Phyllis and her husband David remained in Malaya for many years. It is hoped that the story of his time in captivity and more stories on his life after World War II will be forthcoming in the book on Chartered Surveyors in Malaysia, as only a small part of his career was in Singapore — most of his career was in Johor and Kuala Lumpur working for the PWD

Alexander Niven and the Old Supreme Court building

In this Chapter, there have been a number of references to construction projects for the Armed Forces and public sector in Singapore (and across Malaya) in the 1930s, which saw the Institution establish a Branch in Malaya. However, one project that deserves special mention is the old Supreme Court building constructed between 1937 and 1939, which was the first major project in Singapore to employ the use of Bills of Quantities. This was achieved under the guidance of **Alexander Niven**, a Quantity Surveyor in the PWD

Niven (known professionally as "Alex" but as "Alec" to his family) was born in Hamilton, Scotland, and qualified as a Chartered Quantity Surveyor (PASI) in 1930. A photograph of him is shown in Figure 7.8(a). Upon qualification in 1930, he went to Africa to work for the Kenya and Uganda Railways and arrived in Malaya in 1935, then Singapore. He became a Colonial Fellow in 1936. As one of the first qualified Government Quantity Surveyors, he was a pioneer in the use of Bills of Quantities for Government PWD construction projects. He was also an expert on the Law of Contract. By 1938, in the middle of the Supreme Court development, he was made a Colonial Fellow of the Institution.

Shortly before that, on 5 November 1937, he met Eileen Mills (a native of Seattle who was born in 1910) at a dinner party in Singapore. Eileen was a very intellectual person who had graduated from the University of Washington (in June 1929) in Philosophy at the age of 19 before going on to Radcliffe College (1930–31) and receiving an M.A. in Philosophy in February 1932. Eileen made a decision to travel after her studies and ended up in Singapore in October 1937, where she landed a job with Marsman Investments. Eileen and Niven became romantically involved and made the decision to get married back in her home city of Seattle. Eileen headed back to Seattle in March 1939, closely followed by Niven, who reached Vancouver on 20 August 1939. Eileen's brother brought him from Vancouver to Seattle, and they married on 20 September 1939 at her parents' house.

They made the decision to return to Singapore, arriving in March 1940, where Eileen worked at the American Consulate General, and Niven resumed his former job. However, their lives were changed completely on 8 December 1941 when the first Japanese bombs fell on Singapore, bringing with it the first indication of an imminent Japanese invasion. Niven joined the Straits Settlements Volunteer Force (SSVF) 1st Battalion whilst Eileen managed to get out of Singapore on 30 January 1942, and reached Batavia, Java, several days later, eventually reaching Melbourne on 28 February after a Quantas plane journey from Tilatjap to Broome. The journey from Singapore to Java was largely uneventful, but as Eileen's diaries say, it was purely by chance that the ship missed all the Japanese ships that crossed their path to reach Sumatra. It was some time before she heard of the fate of Niven, and during this time, she grew to resent the universal question: "I suppose you're glad to be out of there?"

Niven left Singapore just before the Japanese invasion on the *H.M.S. Kuala*, which sailed from Singapore on 13 February 1942 with approximately 500 evacuees on board — men, women and children. Japanese planes circled above as they reached the coast of Indonesia and dropped bombs directly on the *H.M.S. Kuala*, hitting the upper bridge, the stokehold and the engine room. The ship burst into flames at once and the whole superstructure was soon ablaze. Few survived but most of those who did, reached Pompong Island. The following day, a small ex-Japanese fishing craft, which had been captured in Singapore and was commanded by an Australian, Captain Bill Reynolds, MBE, late of the Johor Government Service, arrived and rescued 70 of the walking wounded, Niven and the ship's captain Lieutenant Caithness being amongst them. The rescue vessel was the *Krait*, later to distinguish herself in these waters by an attack on Japanese shipping in Singapore Harbour. Tragically, this ship was sunk by the Japanese on 17 February 1942.

After some time travelling through Indonesia, Niven and a handful of other survivors finally reached Colombo by boat before he travelled further to Bombay (now Mumbai). Others who survived the sinking of the *H.M.S. Kuala* and did not perish on Pompong Island ended up in Japanese internment camps in Indonesia, where many of them perished.

Eileen and Alex were reunited after the war ended, and they went to Hamilton, Scotland, to spend time with his mother, after which Eileen returned to Seattle to attend to family business. In 1946, they returned to the region to live in Kuala Lumpur, where he worked for the PWD. By 1947, he was the only Malaya-based R.I.C.S. Branch Committee member, as all the others were in Singapore. Around 1954, Alex and Eileen returned to Seattle, where he retired, and Eileen worked as a secretary to the Classics Department at the University of Washington until her retirement.

Niven suffered 2 or 3 heart attacks (most if not all in Seattle), and the last one took his life on 10 January 1964 at home in Seattle. Knowing that his time might be short, he made a long trip in the early 1960s to see many of his old friends in Scotland and across Asia for the last time. Eileen passed away in Seattle on 4 November 2000.

Niven's manager was Frank Dorrington Ward. Niven was appointed in 1935, and in 1937, he was earning a salary of $7,200 per annum. By 1938, another Quantity Surveyor had joined Niven, no doubt due to the immense size of the Supreme Court project and other ongoing public works. He was Alfred Seller Harris (1913–94), who qualified as a Chartered Surveyor (PASI) in 1935 and started work in 1938 at the PWD, earning $6,000 per annum salary. Niven's salary was then $7,500 per annum by that time, and the annual salary of Ward (Chief Architect of the Public Works Department) was $12,600, plus a car allowance of $720 per annum.

FIGURE 7.8. (a) Alexander Niven, whose Quantity Surveying skills were critical to the success of the Supreme Court construction project. (b) Singapore Old Supreme Court Building, St. Andrew's Road: built between 1937 and 1939.

The Supreme Court (see Figure 7.8(b)) was designed by Ward and modelled after the Old Bailey in London, in the classical style resplendent with colonnades, to complement the adjoining Municipal Building. The building contract was awarded in 1937 to United Engineers through a tender, and its cost was estimated at $1,500,000 (although some sources say $1,750,000). A total of 2,400 tons of steel was required (construction was structural steel rather than reinforced concrete) for a 2.5-year contract. The decision to pre-order the steel at the outset was wise, as steel prices has risen sharply by the end of the project.

Although Frank Ward is remembered as the person who designed the Supreme Court, it was Niven who pioneered the introduction and use of Bills of Quantities on Government construction projects in Singapore with the construction of the Supreme Court.

Chapter 8

The Story of Seah Mong Hee, the First Asian Chartered Quantity Surveyor

In the earlier Chapters, the emergence of Eurasian Chartered Surveyors was referenced, and Singaporeans of Chinese ethnicity had also started working as surveyors in Singapore by the 1920s. However, the world of Chartered Surveyors in Singapore was still completely dominated by white, middle-class British males. Despite the efforts of some to grow local talent, there was general resistance to the idea of "locals" joining the ranks of the elite and enjoying the gracious living that the British and European minority were enjoying.

In his book *A History of Singapore 1819–1988*, C.M. Turnbull states, *"the ruling classes encouraged this aloofness in order to preserve the last vestiges of the mystique of superiority, which had largely been stripped away by exposure through the cinema, Western education and the popular press."*

However, things had started to change with the likes of David Waters and Eric Watson taking on local apprentices and actively encouraging them to become qualified professionals. However, resistance prevailed, and so every Chartered Surveyor in Singapore from the 1900s to the late 1930s was either British or Australian, apart from that small group of Eurasian Chartered Surveyors who took the route to becoming Chartered Surveyors as a result of their Survey School examinations some years earlier.

In 1940, there were seven Probationers of the Institution in Malaya (six in Singapore) awaiting their Final Examination to become Professional Associates. Of those, three were young British surveyors working in the

RAF Works Directorate at Seletar, and another surveyor (Tan Chin Siong) who worked for Charles Frederick Snell as a Government assessor. The other two were apprentices at Waters and Watson — Seah Mong Hee and Joseph Maschoieff.

Encouraged by the Partners to study and take their professional examinations, *The Straits Times*, in October 1940, proudly announced that Seah and Maschoieff had successfully passed their Final Examination (in the Quantity Surveying division) and had become Chartered Surveyors.

Both Seah and Maschoieff were educated in Singapore at St. Andrew's School, and both were also apprenticed at Waters & Watson: Seah was first in 1935, and Maschoieff followed in 1936. Maschoieff took the three professional examinations in three successive years and got a first-class pass in the builder's quantities examination of the City and Guilds Institute. He also found time to be a scoutmaster of the Singapore 19th troop.

Seah Mong (also known as "Bill") — see Figure 8.1 — passed the first two examinations in 1937 and 1938 but wisely decided to take two years to study for the final examination.

FIGURE 8.1. Seah Mong Hee.

Seah Mong Hee (1918–2009) was born in Hainan Island in February 1918, one of five children (two boys and three girls). In 1930, he left Hainan Island to live in Johor with his father, and the rest of his family soon followed. Thanks to the patronage of his father's employer (see Chapter 7), he did his schooling in Johor and Singapore.

He was educated at the English College, Johor Bahru (five years) and then at St. Andrews School in Singapore (two years), where, as the school's top student, he passed the Cambridge School Certificate, leading to his position as an articled pupil with Waters & Watson. He obtained a first-class pass in the builder's quantities examination of the City and Guilds Institute and a second-class pass in the structural engineering examination. Both Waters and Watson encouraged him to take his R.I.C.S. examinations alongside his City and Guilds examinations.

Seah was also proudly and rightly proclaimed in the local newspapers (and by the Institution's Malaya Branch Chair) as the first Chinese person in the world to qualify as a Chartered Quantity Surveyor. See Figure 8.2(a) and 8.2(b) for his achievements.

Seah passed away in Singapore in January 2009 at the age of 92. He will be mentioned again at various points in later Chapters, as he and his family are the pioneers of the Quantity Surveying profession in Singapore and South-East Asia. However, here is a summary of his career from becoming a Chartered Surveyor in 1940 onwards:

- Seah managed the office of Waters & Watson when the Japanese invaded and occupied Singapore, at which point the office closed. By working as a domestic gardener far away from the construction industry, he managed to avoid detection during these years. The last job they handled prior to the Japanese Occupation was the Johor Military Barracks.
- After World War II ended, with David Waters now deceased, Eric Watson and Francis ("Frank") Angell returned to England to work for Horace W. Langdon & Every in London, but Tom Every encouraged them to re-establish the Singapore office, which they did. At this point, Seah re-joined them.
- By 1949, Seah was a partner in the Singapore office.
- By 1956, Seah had become Senior Partner in the Far East.
- In 1963, the firm incorporated his name into the company's name, which became Langdon Every & Seah.

A.R.P. Wardens of Kandang Kerbau Division, Group Post No. 2, taken at an inspection parade held at the New World. In the centre is Mr. A. H. Harrison, Divisional Warden, with the Group Warden, Mr. Sim Boon Eng on his right and the Deputy Group Warden, Mr. Koh Pheow Hong, on his left. On the right of the Group Warden, is the Deputy Group Warden, Mr. Ong Kim Kee.

Mainly About Malayans

—By— The Onlooker

TWO locally educated young men, Messrs. J. B. Maschoieff and Seah Mong Hee, have passed the final examination of the Chartered Surveyors' Institution in the quantity surveying division.

The Chartered Surveyors' Institution was founded in 1878 and incorporated by Royal Charter in 1881. Persons who have passed the three professional examinations of this Institution and have been elected professional associates have the privilege of using the designation P.A.S.I. and may call themselves "chartered surveyors."

Good Effort

MR. Maschoieff became an assistant in the office of Messrs. Waters and Watson, chartered surveyors of Singapore, five years ago, and took the three professional examinations in three successive years, a feat of which anyone would be proud.

He also obtained a first class pass in the builders' quantities examination of the City and Guilds' Institute.

He was educated at St. Andrew's School, Singapore, and his success reflects credit on his ability and the education he received at the school.

Mr. Maschoieff did not allow his time to be monopolised with his examination work but found time to be a sapper in the Singapore Royal Engineers (Volunteers) Field Company and is a scoutmaster of the 19th Singapore Troop.

Chinese Success

MR. Seah Mong Hee had the good fortune of receiving some personal interest from Mr. H. R. Cheeseman, who guided his education and choice of a profession.

He became an articled pupil to Mr. D. R. Waters, F.S.I., M.R.San.I., of Messrs. Waters and Watson, of Singapore, and during this period, passed the first two professional examinations in 1937 and 1938, but wisely studied for two years for the final examination.

Mr. Seah Mong Hee was educated at the English College, Johore Bahru, (five years) and St. Andrew's School, Singapore (two years) and passed the Cambridge school certificate.

During his pupilage he also obtained a first class pass in builders quantities and also a second class pass in structural engineering examinations of the City and Guilds' Institute. Seah Mong Hee is the first Chinese to qualify as a quantity surveyor.

Other local candidates taking the intermediate examinations include L. B. Gomes, M. W. Aeria, L. P. Nazareth, Toh Chee Seng, Seto Kee Yin and Laurence Chin Howe.

Local Branch

THE Chartered Surveyors' Institution is the leading professional body for agricultural and urban property management, valuations, buildings, quantity and mining surveying, and has more than 9,500 members and students.

It is frequently consulted by various departments of His Majesty's Government and many laws of England impose statutory duties on the Institution.

The Compensation (Defence) Act engages chartered surveyors almost to the exclusion of other professions.

There is a Malaya branch of the Institution, having about 30 members, [illegible] airman this year being Mr. C. [illegible], F.S.I.

(a)

MR. SEAH MONG HEE

MR. J. B. MASCHOIEFF

Bangkok Personality

BANGKOK people heard with much regret of the sudden passing of Mr. F. H. L. Peri at the Royal Bangkok Sports Club.

Mr. Peri had addressed the gathering at the annual meeting, and had the satisfaction of knowing that the proposal he had put forward for the re-election of the present Committee, had been carried unanimously—a unique happening and one which had never happened before in all the years of the annual meetings.

Mr. Peri collapsed and died shortly after making his speech. He had been suffering from heart disease for some time and had been under medical treatment.

Chamber Chairman

MR. Peri went to Thailand to join the Siam Forest Co., Ltd., as it then was in September, 1936 and some years later resigned to join the firm of D. Couper Johnston and Co., of which he was the senior partner at the time of his death.

Mr Peri took a full share in the mercantile interests of the port of Bangkok and had been Chairman of the Bangkok International Chamber of Commerce since 1938, and a member through his firm for a much longer period.

He was a life member of the British Association of Thailand, and was Chairman of the Riding and Polo Club for 1938-39. He was 55 and his mother is still alive, as are also a married sister and brother.

* * * *

American Authoress

THE death of Miss Catherine Mayo, the American authoress, last week, calls to mind one of the greatest controversies that modern India has seen when her book "Mother India" was published in 1927 on the eve of the visit of the Simon Commission to India.

Mahatma Gandhi described the book as a "gutter inspector's report" and said that it was a book every Indian should read and no foreigner should see.

Apart from articles in the press seeking to disprove many of her statements about Hindu social customs and manners, many books were written in reply in several Indian languages.

The late Lala Lajpat Rai's "Unhappy India," C. S. Ranga Aiyar's "Father India" and K. L. Gauba's "Uncle Sam" are some of the books written in English.

The late Rev. C. F. Andrews also wrote a book last year—"The True India"—as a sort of reply to "Mother India."

Miss Mayo's two other books on India, "Slaves of the Gods" and "The Face of Mother India" received very little notice in India.

* * * *

Pilot Officer Mudie

PILOT Officer Michael Robert Mudie, the third son of Mr. and Mrs. J. D. Mudie, who spent many years in Malaya, died in July after wounds received in action.

Pilot Officer Mudie was shot down in flames after accounting for a Messerschmitt in one of the air battles over Dover which was broadcast.

This was not his first success against great odds, as Pilot Officer Mudie had a Dornier to his credit in a lone combat in France against nine German planes.

The funeral took place privately at Molesey and the mourners included his mother and two of his brothers.

Pilot Officer Mudie was born in Singapore in 1916 and his three brothers are serving in the forces, two with the Royal Air Force and the other in the Army.

Three were in France during the evacuation and the fourth is in Iraq.

Mr. J. D. Mudie was in the Malayan Civil Service and was for some years a judge.

still."

First Local Trained Quantity Surveyors

TWO old boys of St. Andrew's School, Seah Mong Hee and J. B. Maschoieff, have passed their final examination in quantity surveying for the Surveyors Institute.

They thus become the first successful Malayan candidates in these examinations and when they complete their apprenticeship with Messrs. Waters and Watson, they will be the first Malaya-trained quantity surveyors.

Seah Mong Hee will have the added distinction of being the only Chinese [illegible] chartered as a quantity surveyor [illegible] Surveyors Institute when his period of apprenticeship is completed.

(b)

FIGURE 8.2. (a) A *Straits Times* article on 6 October 1940 celebrating the success of Seah Mong Hee and Joseph Maschoieff in becoming Chartered Surveyors. Source: The Straits Times © SPH Media Limited (b) A *Mainly About Malayans* newspaper publication on 13 October 1940, a full article detailing the achievements of Seah and Maschoieff.

- By the time of his death in 2009, this firm (by that time, it was called Davis Langdon & Seah) had become one of the biggest (if not the biggest) in Asia, providing cost management, project and program management, and specialist consulting services.
- Many of the pioneer Chartered Quantity Surveyors of the 1950s, 1960s and 1970s started their professional careers with Langdon & Every or Langdon Every & Seah, so Seah was the catalyst for the growth of the profession during these years.
- Seah was the last Chair of the R.I.C.S. Malaya Branch in 1958–59. Indeed, he was instrumental in the transition of the R.I.C.S. Malaya Branch in 1959 into the Malayan Institution of Surveyors (which later became RISM), of which he was a Founder Member. In Singapore, with the creation of an R.I.C.S. Singapore Branch in 1961 on the dissolution of the R.I.C.S. Malaya Branch, he was instrumental in the transition from an R.I.C.S. Singapore Branch to local professional bodies that eventually merged in 1982 to become the SISV. More on all of this in Chapter 14.
- He was also instrumental in the creation of the Singapore Institute of Surveyors in 1967/68, later becoming part of SISV on its merger with the Singapore Institute of Valuers in 1982. He was the SIV's inaugural President.

FIGURE 8.3. Joseph Maschoieff (later known as Joe Marsh).

The origin and background of **Joseph Bernard Maschoieff (1917–2006)** (see Figure 8.3) was quite different from that of Seah. He was the son of Maurice Maschoieff, who was born in Safed, Palestine, on 5 December 1884. Maurice went to Switzerland to study to become a civil engineer. On returning to Palestine, he married Rachel Mendelson, and they had five children — Adele (b. 1911), Suzanne ("Susie") (b. 1913), Joseph Bernard (b. 1917, Cairo, Egypt), Isidore David (b.

1922, Heliopolis, Egypt) and Angele (b. 1924, Heliopolis). In about 1921, the family moved from Cairo to Heliopolis.

Maurice was admired for creating parks in Tantah and Alexandria in Egypt. In 1926, the family moved to Singapore, where Maurice worked as a civil engineer and director of Metropole Hotel Ltd. Maurice was an excellent chess player, and in 1933, he and 24 members of the Singapore Chess Club played a simultaneous game of chess against the World Chess Champion, Dr. Alexander Alekhine. The matches lasted for 4 hours, and Dr. Alekhine won every game. Maurice, who was remembered for his great sense of humour, died on 18 July 1936 from cancer. His headstone is located in the Jewish section of the Chua Chu Kang Cemetery in Singapore.

In early 1942, Maschoieff helped to evacuate his mother, Rachel, and sisters Suzie and Angele. They went to Mumbai, India and then Haifa in Palestine. Maschoieff was educated at St. Andrew's School in Singapore and wanted to be a civil engineer like his father, but Maurice's untimely and premature death resulted in Maschoieff having to give up his ambitions to study overseas to be a civil engineer. Instead, he took the opportunity in 1936 to take up employment with Waters & Watson and train to become a Quantity Surveyor.

He was a Scoutmaster of the 19th Singapore Troop and later became a sapper in the Singapore Royal Engineers (Volunteers) Field Company in the build-up to the Japanese Occupation, where he was involved in the construction of an airfield in Malaya, which he was also involved in destroying, as the Japanese Army advanced on to Malaya.

Maschoieff left Singapore the day before the Japanese Occupation. He reached England, where he lived from 1943 to 1946, at No. 50 Cirencester Road, Tetbury.

On 6 June 1944, he gave notice, as a natural-born British subject, in the *London Gazette* of the change of his surname to Marsh. Maschoieff (now Marsh) wrote to Sarah Semple, a family relation, on 11 March 2005 and explained that the reason why he changed his family name was that "Maschoieff was too difficult for everyone, and I thought it best to have a simpler name".

Marsh married Faith Blair in Wandsworth, London, in 1946. Faith was born in 1914 in Glasgow, Scotland. They lived in South-West London for a short time before migrating to Malaya onboard the *Mauretania* in June 1946. However, she sailed back to England in 1949 and spent four months there before returning to Malaya in August 1949 on-board the *Canton*. That same year, Marsh became a Fellow of the Institution. Their son Andrew Charles Marsh was born in Bangsar, Malaya, in July 1950, but his time in Malaya was short as the family decided to return to the UK permanently in 1951.

Upon his return, Marsh set up Jessamine & Marsh from a small office in Red Lion Square, London. After closing the practice after Jessamine's 1956 death and a stint in the local Government in Croydon, Marsh retired to Dunoon, Scotland, in the late 1970s.

By 1980, Marsh and Faith had moved to Innellan, Scotland. He was passionate about his masonic activities and the game of Bridge. Faith died in 1995 in Dunoon, Argyll, Scotland, and he followed on 2 July 2006. Their son Charles and his family still live in South-West England.

There is more about Marsh in the next Chapter: he returned post-war and was the Institution's Malaya Branch Honorary Secretary from 1946 to 1951.

Conclusion on Seah Mong Hee

Seah Mong Hee, his family and their associated businesses have led the industry since the 1950s, and they are still market leaders to this day. Seah is recognised as the father of modern Quantity Surveying in Singapore. Indeed, he was the first Chinese person in the world to become a Chartered Quantity Surveyor. He went beyond the call of duty in his efforts to re-establish the Institution's Malaya Branch after World War II, and this enabled Singapore to benefit from the development of local professional surveying talent.

Prior to this time, the R.I.C.S. Malaya Branch had been dominated by British surveyors, who (apart from Waters and Watson) did not encourage local talent to attain professional qualifications, so it is fair to say that Seah was a true pioneer. He picked up from where Waters had left off in 1941 as the champion for the surveying profession in Malaya and was a strong proponent of professional and technical standards. His influence was a very important factor in the growth (and recognition) of the surveying profession in Singapore from the immediate post-war years right through until the 1970s.

Final Notes

At the same time as Seah Mong Hee and Joseph Maschoieff took their Final Examination in 1940, six local candidates (five of whom were Singapore-based) took their Intermediate Examination.

One of them was **Toh Chee Seng**, also known as **Dick Toh** — there is more on him later in this Book. Another was **Lawrence Chin Yun Howe**, who was then a trainee Quantity Surveyor with the firm of Jones-Davies & Smith. Chin qualified in 1953, and he was an R.I.C.S. Malaya Branch Committee member in the late 1950s. At the same time, he was a founder (and later President) of the ISM, which is now the Royal Institution of Surveyors Malaysia.

Another candidate was **Melville W. Aeria**, born in Singapore in 1922 into a prominent local Eurasian family. Like his father, he was a member of the Straits Settlement Volunteer Force — both were interned by the Japanese in Singapore in 1942. Aeria moved to Australia after World War II and became a prominent member of the Australian Institute of Quantity Surveyors (AIQS). He died in Ryde, NSW, Australia, in 2005.

Chapter

9

1941–46: The War Years and Immediate Aftermath

In the previous Chapter, the lives of the first two locally-trained, locally-qualified Chartered Quantity Surveyors were celebrated.

At that time, in 1940, World War II was starting to have an impact on the lives of those living in the Malay Peninsula and the island of Singapore, but even then, no one had seriously contemplated a Japanese invasion and occupation.

Indeed, for some, things continued as before — as illustrated by a Job Advertisement in the *Singapore Free Press and Mercantile Advertiser*, 2 August 1941, seeking an Assistant Assessor on a 3-year contract. Clearly, no Japanese occupation was envisaged when this role was advertised.

The other interesting thing about this advertisement is that the criteria included being a British subject (with "preference given to candidates born in Singapore") between 24 and 29 years old, unmarried and with a good technical and general education. Also, candidates "should be Members of the Chartered Surveyors' Institution or a similar Professional Association". The Singapore preference could be indicative that times were changing for the better, but it is more likely that due to World War II raging in Europe, the supply of British assessors had stopped completely.

The Loveday Corruption Trial

In 1940, whilst war raged in Europe and the Japanese army advanced through China towards South-East Asia, a Chartered Surveyor working in Singapore was accused of conspiracy and corruption and faced a trial by court-martial (i.e., a military court).

Captain Robert Charles Loveday, a Surveyor of Works, Royal Engineers, later a Senior Surveyor, who worked at Fort Canning and lived at No. 1 Sherwood Road, Tanglin Barracks (and also sometimes a resident of 19 Capitol Flats, Stamford Road), faced trial by court-martial at Tanglin Barracks on 25 charges of conspiracy (to defraud the War Department) and corruption. The trial, and sentencing of Captain Loveday, were covered extensively by Singapore newspapers at the time, with the court-martial verbatim records amounting to over 700 pages. The court-martial was presided over by Brigadier A.D. Curtis (president), Lieutenant-Colonel Mead and five others. The Judge-Advocate was Dr. C.H. Withers-Payne.

Francis George Vaux (1887–1956) of Rodyk and Davidson was Leading Counsel for the Royal Engineers, and E.M. ("Eddie") Tampoe-Phillips, a Singapore lawyer from a prominent Sri Lankan family, was Counsel for Loveday.

Captain Loveday, who on 5 September 1940, was said to be 49 years old and a decorated military man, having been in military service since 1913, was found not guilty of 17 of the 25 charges after 7.5 hours of consideration by the court. However, having not reached a verdict on the other charges, the court reconvened the next day and found Loveday guilty of seven of the remaining charges relating to the acceptance of a total of $169,000 (a significant amount in 1940) for assisting in the award of construction contracts for military buildings in Singapore (and Penang) to various parties. Loveday was cashiered (i.e., "dismissed from the Armed Forces in disgrace") and sentenced to 4.5 years of penal servitude.

The trial highlighted deficiencies in Singapore's Penal Code in relation to Criminal Conspiracy. Had it covered this, the sentence could have been even more severe, and the others involved in this ring of conspiracy could also have been punished. The Code was amended accordingly in 1941. Furthermore, a Commission (Public Inquiry) was set up in 1941 to review the system of awarding contracts in the Government Departments and the Armed Forces.

One of the Prosecution's principal witnesses told the court martial that the plan to form a ring of contractors for War Department work was put together by Loveday, who told the contractors to allow for 10% of the contract sum to be set apart — 5% to Loveday (and his wife) and the other 5% to the members of the ring. Other newspaper reports state that evidence given to the court-martial was that the other 5% was divided amongst the unsuccessful contractors. It was said that the ring would work as a team to ensure that the allocation of work was evenly distributed, albeit recognising the capacity of each contractor to do the work. This could apparently only be achieved through Loveday because of his position to influence the selection process.

The court considered a large amount of evidence pointing to Loveday's guilt. The National Archives of Singapore contains detailed records of the trial, including many stories relating to evidence placed before the court-martial. All the details are not in this book but are definitely worth reading further.

It is believed that Captain Loveday was born in London in 1888. What is known from the newspaper reports of the court martial is that Loveday trained as a surveyor with the Ordnance Corps in Chatham, England (at the School of Military Engineering) and served as a Foreman of Works and Surveyor in various locations with the Armed Forces, including a tour of Egypt from 12 March 1926 to December 1930, during which time he passed the Intermediate and Final Examinations of the Chartered Surveyors' Institution. On 14 December 1930, as Clerk of Works in the Royal Engineers, Sergeant-Major Loveday was promoted to the rank of Lieutenant (Surveyor of Works), and on 27 October 1933, having passed the additional subjects of Laws of Arbitration and Laws of Dilapidations, he became a Chartered Surveyor.

During his trial, Loveday told the court that he was an expert on all contract matters and questions involving contract questions and prices, building, construction, sanitation, building materials, etc. He also told the court that all Surveyors of Works had to pass the Intermediate and Final Examinations of the Chartered Surveyors' Institute; these examinations covered a wide range of technical subjects, including the Laws of Dilapidations and the Laws of Arbitration.

An interesting part of Loveday's defence was that when he arrived at Fort Canning (his main base) in December 1936, he expressed concern that contracts were all "lump sum" (with only six being tendered), which he said was a breach of Army Regulation for Engineer Service (Peace), S.27, para. 275. This stipulated a tender for all jobs above £750, but he was advised that the "lump sum" practice had been adopted because much of the Military construction work was urgent and there was not enough Quantity Surveying staff to prepare Bills of Quantities. By 1939, he had persuaded Brigadier James A.C. Pennycuick (1891–1966) to change to Bills of Quantities — Pennycuick agreed to do so because the change would give him, as Chief Engineer, an opportunity to check and see if the prices were reasonable. Loveday went on to say that the change was not popular with the Chinese contractors, resulting in animosity towards him. In other words, part of his defence was that the witnesses brought forward were not telling the truth, and the reason for this was that they wanted revenge for the shift away from lump-sum contracts.

The case led to a change in the Singapore Penal Code (see above), and a Commission of Enquiry was set up to investigate the allotting of contracts by the Public Works Department (PWD), the Admiralty and the Air Force Services. The Commission comprised Justice Worley (Chair), J. Jeff, E.N.C. Woollerton, R. Burns and Frank Lundon (he is mentioned elsewhere in this Book). The Commission heard that things had tightened up since the Loveday case, and one witness (Mr. Fogden of Fogden, Brisbane & Co. Ltd., contractors) stated that the Loveday case highlighted two main faults: the lack of a proper system governing the selection of firms invited to tender and the lack of any check on estimates put forward by the Surveyor of

Works. He added that unlike "back home", the position "out here" was that there was no system of selecting contractors, and this opened the door wide to corruption.

A further point of note is that Brigadier Pennycuick stated to the Commission that Captain Loveday was in a position of trust and responsibility, and that was part of the problem, but the main problem was that the estimate appeared to be public knowledge. He also agreed under questioning that often, Loveday had driven down the price of materials and that by introducing Bills of Quantities, Loveday had helped them considerably on construction projects.

He also agreed that Loveday had risen through the ranks and was Technical Adviser to the Chief Engineer, and as such, "we depended a great deal upon his advice".

The system had (by the time of the Commission) changed such that another member of the Chief Engineer's staff sent out the tender invitations, not Loveday. However, E.V. Fowler of the Singapore Police (more on him later) was of the view that Loveday was entrusted with too much responsibility and that the same problem did not exist in the RAF or the PWD. Pennycuick had mentioned earlier that they had depended heavily on Loveday, and due to severe shortages of resources available in the Royal Engineers at that time due to the War effort, they had no choice but to place trust in individuals such as Loveday. Furthermore, due to the large sums of money involved in Singapore construction contracts, the temptation was there, and the risk of corruption and conspiracy was therefore higher.

The Hidden Microphone

The court-martial heard that once suspicions had been aroused, Brigadier Pennycuick and Colonel Urquhart, the Deputy Chief Engineer, Royal Engineers, arranged with Lieutenant Frederick Croft to hide a microphone inside his flat on River Valley Road, so that they could listen into a meeting between one of the "ring" members and Croft. In that conversation, they heard the “ring” member offer Croft $10,000 in relation to a contract for hutted camps in Singapore. The allegation was that some or all of the other contractors in the ring would know what price to submit for the tender based on the estimate that Loveday had placed on it (which he had disclosed to some or all of those in the ring) and that once the contract had been awarded, the profit would be divided up among the ring members. In cases where the estimate was high, and the winning tender was just below the estimate, then the ring would get a higher profit to share between them.

Lieutenant Croft gave evidence that he had been asked to handle the tender for the erection of huts and told by a contractor that if he were to send out tenders for the hut contract at the Naval Base — having rejected the first lot of tenders — he (Croft) would receive a reward of 2% of the awarded contract sum. Later, the offer was confirmed as a fixed sum of $10,000. Croft had already informed his superior officer (Pennycuick) of this, and this led to the decision to lay the above mentioned trap.

It should be noted that Loveday was not involved in the hut tender because he was on annual leave, and Croft was instructed to handle the tender in his absence.

Under further cross-examination, Croft admitted that Loveday’s name did not come up in the dialogue with the various members of the ring. He also admitted that he had not suspected Loveday during this time, and he did not talk to Loveday about the matter because he was advised by his superior officer not to do so.

Croft also stated that he had been offered a new car by one of the “ring” members in addition to the $10,000. He had also been offered an extra reward by another “ring” member to get his products into War Department contracts. Croft recalled he told this “ring” member that it was very foolish to try and bribe him.

Despite the huge negative impact that all this had on public trust in the British Military in Singapore, one positive outcome of the Loveday case was the establishment in April 1941 of a special Anti-Corruption Unit in the Detective Branch of Singapore Police, under the leadership of

E.V. Fowler (known as "Big Ted"), later to be Deputy Commissioner and Chief of CID in the Singapore Police Force.

During the Commission of Inquiry, Justice Worley was critical of Pennycuick for bringing a contractor name into the mix, stating that this was outside the terms of reference of the Inquiry and was irrelevant. Pennycuick asked him (in response) why the chief contractor involved had not been brought to justice, and Justice Worley interrupted him to tell him that Pennycuick knew full well the reasons why (the reason is mentioned above, in relation to the Singapore Penal Code). Justice Worley added that by bringing individual names into the Inquiry, Pennycuick was putting the Commission at risk of degenerating into a personal wrangle, which would not serve the purpose of highlighting the dangers of corruption rings.

A final point on the Commission: David Waters of Waters & Watson (whose name features heavily in earlier Chapters and later on in this Chapter) was called upon to provide his expertise to the Commission on the problems within the building industry, as highlighted by the Loveday case. He told the Commission that problems existed because the building industry was not as highly regarded in South-East Asia as it was in Great Britain, so its growth in this region had been haphazard and unsatisfactory. He recommended that the industry be regulated properly through legislation and an Office of Works, and that the process of Malayanisation should start as soon as possible to develop local talent, in particular "secondary schoolboys". He was scathing of Mr. Fodgen's view that the problems lay with Chinese contractors.

Waters also recommended that a (fee) system of the percentage of costs should be introduced despite its unpopularity. However, he was in agreement with Mr. Fogden that no proper system of contractor selection and pre-qualification then existed in Malaya/Singapore. Waters also said that the problem of non-competitive tenders needed to be dealt with, as many contractors felt obliged to tender even if they were not really in a position to do so for whatever reason. These contractors should have first been asked if they wanted to tender or not. Fogden suggested that four pre-qualification criteria be adopted: knowledge of the type of work involved, adequate plant and machinery to do the job, adequate funds (cash flow) to ensure that their business does not experience any financial issues during the project, and finally, they should have sufficient resources with the right skills to complete the project successfully.

The decision to bring Waters in to advise the Commission no doubt helped to maintain the reputation of Chartered Surveyors in Singapore and Malaya, which could have been badly damaged by this court-martial because it was featured extensively in the Singapore press at the time.

At the time of writing this book, the fate of Captain Robert Charles Loveday and his wife and daughter after World War II is not known. What is known is that Loveday was interned by the Japanese in Singapore from 1942 to 1945 but his wife and daughter are not known to have been interned. It is possible that his wife returned to England with her daughter prior to the Japanese Occupation whilst her husband was in jail. Brigadier Pennycuick went on to serve with distinction with the Royal Engineers for the remainder of World War II.

The entire Loveday trial and its consequences could form the subject of a book in its own right. There is so much more to be known about the conviction of Loveday and those who were part of the ring.

Singapore Naval Base Corruption Trial

In March 1941, a case of alleged bribery reached the Singapore criminal district court. At the trial of two contractors Bakhtava and Bickram Singh, an assistant Quantity Surveyor called Henry Moore alleged that he had been offered a share of the profits if he helped a sub-contractor to obtain contracts in relation to (earth) works at the Singapore Naval Base. He also alleged that Bickram Singh had sounded him out on whether his superior, Mr. Lang (John Courtenay Lang, who is also mentioned in Chapter 8), a Chartered Surveyor, could also be induced to help and also take a share of profits. Moore told the court that he had laid a trap for the contractors and provided details of what transpired. In December 1941, both contractors were found guilty of attempting to offer a bribe to Moore and his superior and making a payment of $50 to Moore. Both were imprisoned.

Three months after the contractors were imprisoned, Singapore was under Japanese occupation.

Chartered Surveyors in Singapore During the Japanese Occupation

As mentioned in the previous Chapter, Seah Mong Hee helped to maintain the Waters & Watson office in the months prior to the Japanese invasion in February 1942, at which point their operations in Singapore ceased until the end of the war.

A fortunate number of expatriate Chartered Surveyors and their families managed to get out, but others were not so fortunate, and locals such as Seah had no choice but to remain in Singapore and take what chances they could to survive.

So, from the time of the Japanese occupation of Singapore in February 1942 to September 1945, when Allied forces liberated Singapore, a number of Chartered Surveyors were amongst those imprisoned by the Japanese, including Charles Frederick Snell.

There are many publicised (and untold) stories of those who lived through the Japanese occupation of the Malay Peninsula and Singapore, either as civilians or internees. I have included below some of these stories:

- **John Crickmay** (then Captain J.R. Crickmay), a **Chartered Surveyor**: He was not based in Singapore but was imprisoned for a short time in Singapore in February 1942. He then spent time in Burma and was liberated in September 1945 when back in Singapore as a prisoner-of-war. Prior to his death, he shared his story with the British Imperial War Museum, including details of how his building construction and surveying training came into use when he was forced (along with thousands of others) to work on the construction of railways and railway bridges in Myanmar (then Burma) and Thailand (then Siam), including the infamous Pack of Cards bridge, where conditions were so harsh that for every sleeper laid, a prisoner's life was lost. It was a miracle that he survived.

However, Crickmay narrated that there was little opportunity for him to practice the skills he learnt during his imprisonment in Changi. Instead, textbooks were not available, and despite the strenuous efforts of the Red Cross, neither the Institution nor the College of Estate Management could break the Japanese embargo, although a limited supply of (censored) letters from home did get through. To keep their brains active, Crickmay and other prisoners-of-war set up their own ad hoc courses (mostly based on private tutorials) on matters such as heraldry, architecture, valuation and property management. Crickmay's view was that this activity, and an ability to get on with others of different nationalities and backgrounds, helped towards his survival. As time passed, the number attending these "courses" dropped as prisoners were sent on working parties across Singapore on building repairs tasks (see the section on Tom Kitching below). Some were also sent to Japan, while others were deployed to work on the "Death Railway".

- **Tom Kitching:** Although not a Chartered Surveyor, Kitching was a Chief Surveyor based in the Fullerton Building. He was born in Morecambe, Lancashire, and studied at the Lancaster Royal Grammar School and then at Peterhouse, Cambridge. After graduating in the Natural Science Tripos, he became a surveyor with the Colonial Service and arrived in Malaya in June 1913 at the age of 23. He married while on home leave in Morecambe in 1919 and returned to Malaya. Kitching was not evacuated in February 1942 as the Governor deemed him as not being essential, so he was interned in Changi. Sadly, his wife Nora (who stayed in Singapore to care for those injured and was one of the last to leave) died on or around 13 February 1942, either by drowning as a result of Japanese attacks on the *H.M.S. Kuala* or possibly as one of the wounded who made it to safety to Pompong Island, only to die as a result of further attacks by Japanese planes. She may also have perished as a result of the Japanese attack on the rescue ship *Tanjong Pinang* that had picked up the few survivors who escaped the attack on the *H.M.S. Kuala*.

Kitching soon developed throat problems, and by early 1944, it was discovered that he had a tumour in the oesophagus. His condition worsened, and he died on 14 April 1944. He was buried in Bidadari Cemetery. How his diary survived Changi is a mystery, but it is now held by his son, Brian, who included large diary sections in the book *Life and Death in Changi*. The book is a fascinating read. Kitching had spent time preparing a talk on the "Development of Malayan Surveys", which he could not present himself due to his throat issues, and so "Sworder" (see below) presented it for him on 21 July 1943. It went well, and Kitching was congratulated for it.

- **Gerald Hope Sworder**: He is mentioned in Tom Kitching's diary along with surviving Survey Department staff from Malaya and Singapore that were interned in Changi. As Chief Surveyor of the Government Topographical Survey Department, he had been based in Malaya since the 1920s. His children were at boarding school in the UK, and in 1939, his wife returned to the UK due to the illness of one of their children. To ensure that vital survey maps did not fall into Japanese hands, Sworder remained to deal with the situation and was interned as a result, not seeing his family again until 1946. He had not seen his family for 7 years by the time he got back to the UK, by which time his children did not recognise him.

In September 1945, the Allied forces took Singapore back from the Japanese. Life would never be the same again for those who survived, and it was the beginning of the end of British colonial rule. However, for the immediate future, life returned to how it was in 1941.

Indeed, in January 1946, Eric Watson and Francis ("Frank") Angell (who had both been repatriated to England at the end of World War II) returned to Singapore and announced in *The Straits Times* that as the surviving partners of Waters & Watson, FFSI, Chartered Quantity Surveyors and Loss Assessors, they would shortly reopen for business as Waters, Watson and Angell at temporary offices on the 2nd floor of 181 Battery Road.

Their move back to Singapore arose from a discussion in London with Cecil Tom Every of Horace W. Langdon & Every, and indeed by May 1946, they had changed the name of their firm to Horace W. Langdon & Every (incorporating Waters & Watson). The word "surviving" is poignant, and there was the noticeable absence of David Waters, who had been such an influential industry figure in the pre-war years. Sadly, the answer to Waters' absence lies in Chungkai War Cemetery, Thailand, where Lieutenant David Ross Waters, FSI, PASI, of the 1st Battalion, Singapore (Volunteer) Royal Engineers, was buried after his death on 30 August 1943, aged just 36. See Figure 9.1(a).

Margaret Shennan's *Out in the Midday Sun: The British in Malaya, 1880–1960* tells us that David Waters and Alexander Cullen (who we mentioned in Chapter 5) both died whilst working as forced labour on the "Death Railway" in Thailand.

Furthermore, Shennan tells us that Waters cared for Cullen in the final days of his life. The book describes in some detail those last brutal months of their lives. Cullen (then Captain Cullen of the 1st Singapore Volunteer Corps) died on 26 June 1943, aged 39. He is buried at the Kanchanaburi War Cemetery in Thailand. See Figure 9.1(b).

FIGURE 9.1. Headstones of (a) David Waters in the Chungkai War Cemetery, Thailand, and (b) Alexander Cullen in the Kanchanaburi War Cemetery, Thailand.

Part III

The Work of Surveyors Post-World War II

Chapter

10

1947–51: The Post-War Years

In the aftermath of World War II, only 14 Chartered Surveyors remained in Singapore and 7 in Malaya. It would take time for the profession and for Singapore to recover. However, over the next 15 years, the number of Chartered Surveyors in Singapore increased substantially, including Singaporeans, who would later play a dominant role in the profession in post-independence Singapore.

On 17 August 1947, it was announced in *The Straits Times* that King George VI had made an Order in the (Privy) Council approving the name change of the Surveyors' Institution to the Royal Institution of Chartered Surveyors. Now, Fellows were designated as FRICS instead of FSI and Professional Associates as ARICS instead of PASI

(Note: On 1 January 2001, the title of Professional Associate was changed to Professional Member, and with that change, the ARICS designation, which during the 1990s had become abbreviated to ARICS, became MRICS. Likewise, during the 1990s, the letters F.R.I.C.S. became FRICS.)

The Institution's Malaya Branch re-formed and held its first post-war gathering on 1 November 1946 at the Adelphi Hotel, Singapore, to hear a talk from Singapore's Forest Officer Charles Owen Flemmich on the use and classification of Malayan timbers. A type of lowland rainforest tree from the Dipterocarpaceae family was eventually named after him *(Shorea flemmichi Symington).*

On 28 February 1947, the first post-war General Meeting of the Institution's Malaya Branch was held at the Adelphi Hotel again. The re-formed Malaya Branch Committee had Francis "Frank" Angell as its Chair and Colonel Cornelius Burt as Vice-Chair. The Branch Committee members all had key roles in helping Malaya and Singapore recover from the severe impact of World War II. Aside from Angell and Burt, the other Chartered Surveyors based in Singapore in 1947 were:

- **Eric Clarence Watson** of Horace W. Langdon & Every. He is, of course, mentioned in Chapter 8. Watson continued to work for Horace W. Langdon & Every (incorporating Waters & Watson) from 1946 until 1956. However, his time working on the "Death Railway" during the Japanese Occupation had taken its toll on his health, and he never fully recovered. Indeed, whilst convalescing in Cameron Highlands, he died on 5 February 1956. He was buried in Ipoh, Malaysia. A memorial service was held at St. Andrew's Cathedral in his honour later that month.

John Aspinall

John George Aspinall was **Deputy Assessor at the Assessment Department, Municipal Buildings** (later called City Hall) in 1947. Aspinall's career in Singapore is featured more in Chapter 19. He arrived in early 1947 to take over the role vacated by Charles Frederick Snell, who returned to England permanently in 1946. In his first year in the role, Aspinall announced that would be new assessments in 1948, which would, in the case of properties not assessed since 1941, increase the rates levied by more than three times the current payment. Furthermore, he now had the power to reassess properties where landlords had benefited from rent increases under the Control of Rent Ordinance, i.e., where the Board had fixed an increased rent for a "rent controlled" property. However, he added that the full effect of his task would not come to fruition until the 1949 assessment due to a shortage of professional staff. He also stated that unlike

Snell during the pre-war days, he would, in the case of non-controlled properties, take sublet rents into consideration, as that was more of an indication of true market value and, thus, the assessed value. He would also take into consideration the position where tenants had done repairs and improvements to a property that was the landlord's responsibility. He also announced that Hotels, including the Raffles, Adelphi and Sea View Hotels, would henceforth have to pay higher rates. His vigour and determination to inspect properties and not "assess blindly" was not well received by landlords, who now faced the prospect of having to pay a lot more in rates than they were paying during pre-war days.

Frederick Jones-Davies

Frederick ("Freddie") William Jones-Davies operated **Jones-Davies & Smith from 33 Paterson Road**. Jones-Davies was, in 1947, a Branch Committee member, as he had been in 1939–40. He is mentioned in Chapter 8. He returned briefly to Singapore after the end of World War II to continue Jones-Davies & Smith alone because his partner (Arthur Kenneth Smith) did not return — Smith went to work in Palestine. Even Jones-Davies' return to Singapore was short-lived, as he moved to the United States to be with his cousin Dr. Donald Honeywell, founder of the Honeywell Foundation School in Washington D.C. (see Chapter 8 for more on him). Jones-Davies returned permanently to the UK in the mid-1950s.

The eight other Chartered Surveyors in Singapore in 1947 were:

- **Joe Marsh**, Branch Honorary Secretary and of Jessamine & Marsh, French Bank Buildings, Raffles Place. He is, of course, mentioned in Chapter 8.
- **Lieutenant-Colonel Horace Riley**, Royal Engineers. A Chartered Quantity Surveyor, Horace was on the Branch Committee. He worked in Singapore as a Senior Surveyor of Works for the Allied Land Forces (SEAC).

- **Graham Stuart Robson**, Branch Committee member. He was based at Naval HQ, which was located in Union Buildings, Collyer Quay.
- **Seah Mong Hee**, Branch Committee member and of Horace W. Langdon & Every. He is, of course, mentioned in Chapter 8 and elsewhere in this Book.
- **Charles Eric Side**, Branch Committee member. A Superintendent Civil Engineer at H.M. Naval Base. He is also mentioned in Chapter 8.
- **Keith Alan Suckling**, Branch Committee member. Suckling came to Singapore in 1947 with the Air Ministry Directorate-General of Works, RAF Meyer Chambers, on Raffles Place, and then joined Jessamine & Marsh's Singapore office in 1952. He also worked for Swan and Maclaren for a short time. He is mentioned again in later Chapters as he remained in Singapore for 35 years. He was born in Eton, Berkshire, England, in 1912, where he married in 1939. After retiring in 1982, he returned to England and died in Brighton, England, on 30 March 1992. Suckling played an active role in Branch Committee matters from 1947 right through until the Branch was dissolved and replaced with the R.I.C.S. "Corresponding Secretary" role in 1971. Indeed, he played a key role with Stan Godfrey in securing R.I.C.S. support for the dissolution of the Branch to encourage the growth and advancement of local professional bodies — these bodies eventually merged in 1982 to become the SISV.
- **John Jessamine** of Jessamine & Marsh, French Bank Building, Raffles Place. A Scotsman who spent most of his career in Ipoh, not Singapore, so he will feature more in a future book on Chartered Surveyors in Malaysia. However, he is also mentioned later in this Chapter.
- **Glanville Clayton**, who was also on the Branch Committee. He was in the Claims and Hirings Department of Prudential Assurance, whose offices were at the Mercantile Bank building in the immediate post-war years. The Mercantile Bank building on Raffles Place was demolished in the late 1970s. It was situated where The Providore Coffee House and Raffles MRT Exit E now stands.

1948 AGM of the R.I.C.S. Malaya Branch

On Friday, 9 April 1948, the Royal Institution's Malaya Branch's AGM was held at the Roof Garden of the Adelphi Hotel in Coleman Street. The event was presided over by John Jessamine, who told those present that there was a large number of well-educated local "boys" wishing to make a career as Quantity Surveyors and were studying for the Royal Institution's examinations. Six applications for admission to the class of students had been submitted, but only one had the qualification and was admitted. A record number of 17 applications were received for the examinations, of which five were rejected and four withdrew.

Jessamine also announced that the Branch Committee had approached the War Department to ask for the revision of rates for subsistence allowances included in agreements with private Quantity Surveyors in Malaya — revised rates had been agreed upon and were now being incorporated into agreements.

The Branch Committee at the time of the 1948 AGM consisted of:

- Chair — Eric Watson, FRICS
- Vice-Chair — Robert Cotton Sims, FRICS
- Honorary Secretary — Joe Marsh, ARICS
- Honorary Auditor — Charles Richard ("Dick") Holmes, ARICS
- Lieutenant Colonel W.C. Payne, OBE, FRICS — he was a Quantity Surveyor who had retired back to England by 1949
- John George Aspinall, ARICS
- Thomas F. Stevenson, ARICS
- Major R.J. Widgery, ARICS
- Seah Mong Hee, ARICS
- Keith Alan Suckling, ARICS

There are conflicting messages within John Jessamine's statement at the AGM. On one hand, he said that well-educated local "boys" were seeking to become Quantity Surveyors, but on the other hand, he makes it clear that only one had the right qualification to go forward. The lack of progress was partly due to their careers having been put on hold during the Japanese Occupation during World War II. The tragic death of David Waters did not help, as he was a huge champion for the advancement of locals in Singapore. However, Jessamine's reference to men as "boys" was a reflection of attitudes towards locals during these years, and again, there is no mention of women in the profession.

Following this, in May 1948, it was announced in *The Straits Times* that the Council of the R.I.C.S had raised the entry standard for student membership. Now, students from 1 March 1949 had to possess a School Certificate with credits or higher standard in no less than four subjects, including English and Elementary Mathematics. From the same date, the R.I.C.S. Preliminary Examination and Special Test standards were raised. The *Morning Tribune* of 9 June 1948 invited applicants to ask the Honorary Secretary for application forms for the next round of R.I.C.S. examinations in 1949, with the deadline for applications being 31 July 1948. Applicants were advised to be precise as to which examination they wished to take, and the forms had to be "duly subscribed" by a Fellow of the Institution and the candidate's Principal. The same happened in 1949 for the 1950 examinations. Progress can, in part, be attributed to the efforts of Seah Mong Hee and the Branch Honorary Secretary Joe Marsh in making all efforts to ensure that the invitation was open to everyone. However, given the context of the 1948 AGM speech by Jessamine and the worrying lack of local candidate success, this and subsequent Chapters will devote some time to how this unfolded both before and after the independence of Malaysia and Singapore from colonial rule.

In 1948, Horace W. Langdon paid his first visit to the region, and in February of that year, he was in Singapore prior to visiting Hong Kong and Borneo, with his return to England planned for May 1948. He visited Singapore (see Figures 10.1 and 10.2) for client meetings with Frank Angell and Eric

Watson and for pleasure — he told the *Malaya Tribune* that he was the only Englishman to have played golf on over 650 courses, and he looked forward to adding to that list during his trip. The *Tribune* mentioned that his practice managed the Quantity Surveying needs of the Government in Great Britain, Malaya and Iran, as well as large businesses. At that time, Langdon was Mayor and Alderman of the Metropolitan Borough of Holborn, an R.I.C.S. Fellow and a holder of the London Grand Rank of Freemasons.

FIGURE 10.1. An extract from a 1948 Horace W. Langdon & Every Staff Photograph, with Francis "Frank" Angell (centre) and Horace Langdon sitting to his right, on the occasion of Langdon's first visit to the Singapore office. James Khoo (centre, top row) was then a trainee (he is further mentioned in Chapters 11 and 16) and Charles Holmes (featured towards the end of this Chapter) is in front of Khoo

The year 1950 saw the announcement of a new R.I.C.S. Malaya Branch Committee, with the outgoing Chair Robert Sims (a Quantity Surveyor who was based at H.M. Naval Base, Singapore) announcing at the AGM (held once more at the Adelphi Hotel) that the number of examination applications had doubled — there were 12 applications for student membership and 20 for Fellowship. There was a total of 58 R.I.C.S. qualified professionals now, 8 more than the previous year. This growth tells us that the demand for qualified surveying professionals in the region, in particular, Quantity Surveyors, was increasing rapidly.

FIGURE 10.2. Another extract from the 1948 team photograph showing John Kershaw (who is mentioned elsewhere in this Chapter) (front, left) and Seah Mong Hee (back, right), who features in Chapter 8 as well as in subsequent Chapters.

Officers of the Branch were all Singapore-based despite at least half of the Branch membership being based in the "Federation of Malaya". This "power shift" towards Singapore would no doubt have been a factor in the decision seven years later to create a local professional body in Malaya on the occasion of the country's independence.

Indeed, by this time, even the Honorary Secretary role had moved from Kuala Lumpur (Joe Marsh) to Singapore (Thomas Stevenson). However, given that Marsh's son was born in Kuala Lumpur that year (perhaps his priorities had shifted), and indeed, soon after that, the Marsh family decided to move permanently to England.

The 1950 R.I.C.S. Malaya Branch Committee consisted of:

- Branch Chair — Eric Clarence Watson
- Vice Chair — John Jessamine
- Honorary Secretary — Thomas F. Stevenson
- John George Aspinall
- Aubrey Distin
- Robert Neil
- Ronald H. Ricketts
- Keith Alan Suckling
- Lieutenant-Colonel Alexander Henry Higgs
- Squadron Leader Robert Henry Downer

There were 34 Chartered Surveyors living and working in Singapore between 1949 and 1951, including the aforementioned ten individuals who were on the Branch Committee. The majority of these individuals worked for the UK Military service, either with the Royal Engineers/Far East Land Forces (FARELF), the Civil Engineer's Department at H.M. Naval Base, or the Air Ministry Works (RAF). Others worked in the public sector with the City Council or Public Works Department, while others worked in private Quantity Surveying practices, the most significant one during this period being Waters, Watson & Angell, etc., which, in 1948, was absorbed into Horace W. Langdon & Every.

All 34 Charterer Surveyors were involved in the post-war effort to repair the damage done to both military and non-military buildings, as well as carry out initiatives (both public and private) to rebuild and grow economies that had been so badly affected by the war. Examples of construction activity at this time are the Asia Insurance Company building on Finlayson Green and the China Bank Building on Battery Road — two of several higher-rise offices buildings for which construction started in these years — and also a number of higher-rise public housing provided by the Singapore Improvement Trust and the Singapore City Council in Chinatown and Monks' Hill, respectively. Also, the Municipal Commission's plans for a $90 million power station were unveiled in 1950, with Pasir Panjang Power Station (the "red brick" power station) being completed in 1953.

The rest of this Chapter is devoted to each Chartered Surveyor and the role they played in Singapore during that time:

Fellows

- **Francis ("Frank") Leslie Angell:** He is mentioned in Chapters 7 and 9 as he arrived in Singapore in 1937. The period from the late 1940s to the early 1950s proved to be a turning point in his life. After World War II, he joined Horace W. Langdon & Every in London, but he returned to Singapore in 1946 to form Waters, Watson & Angell, etc., with Eric Watson, which in 1948 was absorbed into Horace W. Langdon & Every. By 1950, he was a Fellow of the Institution. He married Daphne Whyatt at St. Andrew's Cathedral in 1948, and they had a son (Philip) who was born in England in 1950. When Watson died in 1946, he became Senior Partner of the practice in Asia but made the decision to return to England with his family permanently in 1956, still working for the firm in its Southampton office. He came back to Singapore in 1961 with the firm for a short period but returned to England once again and retired from the practice in 1970.

- **John Jessamine (Vice Chair of the Branch Committee in 1950 and Branch Chair in 1951):** A Fellow of the Institution, he was a partner of Jessamine & Marsh, Uniteers Building (where UE Square is now located, at the junction of River Valley Road and Clemenceau Avenue). He is mentioned at various points in this book, and for most of his career, he was a planter and estate manager in Ipoh, where he died in 1956. He will feature more in the future book on Chartered Surveyors in Malaysia.
- **Luis Sigismund Himely (1906–84) (Senior Surveyor in the Singapore Survey Department):** He qualified as a Chartered Land Surveyor in 1950 and was in Singapore for only a short time before returning to Malaya, where he became Deputy Chief Surveyor and Surveyor-General. Indeed, most of his career was in Malaya. He was born in 1906 in Cuba, but in the 1920s, his family moved from Cuba to England. After marrying in Devon, England, in 1934, he came to Malaya and worked alongside Basil Walker-Taylor, whom he was interned with in 1942 by the Japanese in Singapore. Fortunately, his wife and children managed to get to Australia before the Japanese invasion. He retired to England around 1959–60 at the time of the Malayanisation programme (more on that later in the book) and died in 1984 in Devon, England.
- **Eric Clarence Watson (Branch Chair):** He is separately mentioned in Chapters 8 and 9 and also throughout this Chapter.
- **Robert Cotton Sims:** He was the 1949 Chair of the Branch, a longstanding member of the Institution, having qualified in 1929, and a Fellow since 1944. He was in the Civil Engineer-in-Chief's Department, H.M. Naval Base, Singapore.
- **Major Richard John Widgery, Deputy Senior Surveyor of Works in the Chief Engineer's Branch HQ Singapore District:** Born in Devon, England, in 1904, he qualified as a Chartered Quantity Surveyor in 1940 and became a Fellow in 1949. He was a former Branch Committee member.

Colonial Fellows

Cornelius Burt

Lieutenant-Colonel Cornelius George Burt was a Branch Committee member (Vice-Chair in 1947) living at 34 Stevens Road, Singapore, in 1950. Prior to World War II, he was the Command Land Agent based at Fort Canning, and from 1947 to 1951, he was working in the Claims Commission & Hirings team within the FARELF. He was the first Singapore-born person to become a Chartered Surveyor — he was born on 24 May 1892 and baptised at St. Andrew's Cathedral on 9 July 1892, the son of Annie Emma Burt (née Smith) and Cornelius John Burt, who was for many years a Prison Warder at the Singapore Prison.

His family was originally from London and came to Singapore circa 1890. Burt became a Chartered Surveyor prior to World War II, and as a member of the Singapore Volunteer Corps, he was awarded the OBE for his services to the Corps in 1937.

In 1940, he was involved in an interesting case brought before the Rent Assessment Board, which included Charles Frederick Snell as a Board member. The case involved seven squatters on War Development Land. Burt told the Board that they had no jurisdiction over the matter, as the land on which the squatters were located was required for urgent war business. The Board decided to hear the case but, in the end, made no order on the matter, asking instead that Burt ask the Chinese Protectorate to explain to the squatters why they needed to vacate the land. In post-war Singapore, one task that fell to the FARELF was to assess, negotiate and settle compensation claims from contractors for work carried out prior to the Japanese Occupation. Much of this would have been for land and construction work, for which Burt's expertise would have been vital.

Burt does not have any known living descendants, although he was married on 23 September 1939 at the Singapore Presbyterian Church to

Alice Turner of Ayrshire. Newspapers of the time described it as a "military wedding of interest" and a "quiet War Wedding". See Figure 10.3.

MALAYA

Major Weds

Major C. G. Burt, Command Land Agent, Malaya Command, was married at the Singapore Presbyterian Church on Saturday to Miss Alice Turner, daughter of Mr. and Mrs. Peter Turner, of Ayrshire, Scotland. Major and Mrs. Burt are here seen after their wedding.—Free Press picture.

FIGURE 10.3. Photograph of Cornelius Burt and his wife Alice on their wedding day, taken from the *Singapore Free Press and Mercantile Advertiser*, 25 September 1939.

Harry Ward

Harry Laurence Ward (1902–96) worked for the Singapore Municipality (later City Council) as Chief Surveyor. More on him later in this Chapter. He was born in Queensland, Australia, and was an Honorary Fellow of the Institution of Surveyors, but he was also designated as a Colonial Fellow of the R.I.C.S. in 1950. He left Australia to join the Malayan Survey Department in 1925 as a "Surveyor-on-Agreement" after working in private practice as a junior surveyor and then with the Brisbane Water and Sewerage Authority, where he obtained his surveyor's licence in 1925.

Most of the next 14 years were spent mapping the Malayan jungle. During that time, when he was in Taiping with the Topographical Branch of the Survey Department, he came to Singapore in December to marry Jean Smyth in 1928 at St. Andrew's Cathedral. She came to Singapore from her home in Queensland, Australia, via the ship *Niew Zeeland* for the marriage.

He moved from the Malay Peninsula to Singapore in 1939 to become Senior Surveyor and Acting Deputy Chief Surveyor with the Survey Department.

At the time of Singapore's fall to the Japanese, he was a member of the Singapore Royal Artillery Volunteers, part of the Straits Settlement Volunteer Force. His wife managed to escape, and his three children were already at school back in Queensland, but he was interned by the Japanese on 15 February 1941 at Changi. He was told one day that he was going to a holiday camp and that he should bring a musical instrument, but this turned out to be the "Death Railway" in Thailand. Despite being asthmatic, he survived the ordeal, although he suffered very badly at the hands of his captives, and returned to Australia in October 1945 weighing only 7 stones. He would have received preferential treatment had he told his captors that he was a surveyor, but rather than assist them, he kept quiet such was his character.

In 1946, he returned to Malaya, and then moved to Singapore in August 1949 as Chief Surveyor. He retired to Australia in December 1952. After two years of working on his sons' farm in Queensland, the lure of surveying took him to Thiess Brothers as a surveyor in construction, but he left in 1962, when they formed a partnership with Mitsui. In 1965, Ward took a position at the University of Queensland as a tutor in surveying, a role he kept until he finally retired in 1982. Ward passed away in Brisbane in November 1996, aged 94, after a career spanning 75 years.

Professional Associates

- **John George Aspinall:** He became an R.I.C.S. Fellow in 1956 and was Singapore's Municipal (later City) Assessor & Valuer in the Assessment & Estates Department. His career in Singapore is featured in Chapter 20.

Aubrey Distin

Known as "Dizzy" to his Naval colleagues, **Aubrey Distin (1912–99)** had been a Professional Associate of the Institution since 1938. He was a Quantity Surveyor in the Superintending Civil Engineers (SCE) Department at H.M. Naval Base, Singapore, from 1949 until around 1959. He was born in Devon, England, in 1912 and married Adeline Jarvis in 1939 at the start of World War II. Their son Philip was born in 1942. See Figures 10.4 and 10.5.

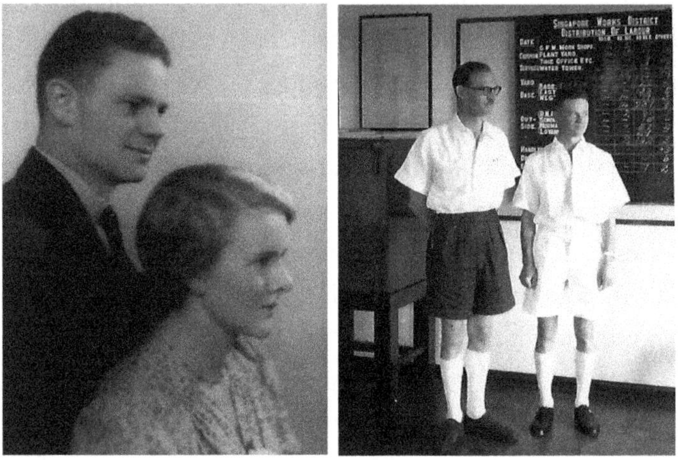

FIGURE 10.4. (a) Engagement photo of Aubrey Distin and his then fiancée Adeline Jarvis. (b) Distin at Singapore Naval Base around 1950–51 with a colleague.

Distin and his wife arrived in Singapore in 1949, and by 1951, he was an R.I.C.S. Malaya Branch Committee member.

FIGURE 10.5. Aubrey Distin (extreme left) with his wife Adeline Jarvis and colleagues at an Admiralty dinner at the Capital, Singapore, in 1950.

During his time in Singapore, Distin would have played a role in the major task of repairing and upgrading H.M. Naval Base facilities. From 1951 onwards, the Naval Dockyard was used to service a wide range of naval vessels involved in various conflicts across Asia, such as the Korean War. Distin and his team must have been involved in the maintenance and upgrade of the Dockyard facilities to cope with this additional demand.

In 1980, they both retired to Canada to be closer to their son, who had emigrated in 1967 — Phil was a Professor at McGill University's Department of Metallurgical Engineering at McGill Montréal until his death in 2016. Distin passed away in Canada in 1999, and his wife died in 1995.

- **Robert Neil** (see Figure 10.6): He was a Quantity Surveyor who qualified as a Professional Associate in 1949 and was a Branch Committee member in 1951. He worked at H.M. Naval Dockyard, Singapore, in the Civil Engineer-in-Chief's Department. He worked with Aubrey Distin. He was a Branch Committee member in 1950. He was also Captain of the successful H.M.S. Dockyard football league team in Singapore during the early 1950s.

FIGURE 10.6. Boxing Day celebrations in 1949 at the Singapore Swimming Club. Robert Neil is third from left while Aubrey Distin is at the centre of the photograph.

- **Squadron Leader Robert Henry Downer:** Downer had been a Chartered Quantity Surveyor since 1935. He was, from 1950, a Quantity Surveyor in the Resident Engineer's Office in the Air Ministry Directorate-General of Works (AMDGW) at RAF Changi, where his role would have been the costing and quantification of building works for the construction of new British Military buildings, the new Lloyd Lees Officers' housing estate at Changi and the school, and the repairs or reconstruction of buildings damaged or destroyed during the Japanese Occupation. He was also a Branch Committee member from 1950 to 1951.
- **First-Lieutenant Henry Green:** A Professional Associate since 1943, he was a Quantity Surveyor and Superintending Engineer, AMDGW, HQ, Malaya Works Area, RAF, GPO, from 1950 to 1952.

Arnold Greig

Arnold Lessel Macmorland Greig was a Land Surveyor who qualified as a Chartered Surveyor in 1949 (later in life, he became a Fellow) and worked at the Survey Department, "E Branch" GHQ, FARELF, GPO, Singapore. He was Surveyor-General to the Malaya Federation for most of his surveying career, and indeed from 1937 to 1963 (the year when he was awarded the CBE), he was in the Malay Peninsula, apart from 1950 to 1951 in Singapore. So, this book will not feature his experiences during World War II — this will remain for the future book on Chartered Surveyors in Malaysia. *Te Ara, The Encyclopedia of New Zealand* (1966) tells us that Arnold Greig was born in South Africa on 24 February 1913 but was educated at Auckland Grammar School and Auckland University College in New Zealand. He was a member of the Institute of Surveyors (NZ) as well as a Chartered Surveyor.

From 1937 to 1941, he was in Malaya in the Topographical Division of the Survey Department, but at the end of 1941, he was attached to the Survey of India for specialist training. After that, he joined the Field Survey Company of the FMS Volunteer Force as a Lieutenant and spent three years as a prisoner-of-war in Singapore, Korea and Manchuria. From 1946 to 1958, he was with the Survey Departments of Perak, Singapore, Kelantan, Kedah, Malacca, and headquarters. He became Deputy Surveyor-General in 1958 and Surveyor-General for the Federation of Malaya from 1960. The Auckland War Memorial Museum holds a range of stories and information on Greig provided to them by his daughter.

- **Donald Musgrave Harding (1923–84)**: He became a Chartered Quantity Surveyor in the 1940s and was working for Horace W. Langdon & Every in Singapore in 1951.
- **Lieutenant-Colonel Alexander Henry Higgs (1896–1962): He was a Quantity Surveyor, a Professional Associate of the Institution since 1931, and a Branch Committee member in 1950.** He worked at "E Branch" GHQ, FARELF, GPO, Singapore. Born in Kent, England, in 1896, he retired to England in the mid-1950s and passed away in Kent in 1962.

Charles Holmes

Charles Richard Holmes (1914–2006) (see Figure 10.1) was a Professional Associate since 1939, and he worked for Horace W. Langdon & Every in Singapore from 1948 until he moved from Singapore to Hong Kong in 1949 to re-open their Hong Kong office due to the potential for war office work. That never materialised, but through his leadership, the Hong Kong office not only survived but thrived — within three years, he had secured Hong Kong Land, Hong Kong Housing Society and Hong Kong Housing Authority as their clients. He stayed in Hong Kong until 1962 as the Resident Partner of what was by then Langdon & Every (Far East).

Holmes is remembered as an old-school professional and a gentleman who was impeccably polite and cheerful. In 1961, he passed through Singapore *en route* to England for a six-month break with his wife and three children and made a passionate plea in the *Singapore Free Press* on 15 May 1961 for the Government to step in and encourage more people to become Quantity Surveyors. He mentioned the massive housing development programme both in Singapore and Hong Kong and stated that the continuation of developments on this scale would rely on more trained Quantity Surveyors coming into the industry. He died in 2006, aged 92.

- **John Alan Kershaw (1908–2003)** (see Figure 10.2). He was a Chartered Quantity Surveyor who worked for Horace W. Langdon & Every, Singapore. Born on 12 June 1908 in Horsham, England, Kershaw qualified as a Professional Associate in 1930, and when in Singapore, he married Frances Wilson (1900–81), a Californian, on 8 May 1953. He was also the best man at Frank Angell's wedding at St. Andrew's Cathedral in 1948. He later moved permanently to California with Frances and died there on 9 August 2003.
- **John (J.R.C.) Winship, a Chartered Surveyor (Fellow)** who came out from Horace W. Langdon & Every's London office to Singapore for a short period in 1949–50 to ensure a smooth transition in the difficult post-war years.

Gordon Lowden

Gordon Michael Lowden (see Figure 10.7) became a Chartered Surveyor in 1948, and from 1950, he was an assessor in the Assessment & Estates Department, Singapore Municipality (later called the Singapore City Council), working under John Aspinall. He was also the Secretary of the Senior Officer's Association of the City Council — the President, Aspinall (featured in Chapter 20), was his manager. In his private life, Gordon was the President of the Royal Singapore Flying Club, based at Paya Lebar in the mid-1950s. Gordon was born in Barrow-in-Furness, England, in 1920, got married in England in 1946 and moved to Singapore in the late 1940s.

He was the R.I.C.S. Malaya Branch Secretary in 1951. Lowden moved to Australia after his years in Public Service in Singapore. He was previously a prisoner-of-war during World War II, and on 4

September 1944, he was one of 1,318 British, Australian and American prisoners-of-war onboard the *Rakuyō Maru* when it set sail as part of a convoy of 13 Japanese ships (HI-72) leaving Singapore that day. The prisoners-of-war had prior to this been working on the Thailand-Burma "Death Railway". The ships also contained bauxite and the ashes of fallen Japanese soldiers. On 12 September, the convoy was attacked by US submarines, and one of the casualties was the *Rakuyō Maru.* A few Japanese crew members were rescued by the rest of the convoy, but the majority of prisoners-of-war perished either during the attack or were killed as they tried to escape in rafts. A total of 1,159 of the 1,318 men perished, but Lowden survived, having been rescued by a US submarine. He went into a period of rehabilitation before returning to England, as the photograph below shows.

FIGURE 10.7. Photograph of the survivors from the *RakuyōMaru* during their rehabilitation. Gordon Lowden is seated on the far right of the photograph.

- **Henry Robert Reed**: A Quantity Surveyor working for the Chief Royal Engineer's Department in Singapore.
- **Ronald Horace Ricketts:** A Chartered Surveyor working at the Singapore PWD He was a Branch Committee member in 1951. An active sportsman, among the sports he played was cricket — he was a

regular in the 1st team for cricket for the PWD circa 1948.

- **Seah Mong Hee:** He was an R.I.C.S. Malaya Branch Committee member in 1950. With the expansion of Horace W. Langdon & Every in Asia, he was appointed as a Partner of the practice in 1949. From 1949 to the early 1950s, the practice grew as it started to attract local clients through the influence of Seah, and its activities grew across Asia. In Singapore, it was the largest practice of its kind. He was involved in 1951 in a project that will be of interest to Singaporeans reading this book — the construction of the former Singapore Badminton Hall Building on Guillemard Road, for which he was a member of the Hall Committee. The former Hall is now marked by the National Heritage Board as a historical site.
- **William Henry Smart:** He was a Chartered Surveyor from 1949 who worked in the Survey Office and then in the Singapore Improvement Trust (SIT). There will be more on him later in this Book in relation to his work with the SIT
- **Charles Oldfield Barnett:** He was a Chartered Land Surveyor from New Zealand who worked at the SIT in the late 1940s and early 1950s. His work as a Land Surveyor in those years was invaluable to the SIT during that time for the development of marshland and reclaimed land into social housing to meet the growing housing needs of the island. In 1952, alongside Denis Komlosy, he was a founding member and office-bearer of the Malayan Branch of the Town Planning Institute (now known as the RTPI).
- **Alexander Ewing Smith:** He was a Chartered Surveyor who worked at H.M. Naval Dockyard, Singapore, in the Civil Engineer-in-Chief's Department.
- **Major Harold Davey, MBE**: He was a Quantity Surveyor who worked for the Royal Engineers. In 1950, he was Chief Engineer, Singapore.
- **Thomas Francis Stevenson:** He became a Chartered Surveyor in 1932 in Singapore and worked for the Singapore Municipality Assessor's Department. He became a Fellow in 1950. As the Malaya R.I.C.S.

Branch Honorary Secretary in 1950–51, he was heavily involved in the Institution's activities in Singapore until he returned to the UK in 1958–59 as a result of the Malayanisation programme.

- **Leslie Hayward Stone:** He was a Chartered Surveyor working in the War Department, Lands Office in Singapore from 1950 to 1952. His role would have centred on the repair and maintenance of military buildings, many of which had been badly neglected during the Japanese Occupation or badly damaged in the build-up to it. He may also have been involved in the handing back of (and compensation for) previously requisitioned land and buildings to their property owners and the sale (or transfer to the Municipality) of buildings no longer needed for military use.
- **Keith Alan Suckling:** He was a Chartered Quantity Surveyor since 1946, who also worked for Jessamine & Marsh, Singapore. He was a Branch Committee member from 1950 to 1951. He remained in Singapore until he retired to live in East Sussex, England. With Johan McDonald, he founded Suckling McDonald Quantity Surveyors (later Suckling, McDonald & Mohd Ishak) — see Chapter 17.

Basil Thompson

Basil Ernest Thompson (1913–92) (see Figure 10.8), also known as "Tommy", was born in Surrey, England, in 1913 and qualified as a Chartered Surveyor in the pre-war days. He was a Captain in the British Royal Marines and served in Asia after graduating from Sandhurst Military Academy before World War II. He arrived in 1945 with the liberating forces, and by 1946, he was an Assistant Engineer of the Sewage Works for the Municipal Commissioners. He later became Chief Sewerage Engineer and Deputy City Engineer in 1962, followed by a promotion to City Engineer in 1956. See Figure 10.8 for a series of photographs of him taken over the period of his career and into his retirement.

He lived in the late 1940s at 22 Chatsworth Road, Singapore, which still stands today. In Thompson's time, it was a colonial bungalow set amongst the Black and Whites houses on Chatsworth Road reserved for senior Government employees. However, he and his wife, whom he married in 1955, later moved to Tanglin Road. In 1951, he gave evidence at a Coroner's Inquest into the death of a labourer who suffocated in a 13-foot manhole on Kim Chuan Road as a result of not wearing a life belt or a gas mask. Thompson gave evidence to confirm that the deceased had not followed procedure. Based on his evidence, the Municipal Commissioners were cleared of negligence.

FIGURE 10.8. Basil Thompson, before, during and after his time in Singapore.

It was announced in March 1958 that upon Thompson's retirement, his role would be subject to Malayanisation, with preference to be given to a "Malayan person" to replace him. This created a long debate amongst Councillors — some argued that Thompson's Deputy should be given the role due to his abilities and what they saw as "natural succession", but other Councillors felt that under the existing system of Europeans being automatically given promotion upon the retirement of their managers, locally born staff (referred to as "Malayans" in the Council Chamber) stood no chance of getting promoted. In the end, the decision was made to appoint a locally born staff member to the role.

Thompson did indeed retire from the City Council's employment in 1958 and migrated that year to Australia, flying (via Air India) to Sydney where he settled down. He passed away in Sydney in 1992.

George White

FIGURE 10.9. George White.

Wing Commander George Morham White (1913–97) (also known as "Chalkie") was born in Edinburgh, Scotland, and qualified in 1935 as a Quantity Surveyor. He was with the AMDGW, FARELF, and based at Changi. His photograph is at Figure 10.9.

His role would have been as a Quantity Surveyor involved in construction of new British military buildings, including the Lloyd Lees Officers' housing estate at Changi as well as the new school, and in the repair or reconstruction of buildings damaged or destroyed during the Japanese Occupation. During their time in Singapore, George, his wife and young son lived in a semi-detached sea-facing chalet bungalow in Changi. They arrived in 1950/51 and returned to the UK in 1953.

On his return to the UK, he worked in a variety of different locations in England for the Air Ministry and then with the Ministry of Public Building & Works before retiring at the age of 65. He then worked part-time in Bracknell, England, in private practice, until his 75th birthday. He and his wife moved to Brighton, where they both died in 1997.

- **Ernest George Verdun Cooke**: A Quantity Surveyor in 1950 in Singapore with the Air Ministry Works Department, RAF
- **Major Wesley Frank Willoughby, Corps of Royal Engineers:** He was a Quantity Surveyor who qualified in 1947, and he worked from 1950 to 1952 at the CRE, 162, Works Area, GPO, Singapore. In 1955, he was awarded an MBE, at which point he was a Lieutenant-Colonel (temporary). The story of his time and his experiences in prisoner-of-war camps in Singapore, Taiwan and Japan from 1942 to 1945 during World War II is held by the Imperial War Museum in London.

New Chartered Surveyors are Needed

Stepping back to April 1951, the new R.I.C.S. Malaya Branch was elected with Quantity Surveyor Lieutenant-Colonel C.F. Atkinson (later, temporary Brigadier) of the Royal Engineers as Chair, Harry Ward (Chief Surveyor, Singapore) as Vice-Chair and Gordon Lowden as Secretary. The Branch meeting-cum-AGM was attended by Dr. G. Etzel Pearcy (1905–80) of the United States (US) Geographic Society, who gave a talk on map-making, which gave the Chair an opportunity to welcome recently enrolled Land Surveyors from the Federation of Malaya and from Singapore.

Dr. Pearcy was a former director of the US State Department's Office of the Geographer and an expert in urban geography who served on US Government Committees dealing with maritime jurisdiction, the law of the sea and the codification of geographic names. By 1950, he had joined the US Foreign Service. He was famous for his radical proposal to re-map the US and (in so doing) reduce the number of States from 50 to 38, giving new names based on geographical features to the 38 realigned states. His idea was for a more rational system of state boundaries and improved efficiency in government, transit, planning and public utilities, which would "save billions of dollars".

In the same year, a Singapore City Council advertisement for an Assistant Assessor stated that the successful applicant should be a Professional Associate of the Royal Institution of Chartered Surveyors, the Chartered Institute of Auctioneers and Estate Agents or the Association of Rating & Valuation Officers. Experience in rating and compulsory purchase was required, with a salary per month in the $500 to $1,000 range depending on experience, plus a cost-of-living and housing allowance "if not provided with quarters". The last remark and requirement for a fully qualified surveyor are clear indications that the colonial administration still showed a preference for Expatriates over local talent. This must surely have added to the growing frustration of talented locals, who were continually being overlooked for public sector professional roles.

However, things were about to change.

Chapter

11

The Road Towards Malayanisation

In 1952, in what was to be a boom year for Singapore, the Chief Surveyor in the Survey Department, Harry Ward, retired to Brisbane after 27 years of service. A survivor of three years working on the "Death Railway" in Thailand in World War II, he proudly shared with *The Straits Times* that one of his biggest achievements was helping M.N. Menon and Nagalingam Rameswaran towards their goal of becoming professional surveyors. He stated that the standard of surveying had risen since its pre-war days, and that his team was now working on a project to survey all the low-lying lands in Singapore, to find land suitable for food production. A sign of the times was his additional remarks that the experiment in bringing women into surveying roles originated due to a shortage of men, but they had proved their worth, so the Survey Department would continue to employ women. Indeed, there was a *Straits Times* report on his "experiment" in September 1950 under the heading "Five Bright Girls". Ward explained that not only were the "girls" (Patsy, Priscilla, Lily and Katherine were around the age of 21 and Pauline was 18 — all locals) doing the same job as the "boys", but they were doing it equally as well and getting equal pay too, which was not the case in other Departments.

Their work as Junior Technical Assistants involved preparing permanent Survey Office records, tracing for the public, and preparing titles for the land office. They enjoyed their work but told *The Straits Times* that the boys and men in the office were "dull and solemn looking". They each earned a salary of $50 per month, with a yearly increment of $5 until they passed their examinations at the end of their third year. With housing and other allowances, they each got around $100 per month, which they said was not enough, and how long they could stay on would depend upon their salary scale. They all said that marriage would not prevent them from continuing in the job if the salary was right. Ward was certainly way ahead of others in that era with his equal pay commitment, and indeed, after his return to Queensland, he actively encouraged his former staff to come to Brisbane and study there in order to advance their careers, which brought him into connection again with them for many years to come.

In 1954, there were some further signs of positive change, albeit limited. The Singapore City Council made plans to give two scholarships to local "boys" to train as Chartered Surveyors in Great Britain. It was open to those under 21 with Grade 1 School Certificates, and selected candidates would have to "serve" in the City Assessors' Department for 6–12 months before going to Great Britain. During this period, they would get a subsistence allowance of $85 per month.

Unfortunately, there was still a widely held view within the British Expatriate community that local talents were not willing to get their hands dirty and apply their technical "brains" to the task at hand. This was despite the fact that when the Expatriate staff took their annual home leave (often for up to six months), local staff would step in and cover their duties without fail. In February 1954, a letter to *The Straits Times* argued that young Asian men and women were happy to go for technical study and "stuff" their brains with "theory", but when they entered the industry or public service, they were unable to "ensure that bearings are correctly bedded". The only positive to take from this letter was that Singaporeans were no longer being called "boys" and "girls"!

Fortunately, the large number of Singaporeans who got a professional qualification (such as the R.I.C.S. qualification) from the late 1950s onwards disproved this "theory". Perhaps this negative view was a contributing factor as to why no locally born surveyors successfully navigated the final hurdle to pass their final examination to become a

Chartered Surveyor under the Colonial rule from 1945 to 1957. This was despite the earlier success with Seah Mong Hee and several surveyors born in Malaya and Singapore who passed the first R.I.C.S. examinations in the 1940s.

It is quite clear that the Colonial Government's priority during this time was to bring in qualified and experienced Quantity Surveyors, Land Surveyors and valuers (Government assessors). They did not see it as a priority to train up the next generation of locally born surveyors. As a result, local surveyors were not encouraged to take their final R.I.C.S. Examination, although Expatriate surveyors were encouraged to progress. Another example of how this worked was when in 1955, the City Council was looking for assessors in the \$630–\$1,200 monthly salary band (plus allowances). Again, there was a requirement for a fully qualified surveyor (R.I.C.S., Chartered Institute of Auctioneers and Estate Agents or the Association of Rating & Valuation Officers) well versed in the rating of all property types, which meant that only Expatriates could apply with any chance of success.

There was at least a glimmer of hope in June 1957 when **Robert Lim Geok Eng** (an assessor in the Singapore City Council Assessors Department) passed the first R.I.C.S. Professional Examination. However, progress beyond the first examination was extremely rare.

However, with the acceleration of the Malayanisation process, the favourable environment that Expatriate Chartered Surveyors previously enjoyed was disappearing fast, especially within the Singapore City Council, so they either had to leave or "localise", which included adjusting to localised contract terms.

The Malayanisation process was a process (headed by the Malayanisation Commission) created to staff Government (civil) service positions with local people. It was generally acknowledged that progress towards self-government could only be realised if Expatriates in the Government service were replaced by local people and phased out over a period of eight years, subject to payment of compensation, a lump-sum gratuity and monthly pension based upon the period of service. To build up the local talent pool and enable the eight-year target to be met, the

Singapore State scholarship was created to attract and train officers for the public service.*

Of the Expatriate Chartered Surveyors working in the City Council's Assessors Department in 1958, three — all chartered surveyors — resigned: Thomas Stevenson (who had just only taken over as the City Assessor upon the official retirement of John Aspinall; see Chapter 20 for more on him), Gordon Lowden and Selwyn Lewis — the latter two were both Assistant Assessors. They were not the only Expatriate employees to resign. The Governor, Sir Robert Black, had been conscious of the dire shortage of qualified, experienced locals and the continued domination by Europeans, and so he accelerated the Malayanisation process through inducements for the Expatriates to retire. He did this in full cooperation with prominent local political and administrative figures, and in managing the transition, he dealt with their concerns and helped the Expatriate Europeans prepare for new careers.

By July 1958, there were further signs of progress, with locally trained candidates taking the R.I.C.S. Examinations, due in no small part to the efforts of **Langdon & Every (Far East)** and Expatriates in Government services such as **Derrick Morris** and **Selwyn Lewis**, who had started teaching local candidates part-time at Singapore Polytechnic. Of the five candidates who passed the examinations, three are noteworthy:

<u>Valuation Section, First Examination Passed</u>

Teo Hee Cher — He took the correspondence courses provided by the College of Estate Management whilst in Singapore and qualified as a Chartered Auctioneer and Chartered Real Estate Agent. Upon its merger with the R.I.C.S. in 1970, he finally became a Chartered Surveyor, at which

* The Malayanisation process was a process (headed by the Malayanisation Commission) created to staff Government (civil) service positions with local people. It was generally acknowledged that progress towards self-government could only be realised if Expatriates in the Government service were replaced by local people and phased out over a period of eight years, subject to payment of compensation, a lump-sum gratuity and monthly pension based upon the period of service. To build up the local talent pool and enable the eight-year target to be met, the Singapore State scholarship was created to attract and train officers for the public service.

point his designation became "ARICS". He worked for the Housing & Development Board (HDB) from the early 1960s to the 1980s in various roles, e.g., Principal Resettlement Officer based at Princess House and an Estates Officer based in Toa Payoh.

It is understood that Teo later migrated to Victoria, Canada, where he passed away in 2015.

Quantity Surveying, First Examination Passed

Lau Fook Sing — He was later a founding Partner in Lau Teo & Yong Quantity Surveyors, but at that time, he was working for Langdon & Every (Far East). There will be more on Lau in later Chapters. He is shown in the photograph in Figure 16.2.

Quantity Surveying, Final Examination Passed

Jack LeRoy — He was a Chartered Quantity Surveyor from London who first started working as a surveyor in 1949 and whose career led to professional activity in more than 20 countries over a 70-year career. He retired from Davis LeRoy & Partners in May 2019. From 1957 to 1958, LeRoy spent a year working as a Quantity Surveyor in South-East Asia, his first overseas job. He spent most of that time in Burma (now Myanmar), with intervening trips to Singapore. During a 3-month stay in Singapore in 1958, he took the opportunity to sit for his Final Examination here and passed. Several months later, he collected his R.I.C.S. Diploma in London on his return to England.

Champions for the Emergence of Local Talent

Some Chartered Surveyors, notably Harry Ward (see above and in the previous Chapter) and Basil Walker-Taylor (see next Chapter) and those who worked for the Singapore Improvement Trust took active steps to grow local talent in the face of adversity. Another person who helped to change things for the better for talented Singaporeans was Derrick Morris.

Derrick Morris (see Figures 11.1 and 11.2))

Derrick Morris (1933–2000) was a Chartered Quantity Surveyor who arrived in Singapore during this period to work for the Public Works Department (PWD) out of its High Street offices. Morris was born in Devon, England, in 1933. He qualified as a Chartered Quantity Surveyor in Bristol, England, in 1955, and after a period of National Service with the British RAF in Malta, he arrived in Singapore to work for the PWD on a 3-year contract. He arrived with his wife, June, in 1958 and daughter Susan, who was born in Devon in 1956.

Morris was highly regarded by everyone who knew him, and was seen to be a man of the highest integrity by those he trained to be Chartered Surveyors. His time in public service was spent in the costing, quantification and cost management of many public sector construction projects, including a major programme of school construction. He also lectured on Quantity Surveying at Singapore Polytechnic in the early 1960s.

The National Archives of Singapore (NAS) contains a wealth of Oral History Interviews, one of which is a series of interviews with Morris, which covers his life and career in Singapore, including his recollections when the Singapore City Council and the PWD amalgamated. For those with an interest in the events of this period and a desire to listen to a first-hand account of what it was like to work as a Chartered Surveyor in Singapore from the late 1950s until the early 1980s, a visit to the online NAS is well advised, to listen to this fascinating series of recorded interviews.

In 1961, June returned to England, where their second daughter, Kim, was born, and in 1962, Morris signed a 3-year contract extension with the PWD. However, in early 1963, he left the public service to set up a Frank & Vargeson office in Jesselton (now known as Kota Kinabalu). Morris and the family then moved to London in 1964 to work in the Frank & Vargeson office there, and after a further spell with them in Jesselton covering the whole of North Borneo, he returned to Singapore in 1967, at which point he took over the management of the offices of Frank & Vargeson in Singapore, as well remotely managing their Brunei office.

In 1967, the Singapore office of Frank & Vargeson was on the 2nd floor of Stamford House, Stamford Road, but in 1971, Morris moved the office to the Golden Mile Complex where he continued to be a champion for local surveyors, by employing and then partnering with local Quantity Surveyors who were both Singapore Institute of Surveyors (SIS) and R.I.C.S. qualified. By the late 1960s, he had become a Fellow of the R.I.C.S., and at this time, he and the family moved to Brunei for a year. On their return in 1971, the development of the Golden Mile (Woh Hup) complex was well underway — Frank & Vargeson was the appointed Quantity Surveyor for the project. The shopping centre was completed at the end of 1971, followed in 1972 by the office tower, and in 1973, by the apartments.

FIGURE 11.1. (a) Derrick (second from right) and June (third from left) Morris with Yong Nam Seng (Chairman of Woh Hup Pte. Ltd.) and Kwik Poh Lian (Yong's wife) at the Frank & Vargeson Christmas Party, 1963, in Jesselton, Sabah. (b) Derrick and June resurrect the PWD Family Meeting Place in Singapore, 1961.

FIGURE 11.2. The opening of the Golden Mile Woh Hup Complex, later renamed the Golden Mile Complex. Morris' firm Frank & Vargeson was appointed as Quantity Surveyor for many of Woh Hup construction projects around the region during this time, including this one. Left to Right: June and Derrick Morris and Kwik Poh Lian (wife of Yong Nam Seng, Chairman of Woh Hup Pte. Ltd.).

This period was a busy one for Morris as he was also the Quantity Surveyor for the development of the Westin Stamford Hotel and the Equatorial Hotel in Singapore. In 1974, he set up Frank & Vargeson, Hong Kong. Derrick and June went to Hong Kong to cover periods when the Country Manager, Michael Charlton (later with Morris in James R. Knowles), was on leave. In 1979, Morris left Frank & Vargeson to pursue his lifelong ambition of studying construction and engineering law (at Cardiff University) and obtained a degree in Building Law in 1981. In 1982, he studied for the Bar Examinations at Lincoln's Inn, London, then after, a year's pupillage at James R. Knowles in their Manchester, England office, in 1984. The lure of Singapore meant that Morris decided to study for the Singapore Bar (which he passed) — his London Bar qualification did not entitle him to an automatic Singapore Bar entry because he had given up his Permanent Residency. Whilst studying, Morris did some consultancy work.

In 1988, he was offered a Door Tenancy to practice as a Barrister at Lincoln's Inn, London, and then in 1990, he finally got to set up a James R. Knowles office in Singapore. There, he acted as lawyer for James R. Knowles in arbitration cases pertaining to Building Law. One of Morris' first of several successful arbitration cases for James R. Knowles was against the Hong Kong Government. Morris travelled 4 to 5 times a year to Singapore and then onto Hong Kong, Bangkok and Jakarta, acting as James R. Knowles' arbitration lawyer, so his location depended upon his case work. He also did arbitrations in Dubai, UAE and Oman. At the same time, he continued to give lectures and talks. When Morris obtained his law degree, it encouraged Lim Seng Lim and Low Ah Hai (who he trained as Quantity Surveyors and who worked for James R. Knowles) to take up law.

Morris' ties to Singapore remained until his death in England after a short illness on 18 December 2000.

Morris also brought on the career of **John Hughes**, who qualified as a Chartered Quantity Surveyor whilst with the PWD in Singapore, and who went on to become a key member of his Frank & Vargeson Singapore team.

Aside from helping **Chelliah Tiruchelvarayan** and **Loh Kum Seng**, another Singaporean helped and encouraged by Derrick Morris in his career was **James Khoo**.

James Khoo

James Khoo Cheok Chuan has already been mentioned in Chapter 10 as a young man, working as a Quantity Surveyor in the post-War office of Horace W. Langdon & Every, where in 1948, he passed the City & Guilds of London Institute's Intermediate Examination in Builder's Quantities.

Along with Lau Fook Sing and Toh Chee Seng, Khoo took the College of Estate Management correspondence courses and then his Final R.I.C.S. Examination.

He was the first of this particular group of pioneers to pass the Final Examination and become a Chartered Surveyor, which he did in London in 1959. He later became an R.I.C.S. Fellow.

Khoo joined the PWD in the 1950s, and after a period in London when he studied Quantity Surveying at the then College of Estate Management (now the University College of Estate Management [UCEM]), and with the benefit of a Government scholarship, he returned to the PWD in 1960 and was based at the PWD offices in High Street, Singapore. He later moved to the PWD office on the 11th floor of the new National Development Building when its construction was completed at the end of 1968 on Erskine Hill. The decision for a new National Development Building (which would also be the new Headquarters of the HDB) was made with a view to consolidating the various functions of the Ministry, such as the PWD (then in High Street), the Ministry Headquarters itself (which was then on Upper Pickering Street) and the Planning Department (in the City Hall). This Building is, of course, still there today.

Khoo worked with Derrick Morris on many public sector construction projects in Singapore from 1960 and later took over Morris' responsibilities when Morris moved into private practice in 1963.

Khoo remained with the PWD for the rest of his career apart from a spell teaching at Singapore Polytechnic in the 1980s and retired long before the PWD was corporatized into the PWD Corporation (in 1999), before being renamed as CPG Corporation in 2002. After retirement, he migrated to Western Australia but passed away in Thailand in 2007, aged 82.

A photograph of him at the 1966 R.I.C.S. dinner is shown in Chapter 16.

Langdon & Every (Far East)

In 1954, Horace W. Langdon & Every was renamed Langdon & Every (Far East) after Horace W. Langdon's death, aged 75; its name was changed to distinguish itself from the UK practice. They were now at the forefront of the development and growth of Singapore in the post-war years. Two excellent examples of this are the construction of the new Bank of China building at No. 4 Battery Road between 1951 and 1954, and the Asia Insurance Company Building, Finlayson Green, completed in 1955.

Langdon & Every (Far East) was the Quantity Surveyor for both projects. Langdon & Every (Far East) was at this time based at 14 Laidlaw Building (see Figure 11.4(a)), which is long gone from the Singapore landscape. On its site is Maybank Tower, close to the Fullerton Hotel on the Singapore riverfront.

The Chartered Surveyors within the firm's Singapore office during this period were **Seah Mong Hee, Charles Holmes** and **Peter Sanderson**. Of these individuals, Seah and Holmes have already been discussed in previous Chapters.

Peter Sanderson

Peter W. Sanderson (see Figure 11.3.) was a new arrival in Singapore in September 1956 to join Seah as a joint resident partner of Langdon & Every (Far East) in Singapore.

FIGURE 11.3. Peter Sanderson.

Sanderson joined Horace W. Langdon & Every in 1949, in their London, Persia (Iran) and Kuwait offices. He moved to Hong Kong in 1962 and took over as the Company's Hong Kong Partner on the retirement of Holmes, and he remained in that position until his retirement in 1978.

Singaporeans with the practice in the late 1950s included Toh Chee Seng and Lau Fook Sing, who both went on to become Chartered Surveyors. Both are featured later in this book.

As mentioned previously, Langdon & Every (Far East) was involved in the Bank of China Building construction (see Figure 11.4(b)). In November 1950, *The Straits Times* reported that after 13 years of delay, the 14-storey 200-feet tall Bank of China building would be constructed on a site at the junction of Battery Road, Flint Road and Boat Quay. The Bank of China had originally leased space in Cecil Street from the Great Eastern Life Insurance Co. Ltd., but in anticipation of that lease expiring, they had in May 1937 acquired office premises on the corner of Battery

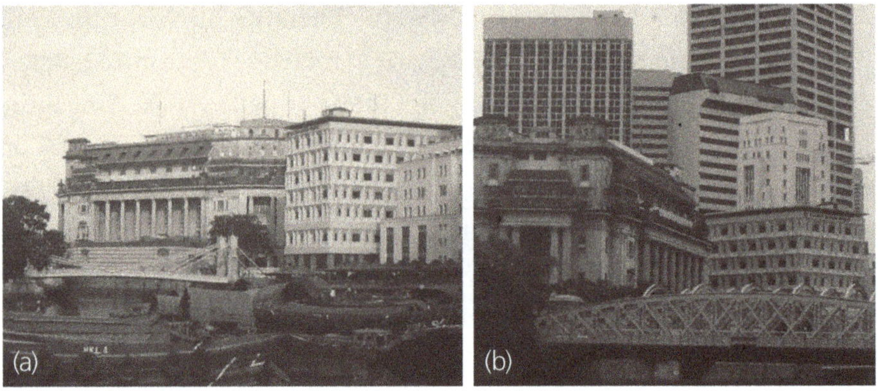

FIGURE 11.4. Fullerton Building with Laidlaw Building: (a) Langdon & Every (Far East)'s offices were in Laidlaw Building, now the site of the Maybank Tower, from 1947 until the early 1960s, and (b) the 14-storey Bank of China Building, completed in 1954, seen next to and behind the Laidlaw Building.

Road and Flint Street from Guthrie & Co. for $175,000, with a view to demolish and construct a substantial new office building on its site. The problem they encountered was in gaining possession of the premises from the various tenants in occupation, some had been in the building for many years. The process of gaining possession was interrupted by World War II and the Japanese Occupation, but they eventually succeeded in getting possession in 1947.

The 1950 newspaper article reported from the Bank of China Branch Manager (and architects Palmer and Turner) that rumours of the Bank's lack of desire to put up the $3,000,000 to build it were untrue, and the unforeseen construction delays had now been resolved, with the foundations laid, the concrete to the ground floor level having been completed and the plans for the upper floor having been handed over to Langdon & Every (Far East), who were the appointed Quantity Surveyors for the development. The target date for completion was 1952 but it was eventually opened in 1954, with Langdon & Every (Far East) managing the tenders for the construction works. The newspaper reported that the development site (21,270 square feet) would provide 90,000 square feet of office space (the first seven floors would be air-conditioned), with the Bank on the Ground and Mezzanine floors.

This was a landmark project for Langdon & Every (Far East), as this was an ambitious project in the immediate post-war years and was the start of the construction of a number of "tall" corporate buildings in Singapore, given its position as a key trading port in Asia. Singapore was fast becoming the HQ location for a whole range of organisations such as Banks, Insurance Companies, Import and Export firms and Shipping Companies — these organisations wanted to occupy prestigious buildings, so their role in the construction of these new landmark buildings was extremely important.

Early planning discussions for the construction of the Asia Insurance Building on Finlayson Green, (see Figure 11.5.) now known as the Ascott Raffles Place, started in 1948, with construction eventually starting in 1950–51. The architect was the renowned Ng Keng Siang (1908–67).

However, a series of problems bedevilled the construction, the main two being the requirement to store structural steel away from the site rather than the preferred location (which was on the "green" traffic island next to the site) and the very late discovery that there was no solid rock at the level that earlier surveys had indicated. The latter resulted in Ng having to seek specialist survey advice and the Municipal Architect (W.I. Watson) then asking for further soil testing expertise to be provided by Soil Mechanics Limited of London. This resulted in the need for deeper foundations as well as specialist construction machinery for the foundations (air compressors) to be brought over from England — experts in this type of foundation construction were also brought in to train up the local teams to perform the task. Finally, experts from the United States were brought in

FIGURE 11.5 The Asia Insurance Building on Finlayson Green, completed in 1955. The building has since been converted to serviced apartments known as Ascott Raffles Place.

to supervise and provide expertise on the steel frame construction. All this resulted in delays to the completion of the new 19-storey structure. It was finally completed in December 1955. Total costs were originally estimated at $3 million but the final figure reached $8 million.

Until 1971, it was the tallest building in Singapore.

Chapter

12

The Critical Transition Years: 1957 and 1958

This period of time encompassed the granting of independence to the Federation of Malaya in August 1957, as well as the British Parliament's passing of the State of Singapore Act in August 1958, which saw the transformation of Singapore from a colony to a state with control over all domestic affairs, including finance.

During these years, Langdon & Every (Far East) played a key role in Singapore's development. Their contributions and those of the Chartered Surveyors working for them during these years have already been covered in Chapter 11. It is worth mentioning that in these years, a frequent visitor to the Singapore and Hong Kong offices was **Cecil Tom Every (1902–87)**, who co-founded Horace W. Langdon & Every with **Horace William Langdon** in 1928, having first joined Langdon's own practice in 1921 in England. In 1958, Every was the sole UK Partner and held all the shares in the "Far East" operations in Asia. Prior to this, Langdon had been a more frequent visitor to Singapore. Every was interviewed by *The Straits Times* in January 1958, where he stated that they employed 35 local staff who were trained by the firm and had proved to be "capable quantity surveyors". Furthermore, they looked forward to training up more locals. Every's son, Giles Every (1931–84), became a partner of the firm in the early 1960s, and he too, became a frequent visitor to Singapore to support the progress of the firm until his untimely death in 1984.

At this time, Langdon & Every (Far East) continued to face competition for work from architectural firms. Indeed one such firm, Raglan Squire

& Partners (now known as RSP), even employed an English Chartered Quantity Surveyor for a while in their Singapore office (his name was **Robert Richard Bye**), but despite the competition from the mid-1950s onwards, Langdon & Every (Far East) continued to be successful, and it grew and diversified into other areas such as advisory and project management. This continued even when, in the 1960s, some employees of Langdon & Every (Far East), having become Chartered Surveyors, decided to set up their own quantity surveying practices in Singapore. More on them in later Chapters.

Some of the Chartered Surveyors mentioned in previous Chapters covering the late 1940s to mid-1950s continued to work in Singapore as before. They include Aubrey Distin, who continued his work as a Quantity Surveyor in the Civil Engineer-In-Chief's Department at the Royal Naval Base at Admiralty, as well as Keith Suckling. Also, during the 1950s, some Chartered Surveyors worked for the Singapore Improvement Trust — see Chapter 14 for their story.

In total, during the late 1950s, the number of Chartered Surveyors in Singapore was around 40 individuals at any one time. Aside from those already mentioned at the start of this Chapter and in Chapter 10, the rest of this Chapter features others who made a noteworthy contribution to Singapore's development in the late 1950s.

Chartered Surveyors in the Directorate-General of Works (the Air Ministry Works Department)

These Air Ministry Works Department (AMWD) surveyors were involved in the construction, improvement and extension of RAF Changi buildings, as well as maintenance of existing RAF buildings at other bases around the island, e.g., Tengah and Sembawang.

A 1968 *Straits Times* report tells the story of all the construction activities at Changi in those years under the article title "The RAF property that Singapore will inherit". It was stated that Changi was the largest of the three RAF Air Bases, standing at 3,500 acres. In 1968, the total number of RAF officer personnel at all three bases was 7,500, with the grand total, including families, at 13,000, which necessitated significant living facilities.

The diverse range of buildings that the AMWD built and maintained in the post-war years included a passenger terminal, Changi Creek Hotel (mainly for personnel in transit, and now part of the Civil Service Club complex in Changi Village), a 115-bed Hospital on the hilltop (a fully functional general hospital, including a maternity ward), plus a primary school and secondary grammar school. Indeed this was the only RAF base in the world possessing a secondary school — it hosted 500 children from all three RAF bases in Singapore.

The school buildings have since been demolished, but the hospital remains, albeit derelict and unused for some years. On the beach side of RAF Changi, there was also the officers' club, swimming pools and the China Sea Beach Club, along with a wide range of recreational facilities. At that time, the future of the Changi base was uncertain, but with the handover planned for 1971, the newspaper article points to the future as a potential second commercial/civil airport, given its size and advantageous approach for aircraft. In reality, the handover of RAF Changi was extended beyond 1971 and was actually completed a few years later.

At that time, Paya Lebar Airport was Singapore's civil and commercial airport, but by 1981, it had shifted to Changi, after which Paya Lebar evolved into a military airport.

Five Chartered Surveyors worked on all these projects. **Wing Commander Leonard A. Balsom** (the Superintending Quantity Surveyor) led the team of four namely, **Flight Lieutenant W.J. Sampson, Flight Lieutenant Donald Russell-Croucher (1925–2016), David G. Hollingsworth** and **Ian Whyte** (see Figure 12.1), who was an Assistant Lands Surveyor.

FIGURE 12.1. Post-war photograph of Ian Whyte prior to his time in Singapore.

Flight Lieutenant Russell-Croucher was the father of Martin Russell-Croucher, a Chartered Surveyor who was the RICS' Professional Standards & Special Projects Director from 2008 to 2014.

Chartered Surveyors at the Singapore Ministry of Local Government, Lands and Housing

Walter Reeve

Born in Middlesex, England, in 1898, **Walter Esmond Reeve** (see Figures 12.2 to 12.4) qualified as a Professional Associate of the R.I.C.S. in 1923 and became a Fellow in 1932. In the 1930s, he was in business with his family as Heaps Son & Reeve in London, a practice that continued in name long after he had left London to work and live in Malaya.

He came to Malaya in 1947 to handle the large volume of war damage claims for the War Damage Commission (in his capacity as a superintending building assessor), and he moved to Singapore in 1951 as a Government valuer in the Valuation Office, based at City Hall. During his time in Government service, he and his family lived in one of the Government-owned Black and White houses at Ridley Park, Singapore. He was able to continue in his Government role after the Malayanisation programme because he, his wife, June, and his children had taken Singapore citizenship.

FIGURE 12.2. (a) Walter Reeve at his home in Ridley Park Singapore. (b) The Reeves at the Padang attending a Land Office vs. Rural Board football match.

Walter Reeve's service to the R.I.C.S. includes being the Malaya Branch Vice-Chair in 1953 and Chair in 1955. In a voluntary capacity, by 1954, he had become a Column Officer in the Auxiliary Fire Service, and their Communications Branch held a "farewell" dinner for him and his wife on the occasion of their departure to the UK in June 1954 for three months on leave. He had previously served in the London Auxiliary Fire Service during World War II, which no doubt provided him with an excellent grounding for this new voluntary role.

He left Government service in 1964 to go into private practice. He was a founder member of the Real Estate Valuers Association (Singapore) (REVA(S)) and he was its 1965/66 President. In 1977, REVA(S) changed its name to the Singapore Institute of Valuers (SIV). In July 1966, it was announced in *The Straits Times* that "W.E. Reeve, FRICS, Chartered Surveyor, Licensed Auctioneer, Valuer and Estate Agent would leave the

FIGURE 12.3. (a) Singapore Auxiliary Fire Brigade (Walter Reeve is fourth from right, seated). (b) Walter and June with Percival Scott ("Scotty") Leathart, who was the Police Superintendent in charge of the Gurkha Contingent of the Singapore Police Force in Singapore from 1949 to 1959.

practice of Cheong Koon Seng & Co. Ltd. from 31 July 1966 and would from 1 August 1966 practice with Messrs. Victor & Mendez Ltd., Auctioneers, Valuers and Estate Agents, at 39A Telok Ayer Street Singapore".*

By 1970, Reeve was practising from the same address but under his own name and not in association with another firm. He was a critic of Section 33 (2) (C) of the Land Acquisition Act 1966. Its effect was that in determining the amount of compensation to be awarded for land acquired under this Act, the Board would not take into consideration the appreciation that accrued to the land as the result of development and the provision of roads, electricity and water by any public or statutory body seven years preceding the date of acquisition. The difficulty facing the valuer, as he explained, was the challenge of distinguishing between this

FIGURE 12.4 (a) The Reeves with friends at Singapore Cricket Club in the 1950s. (b) Walter and June Reeve at a Malaya Society of Architects' dinner in 1954 in Kuala Lumpur.

* Cheong Koon Seng was one of Singapore's earliest auctioneers and estate agents. Koon Seng Road is named after him. He founded the practice in the 1920s, and it was renamed CKS Property Consultants in 1999 — CKS exists to this day.

appreciation and the "natural increase in value due to other circumstances". His concern was that values in the boom years increased significantly over short periods, so these points would be incredibly hard to determine given the likely changes in value if land values rose at rapid rates. Nonetheless, things went full steam ahead and aside from the HDB developments, the industrial sector saw major investment with Jurong, a prime example. Indeed, the success of the HDB (formed in 1960) in developing large tracts of land for public housing was certainly helped by its ability from 1966 onwards to acquire land for this purpose under the Act. The debate continued after Reeve's retirement, and the Act has been amended, notably in 1973 and in 2007, when Section 33 (2) (C) was repealed.

The Reeves retired to Northland, New Zealand, where Walter died in 1979.

Chartered Surveyors at the Public Works Department, 6 High Street

Aside from **Derrick Morris** and **James Khoo**, there were five other Chartered Quantity Surveyors working in the PWD in the late 1950s, including **Stuart Duncan** (Contracts Officer), **Kenneth Goodbourn** and **Chelliah Tiruchelvarayan.** Stuart Duncan is mentioned later in Chapter 15.

FIGURE 12.5. Ken Goodbourn.

- **Kenneth ("Ken") Edgar Norman Goodbourn (1925–2008)** — see Figure 12.5 — arrived in Singapore aged 27 in 1952 and was only with the PWD for a short time because shortly afterward, he joined Langdon & Every (Far East). In 1958, he relocated to their Hong Kong office, and from 1979 to 1982, he was Senior Partner in their Hong Kong office. He retired in 1982. Goodbourn was well known within and outside the surveying profession in Hong Kong, and his contribution to quantity surveying in Hong Kong was mainly through his involvement with the Hong Kong Standard Method of Measurement (HKSMM). His meticulous concern for detail and ability to ensure consistent wording and intent resulted in the SMM being created and used throughout the construction industry in Hong Kong. Goodbourn was on the drafting committee for HKSMM1, 2 and 3 and was Chair of the Committee for HKSMM4.

- **Chelliah Tiruchelvarayan (sometimes known as Tiru) (1927–2014)** — see Figures 12.6 and 16.3 — was a popular figure in Singapore and was a Chartered Surveyor until his death in Singapore in 2014. He also played a prominent role in the creation of the Singapore Institute of Surveyors in the 1960s.

  Tiruchelvarayan was born in Kuala Lumpur on 28 October 1927. He completed his education in Singapore and graduated in 1960 from the London University. He came back to Singapore and re-joined the PWD, where he had started off his career in 1949 prior to his studies in London. During his time with the PWD, he became a chartered surveyor (in 1963) after taking his Final Examination and was appointed as the Quantity Surveyor for all projects undertaken by the Government. He was also placed in charge of all MINDEF Construction Projects.

  He later became a Fellow of the R.I.C.S., a Fellow of the Institute of Arbitrators of London, and a Fellow of the Singapore Institute of Surveyors and Valuers (SISV).

  In the 1960s, he was an active R.I.C.S. member and mentor to young aspiring surveyors. He was also instrumental in organising the 1965 R.I.C.S. Singapore dinner, which was attended by then-Prime Minister Lee Kuan Yew, and the 1966 R.I.C.S. Singapore dinner, in the presence of the R.I.C.S. President Guy Biscoe. These dinners are featured in later Chapters.

FIGURE 12.6. Chelliah Tiruchelvarayan (centre) at a PWD Christmas Party held for the children of PWD staff on Sunday, 10 December 1961.

Chartered Surveyors at the Singapore City Council, City Hall

Chartered Surveyors at the City Council were Basil Thompson (City Engineer) (see Chapter 10 and Figure 10.9.), and in the City Assessment & Estates Department, there was John Aspinall (City Assessor and Valuer) (more on him and the period of transition in Chapter 20), Thomas Stevenson (Deputy City Assessor), as well as Selwyn Lewis (see Figure 12.7) and Gordon Lowden (see Chapter 10 and Figure 10.8) — both Assistant City Assessors.

Aspinall, Stevenson and Lowden are mentioned elsewhere in this book. All four assessors (Aspinall, Stevenson, Lewis and Lowden) resigned in November 1958 despite being offered five more years of service under the Malayanisation scheme.

Basil Thompson, City Engineer, also retired at this time under the scheme, and at the same time, a local replacement was found for him under the Malayanisation programme. The replacement was Dr. Ng Wah Ling, with a salary of $2,500–$2,900 per month.

Selwyn Lewis

Horwood Selwyn Lewis (see Figure 12.7) first came to Singapore in the early 1950s to do his British National Service. Whilst here, he was heavily involved in building a Mess Hall on Pulau Brani (see Figure 12.8(a)). In mid-1955, at the end of his National Service, he secured a quick return to Singapore through a job offer from John Aspinall to work as an assessor in the Assessment & Estates Department on a 3-year contract, working at the City Hall offices. His monthly salary was $785, rising to $875 in year 3, plus Expatriate Allowances.

During this 3-year contract, Lewis and his wife lived in apartments on Marang Road, Mount Faber, and they also celebrated the arrival of their two daughters at the Kandang Kerbau (KK) Hospital.

FIGURE 12.7. (a) Selwyn Lewis in Singapore during his National Service, circa 1953–54. (b) This photograph was taken on Pulau Brani.

Lewis was also permitted by the City Council to teach at the Singapore Polytechnic (see Chapter 19) alongside the likes of Derrick Morris — this teaching activity was pivotal in helping the pioneer generation of Singaporeans take their first academic learning steps on their journey to becoming Chartered Surveyors. See Figure 12.8(b).

FIGURE 12.8. (a) The Mess Hall at Pulau Brani, Singapore, circa 1953–54. (b) Selwyn Lewis' Letter of Reference, dated 19 August 1958, from David Vickery, Head of Department at the Singapore Polytechnic.

Chartered Surveyors at the Singapore Survey Department, Fullerton Building

Basil Walker-Taylor (see Figures 12.9 and 12.10)

Basil Patteson Walker-Taylor (1901–83) was born on 14 November in Orange and Cremorne, New South Wales (NSW), Australia, and died in Pangbourne, Berkshire, England. He was the son of a clergyman (Canon Hugh Walker-Taylor) and Annie Walker-Taylor (nee McWilliam) of Orange & Cremorne, NSW, Australia. He did his Pupillage under Staff Surveyor F.J. Bootle, Lands Department, Orange, between 1918 and 1924. In 1925, he was appointed Native Lands Commission Surveyor in Suva, Fiji, after which he moved to Malaya in 1927 and was the District Surveyor in the various states of Pontian, Muar, Pahang and Kedah (Alor Star) until World War II.

It was in Pahang where he first befriended Sir Robert Black (see the 1957 R.I.C.S. Malaya Branch Dinner in the next Chapter). On 23 April 1938, he and Patricia ("Pat") Barr got married at St. Andrews' Cathedral, Singapore. Pat was the daughter of F.J. Barr and Mrs. Barr of

Ipoh and Singapore. They honeymooned at Fraser's Hill. The ceremony was conducted by Venerable Archdeacon Graham White, the bride was given away by her brother Roy Barr, and Jack Crosse was best man. The reception was held at Eden Hall, the home of Ezekiel ("Eze") Manasseh and his family at that time. There was also a family link: the bride's aunt was E.S. ("Trilby") Manasseh.

Going back to his early career, *The Straits Times* (at the time of his retirement) states that Walker-Taylor left Australia in 1927 to come to Johor. Indeed by 1928, he was playing in the Johor "A" Rugby team. He moved from Johor back to Singapore in 1932, but from 1934 to 1941, he was back in Malaya at various locations in the Survey Department.

The future book on Chartered Surveyors in Malaysia will tell us more about his Survey Department career in Malaya from 1927 to 1956, during which time he applied several times to become a Chartered Surveyor, eventually succeeding after World War II. His World War II experiences have already been documented (including his diary for the three months up to the fall of Singapore) and kept in the Imperial War Museum in London — the diaries (and the book, edited by his son Paddy Walker-Taylor) are an incredible record of that 3-month period. He joined the Federated Malay States (FMS) Volunteer Force whilst based in Alor Setar (then called Alor Star) prior to the fall of Singapore, and having reached Johor on 31 January 1942 he headed for Mount Sophia in Singapore where the Survey Department was temporarily based in a school. Shortly after that, on 17 February 1942, after the surrender of Singapore to the Japanese, Walker-Taylor was interned in Changi for the duration of the war.

Pat had moved to Singapore to live in Eden Hall at the time when Walker-Taylor volunteered. Her mother's sister (Trilby) had married Eze Mannasseh, and immediately prior to the Japanese Occupation, Eze and Trilby had allowed the wider family to stay at Eden Hall. Luckily, Pat managed to escape on the *Gorgon* to Adelaide just days before the surrender. Sadly, Eze died in Changi in 1944 during his internment. After the war, the family reclaimed Eden Hall (which had been kept in remarkably good condition by the Japanese who had occupied it since

February 1942), and it was sold by the family to the British Government in 1957 for a nominal sum subject to the proviso that a plaque be installed at the foot of the flagpole in the garden, stating "May the Union Jack fly here forever". There is, of course, more on Eden Hall in Chapter 3, in relation to Regent Bidwell.

After the war, Walker-Taylor was reunited with Pat in Australia, but they returned to Malaya in 1946. He resumed his survey work in Malaya after the war and became Chief Surveyor for Kedah and Perlis in 1951, then in Penang. In 1956, he was appointed Chief Surveyor of Singapore (see Figure 12.9).

FIGURE 12.9. Basil Walker-Taylor in his Survey Department office in Singapore, circa 1957–58.

Eventually, in 1956, they came back to Singapore until Walker-Taylor's retirement in 1959. One key member of his team (Ho Khye Soo, also a Chartered Surveyor) remained with the Government service in Kuala Lumpur and Penang, and the two remained friends.

He retired on 18 March 1959 and headed for England with his family the next day. *The Straits Times* reported that this completed the Malayanisation scheme in the Survey Department, given that his replacement was to be an Asian. He praised his staff and pointed to their efforts in keeping the survey maps up to date, including sub-divisions, which was not an easy task, with rapid development in Malaya and Singapore continuing to happen and the constant change in boundaries. He stressed the need for the Survey

THE SINGAPORE CIVIL SERVICE JOINT COUNCIL 1957–8

Seated left to right: Messrs. P. Govindasamy, B. P. Walker-Taylor, Baseron bin Jumat, Oon Khye Kiang, K. Thilliampalam (Vice-Chairman), E. B. David, C.M.G. (Chairman), R. C. Hoffman, Goh Teck Phuan, M. Doraisingham, O.B.E., A. K. Karuppiah, R. S. Tufnell.

Standing left to right: Messrs. R. F. Knot, Lee Keng Tuan (Secretary, Official Side), R. Joseph, S. Muthucumaru (Secretary, Staff Side), Lim Joo Hock, W. K. Almushayah, H. H. Beamish, M. Nadarajah, J. H. McLeod, S. Katang Pillai, G. P. Darke.

FIGURE 12.10 Singapore Civil Service Joint Council 1957–58. Basil Walker-Taylor is seated second from left.

Department to continue to keep pace with development to ensure that the maps did not quickly become out of date, though he stressed that the Survey Department was at that time in good shape.

He was quoted as saying:

> Intensive urban and rural development tends to put our maps out of date quickly. It is, therefore, essential for the Survey department to keep pace with this development by providing up-to-date maps. It is also essential to carry out surveys of subdivisions of land and to provide records embodying changes in boundaries as they occur. The Government has an efficient Survey Department, and these requirements are being met as far as possible by the means available.

At that point, the Singapore Survey Department went its separate ways from the Federation of Malaya Survey Department from 1959 to 1963, but during these four years, both Departments still worked very closely together. Before his departure, he was instrumental in preparing the Singapore Guide and Street Directory (Fifth edition, 1958) and also the Annual Report of the Survey Department Survey Department (Colony of Singapore) for the years 1957, 1958 and 1959.

The Annual Report was a fascinating piece of work. The job scope of Land Surveyors in Singapore (and Malaya) has already been mentioned in this book, and the cadastral survey and mapping work they did (from the mid-1800s through to Walker-Taylor's time at the helm) not only enabled the development of Singapore as we now know it, but the property records that it held on land titles (and boundaries) helped to shape Singapore and assist the task of Denis Komlosy and his team on the Master Plan during the 1950s.

Indeed, the Report tells us that the principal functions of the Survey Department were to:

(i) Provide evidence that will completely and permanently identify the land covered by any title issued by the Government
(ii) Compile records of alienation necessary for systematic land administration and public information
(iii) Execute surveys required under the Registration of Deeds Ordinance, the Singapore Improvement Ordinance, the Boundaries and Survey Maps Ordinance
(iv) Compile and maintain records and maps as required by the Boundaries and Survey Maps Ordinance;
(v) Make and maintain reliable maps of the State
(vi) Undertake special surveys and tasks for which the Department is particularly fitted

The Report also explained the process involved in the precise mapping of every inch of Singapore, including the importance of accurate field work, which was no small task given that in 1959, Singapore was 209.5 square miles in area. See Figures 12.11 to 12.12 showing images of the Survey Department staff at work during this time.

Walker-Taylor's positive and strong relationship with the staff in the Survey Department would have benefited his successor, Chief Surveyor Mr. Menon, who inherited a loyal and hardworking team. Menon's Deputy was

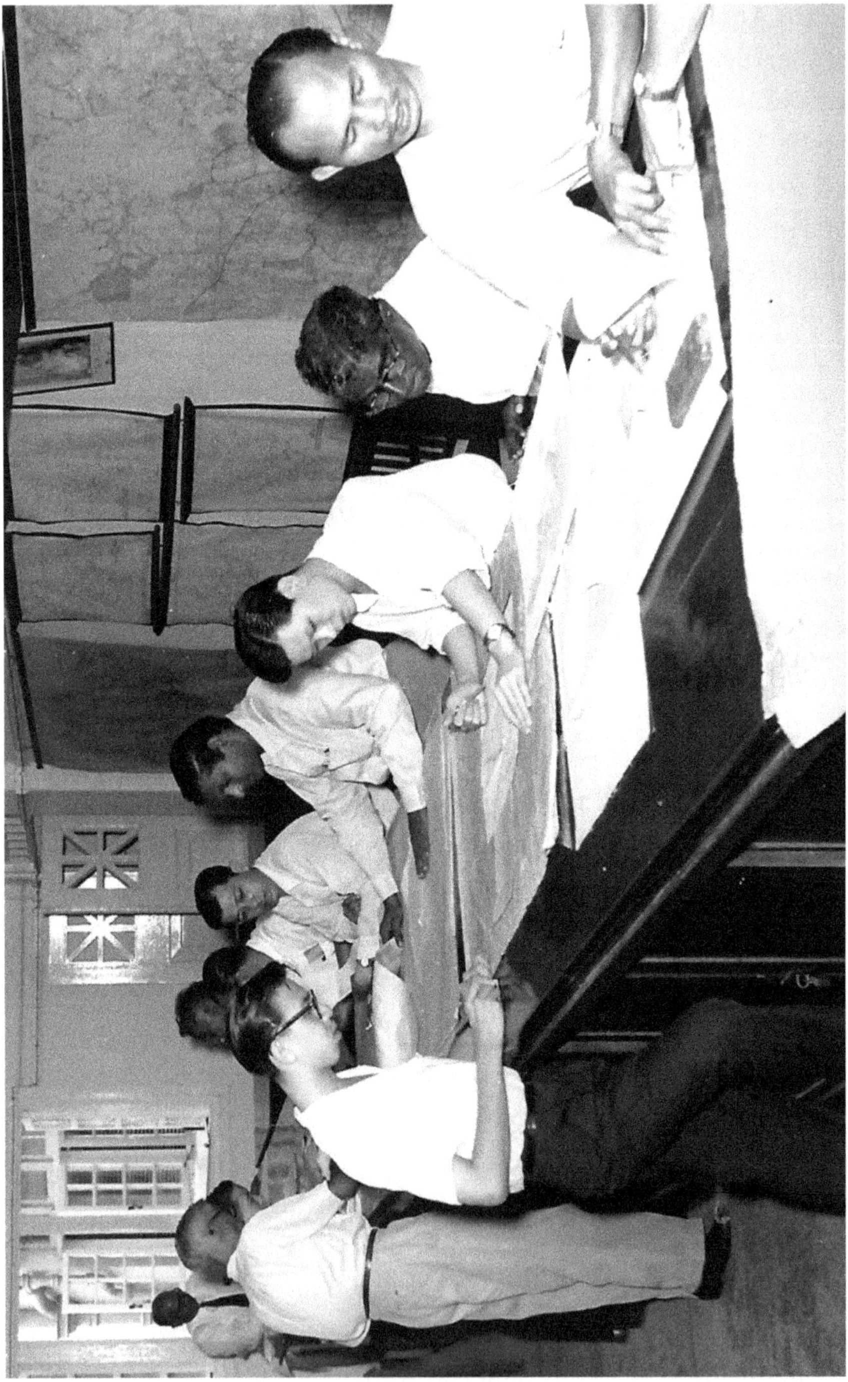

FIGURE 12.11. (a) Draughtsmen in the Singapore Survey Department Drawing Office.

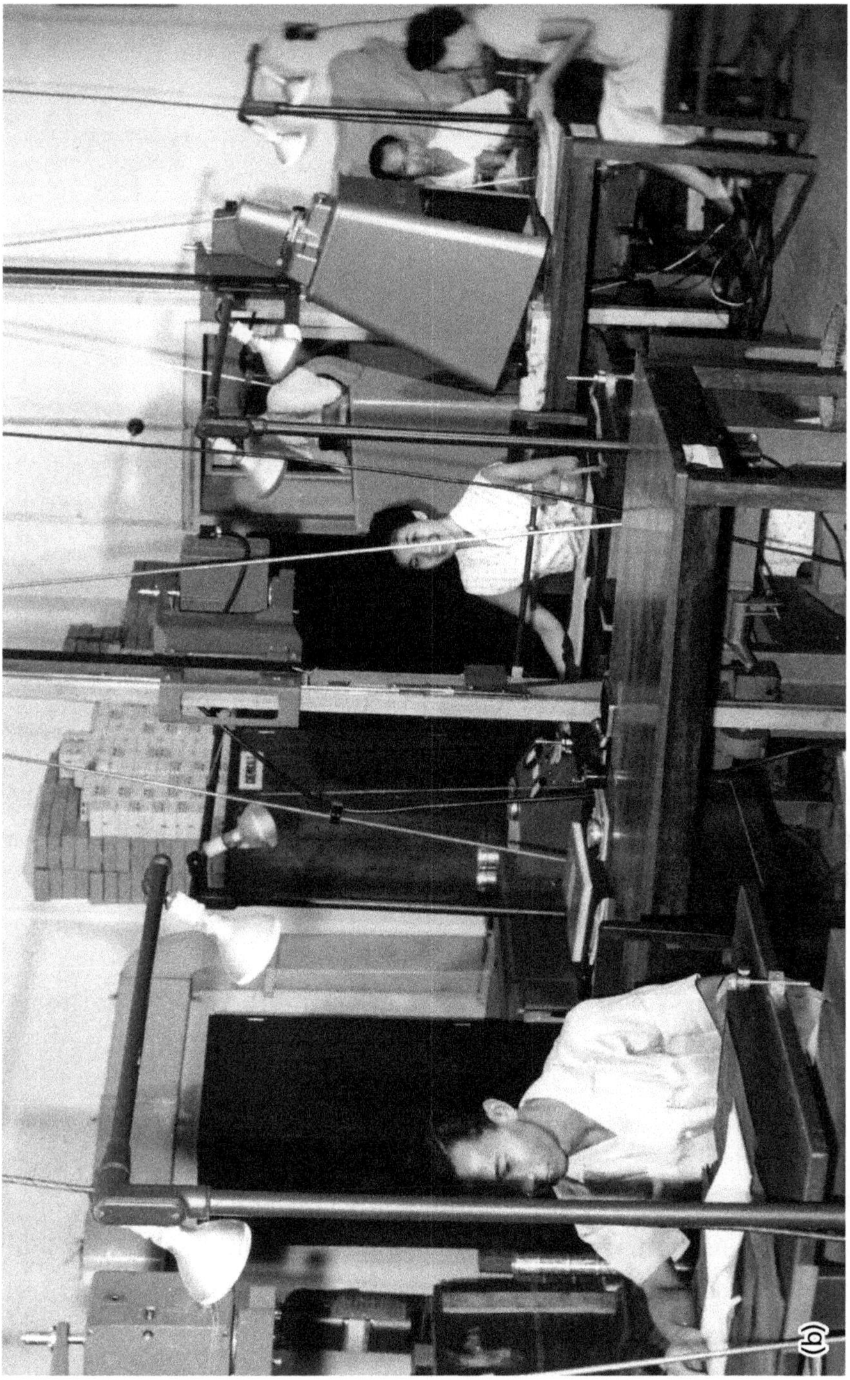

FIGURE 12.11. (b) Workers performing scanning activities in the Singapore Survey Microfilm Department. Both photos were taken in 1957–58.

FIGURE 12.12. Field survey work to measure and map land identified for future development. The location in the image is in Singapore, exact location unknown.

Rameswaram Nagalingam, who took over as Chief in 1963, and who later on was instrumental in the creation of the Singapore Land Surveyors Board.

On retirement in 1959, Walker-Taylor and Pat decided to live in Dublin, Ireland, where their son, Paddy, was schooling, but later in life, they relocated to County Wicklow, followed by a move to Maidenhead, England, and finally Pangbourne, England, where Walker-Peter died on 14 August 1983 at the age of 81.

Chartered Surveyors at the War Department, Chief Engineers' Department, GHQ, FARELF C/O GPO

There were eight qualified R.I.C.S. professionals in this Department during the mid to late 1950s. They were all based at Sime Road. They were led by **Lieutenant-Colonel Frederick Croft**, Senior Quantity Surveyor. Amongst the other seven Quantity Surveyors supporting him were Major John Basil N. Heather (1920–2003) and Lieutenant L.W.S. Rix (also known as "Stuart Rix") (1930–2020). Rix provided some of the photographs in Chapters 13 and 14.

Frederick Croft

FIGURE 12.13. Lieutenant-Colonel Frederick Croft.

Lieutenant-Colonel Frederick Norman Croft (1906–97) — see Figure 12.13 — is also mentioned in Chapter 8 as a key witness in the Loveday Court-Martial. Born in South Africa in June 1906, his family moved to England in 1911. A Quantity Surveyor, he married Daisy in 1934, qualified as a Chartered Surveyor in 1937, and arrived in Singapore around 1939 to work in the Chief Royal Engineer's office at Changi.

The Imperial War Museum in England contains his personal diary for the period of time he spent as a prisoner-of-war of the Japanese in Changi Camp, Singapore (February to August 1942), Taiwan (August to November 1942), Tokyo Camp, Yokohama, Japan (November 1942 to May 1945), his stay at Shinagawa Hospital Camp (August to October 1943), in hospital at Kamanishi (July to September 1944), and finally Kosaka, Northern Japan (May to September 1945), before his liberation by American troops.

Croft managed to write a detailed diary of his experiences as a Japanese prisoner-of-war. The diary was used to help convict war criminals in the trials that took place in the Hague after the war in 1945. He returned to England from Singapore circa 1962 and died in Hampshire in 1997.

Chartered Surveyors at the Directorate of Disposals and Surveys, War Department

There were five R.I.C.S. professionals at the War Department involved in the management and disposal of British War Department assets in Singapore. The Directorate was headed by a Chartered Quantity Surveyor, Lieutenant-Colonel W.H. Johns.

Chartered Surveyors at the Hong Kong & Shanghai Banking Corporation, Collyer Quay

Malcolm Burrells (see Figures 15.17, 15.18 and 16.3)

Malcolm Sidney Burrells was born in Bromley, Kent, England, in 1929 and was married there in 1951, after which he and his family moved to Singapore. After leaving the Hong Kong & Shanghai Bank (HSBC), he became the first-ever General Manager of City Developments Limited (CDL) in 1963. CDL's first office was in Amber Mansions, a three-storey shopping and office building that stood where Dhoby Ghaut MRT station is now situated. Burrells oversaw the first-ever CDL property development, Fresh Breezes in Johor Bahru, and subsequent developments in Singapore, taking his inspiration from initiatives in Australia after a six-week fact-finding tour there. In the 1980s, he moved to Sydney, where he was Head of Property for the State Bank of New South Wales, and by 1990, he had joined Chesterton International in Sydney.

There will be more about Malcolm Burrells and his career and service to R.I.C.S. in Chapter 15.

Chartered Surveyors at Malay-Borneo Building Society Limited, Macdonald House, Orchard Road

Richard Compton-Hall and H.C. Thomas

Malaya Borneo Building Society Limited (MBBS), now known as MBSB Bank, dates from 1950, and at the time when **Richard Compton-Hall** and **H.C. Thomas** were employed as Surveyors there, the Malaysian Government was its major shareholder. The Building Society was responsible for the financing of many new housing developments in Malaya and Singapore during these years, and so Compton-Hall and Thomas would have been heavily involved in assessing land values and development appraisal work to support the financing activities of the Building Society. They would also have managed its offices in the region.

FIGURE 12.14. Richard Compton-Hall, OBE, probably during the late 1930s.

Richard William Compton-Hall (1898–1980) — see Figure 12.14 — was born on 4 September in Croydon, England, and died in Worthing, West Sussex, England, on 20 December. He was the son of Joseph Compton Hall and Ethelind Hall née Barnes-Austin. He first married Gwynedd H.A. Goode in June 1924 in Reigate, secondly to Gwendolen G. Clifford in February 1939 in Paddington, London, and thirdly to Theresa (Tess) Catherine, who died in Worthing in 1979. At the start of World War II, he was an RAF Volunteer Reserve (General Duties Branch), but by July 1943, he was a Squadron Leader in Cairo (Middle East Command), and by 1944, he was Squadron Leader 72212 in the RAF Volunteer Reserve. In the 1949 New Year Honours List,** Wing Commander Richard William Compton-Hall, Assistant Director of Disposals, Middle East Land Forces, was awarded the OBE for services in Palestine prior to 30 June 1948.

Compton-Hall had one son — Commander Patrick Richard Compton-Hall (1929–2001), RN, MBE, who was the author of a number of books on the subject of submarines. Commander Richard was the curator of the submarine museum in Gosport, Hampshire, after retirement. He had two sons; the younger (Simon) tragically died at an early age after being mauled by tigers at the Safari Park (now known as Lüneburg Heath Wildlife Park) near Hamburg, Germany, in August 1982, where he worked as a gamekeeper.

**The 1949 New Year Honours were appointments by the Commonwealth realms of King George VI to reward and highlight good works by citizens of those countries. They were announced in supplements to the *London Gazette* on 31 December 1948 for the British Empire, New Zealand, India and Ceylon to celebrate the past year and mark the beginning of 1949.

Chapter

13

The Separation of R.I.C.S. Malaya and R.I.C.S. Singapore, 1958–61

1957 R.I.C.S. Malaya Branch Dinner (see Figures 13.1 and 13.2)

The 1957 R.I.C.S. Malaya Branch dinner was held at the Sea View Hotel on Wednesday, 19 June. It was co-hosted by Seah Mong Hee and William Henry Smart (the 1957 Branch Chair), and the Guest of Honour was Sir Robert ("Robin") Black. Born in Edinburgh, Scotland, in 1906, Sir Robert was the Governor and commander-in-chief of Singapore from 1955 to 1957. As the Governor, he was representing the Queen in her position as Patron of the R.I.C.S. Before that, from 1930 to 1939, he held administrative posts in the Civil Service in Singapore, Pahang, Perak and Johor. He retired in 1964 to England, where he passed away in 1999. The speech that he made that night was released as a Press Statement. Fortunately, it has been preserved in the National Archives of Singapore. He talked about his knowledge of the profession dating back to the days in Pahang when he worked with Basil Walker-Taylor of the Survey Department in Singapore — Walker-Taylor was at the dinner that evening with his wife, Pat, and Sir Robert acknowledged them both.

Sir Robert recalled that the origins of surveying in Great Britain go back to the Domesday Book, but due to the level of interrogation and taxation that accompanied the survey, it was not regarded highly by the public, so it was not until the Industrial Revolution that surveying became a profession, through the need for land to be acquired for railways, building towns and the resulting need for land and property prices to be negotiated and agreed. He also stated that setting up a professional body of surveyors enabled the R.I.C.S. to maintain the highest standard of integrity and professional conduct that rendered a great service to the public. He mentioned that the Malaya Branch had set up a technical library for its members and students and was also supplying lecturers for a course for surveyors at the Singapore Polytechnic so that the Institution's examinations could be taken there. This, in itself, was a very strong indicator that things were changing, and local surveyors were getting the chance to advance to a full R.I.C.S. qualification. The Final Examination was, for some reason, taken at the British Council premises at this time and not at the Polytechnic.

The Chairman and Members
of the
Royal Institution of Chartered Surveyors
(Malaya Branch)
request the pleasure of the company of
Mrs L.W.S. Rix
at the Annual Dinner to be held at the
Ballroom, Seaview Hotel
on Wednesday, 19th June 1957

8.00 p.m.
To Dinner at 8.30 p.m.

Host L.W.S. Rix Esq.
R.E., ARICS

FIGURE 13.1. A formal invitation to the 1957 R.I.C.S. Malaya Branch dinner at the Seaview Hotel on Wednesday, 19 June 1957.

Prior to the dinner, on 14 June 1957, the R.I.C.S. Malaya Branch Chair issued his annual report, and this was reported in *The Straits Times*. William Smart of the Singapore Improvement Trust was quoted as saying that a bright future lay ahead for the development of Malaya and Singapore — out of its 131 members, 90 were Corporate members (FRICS and ARICS), with 41 being Probationer and Student members, a mixture of Europeans and

Asians. Another strong indication of positive change was that 20 students were attending the Singapore Polytechnic course covering certain aspects of the R.I.C.S. First Examination, and it was noted that examinations at all levels in the training of a Chartered Surveyor were held in Singapore and Kuala Lumpur each year. The final important point of note in the 1957 report was that several students employed by the Government were studying in England, signs of change that were touched upon in Chapter 12 and will be covered in individual stories in the remaining Chapters of this book.

FIGURE 13.2. A photograph taken at the 1957 R.I.C.S. Malaya Branch dinner. L.W.S. ("Stuart") Rix is second from the left.

1958 R.I.C.S. Malaya Branch Dinner

The 1958 R.I.C.S. Annual Dinner was again held at the Adelphi Hotel. This dinner was notable because the Branch Chair was Seah Mong Hee, the first non-British Chair of the Malaya Branch. The Guest of Honour at the dinner was the Honourable Francis Thomas, Minister of Communications and Works. Thomas was a member of the Labour Party of Singapore in

1948, President of the Labour Front (1950) and Minister of Works and Communications (1955–58) in the Labour Front Government under David Marshall and later Lim Yew Hock. He resigned in 1958 and left politics to become a teacher in Singapore.

Toast List

Her Majesty The Queen
(Patron of the Royal Institution of Chartered Surveyors)
Proposer — M. H. SEAH, Esq., F.R.I.C.S.
Chairman: Malaya Branch

The Royal Institution of Chartered Surveyors
Proposer — M. H. SEAH, Esq., F.R.I.C.S.
Chairman: Malaya Branch

The Guests
Proposer — T. F. STEVENSON, Esq., F.R.I.C.S.
Reply by The Honourable FRANCIS THOMAS, M.A. HONS. (CANTAB.)
Minister of Communications and Works for Singapore

Toastmaster — M. S. BURRELLS, Esq., A.R.I.C.S.

Menu

Lobster Cocktail
or
Grapefruit

Consomme Julienne

Roast Duck
Orange Garnish — Apple Sauce
Fresh Peas — Roast New Potatoes
Creamed Cauliflower

Lemon Meringue Pie

Mushrooms on Toast

Coffee

FIGURE 13.3. The Toast List and Menu inside the Invitation to the 1958 R.I.C.S. Malaya Branch Dinner.

1959 R.I.C.S. Malaya Branch Dinner

On 20 April 1959, the final R.I.C.S. Malaya Annual Dinner was held, and once again, it was held at the Adelphi Hotel in Singapore (see Figure 13.4).

Separation of R.I.C.S. Malaya and Singapore Branches

The 1959 R.I.C.S. Malaya dinner was its last official event prior to the amicable split of the Malaya and Singapore Branches into separate Branches.

The creation of the Federation of Malaya on 31 August 1957, with the exclusion of Singapore, meant that whilst the R.I.C.S. Malaya Branch

FIGURE 13.4. Gerald Chilton and his wife, Barbara, with Alan T. Tinkler and Mrs. Tinkler at the 1959 R.I.C.S. Malaya dinner. More on Chilton in Chapter 15.

continued to operate, those within it were now working in completely different markets, particularly in the public and military sectors, whereas previously they had covered the entire Malay Peninsula, including Singapore, at the foot of the peninsula. This led to an amicable split — facilitated and encouraged by R.I.C.S. — to form separate branches for Malaya and Singapore.

The transition to this started in 1958–59 with the creation of a separate professional body in Malaya, known as the Malayan Institution of Surveyors, which was — to all intents and purposes — a "transitional R.I.C.S. Malaya Branch". This new professional body merged with the Institution of Land Surveyors in March 1961 to create the Institution of Surveyors (Federation of Malaya) — since 2011, this has been known as the Royal Institution of Surveyors Malaysia (RISM). The cessation of the

R.I.C.S. Malaya Branch in 1958–59 enabled the formation of the Institution of Surveyors (Federation of Malaya), but it had left a void in Singapore; that void was filled in 1961 with the creation of a new R.I.C.S. Branch in Singapore.

As background, the Malayan Union was the outcome of British efforts to improve administrative efficiency through the creation of a unitary state as the first step towards the eventual self-government of Malaya, but due to objections, a decision was made to replace the Malayan Union with a Federation by the British Colonial Office in July 1946, and an Anglo-Malay Working Committee was convened, which proposed a Federation of the peninsular states and settlements, with increased safeguards for the special position of the Malays and the sovereignty of the Malay rulers. The Federation of Malaya, which comprised the nine Malay states and the Straits Settlements of Penang and Malacca, was inaugurated on 1 February 1948 to replace the Malayan Union, and Singapore was excluded from it, remaining a separate crown colony — this was the sole feature of the Malayan Union that was retained in the new Federation. In Singapore, the colonial authorities took steps towards internal self-governance with the creation of Executive and Legislative Councils in 1948 and the widening of representation, although the Colonial Governor still retained overall power and control. In 1959, Singapore attained self-governance, but the British colonial administration still controlled external relations and key internal policies such as internal security.

It should also be noted that the R.I.C.S. Singapore Branch continued to operate independently between 1963 and 1965, at which time Singapore became one of the 14 states of Malaysia. It is understood that this was because there was no R.I.C.S. Malaysia Branch (see above). However, the creation of a new professional body in Malaysia sowed the seeds for the creation of local professional bodies in Singapore, which will be mentioned again in later Chapters.

Chapter

14

The Work of Chartered Surveyors in the Singapore Improvement Trust, 1952–59

This Chapter is a celebration of the work of five Chartered Surveyors in the Singapore Improvement Trust (SIT), namely, Denis Komlosy, William Smart, John Carter, Max Parker and John Hill. Several other Chartered Surveyors came to Singapore to work for the SIT in the 1950s for a short time, but it is the work of these five individuals that will be remembered.

Denis Komlosy (see Figures 14.1 to 14.4)

The work of **Denis Herbert Komlosy (1912–95)** is particularly significant, given its major, positive impact on Singapore that is still felt to this day.

Stepping back to 1949, Lincoln Page, later to become the Chief Planner of the SIT, attended the National Housing and Planning Conference in Brighton, England, and wrote back to his Manager Jim Milner Fraser, FRIBA (1905–78), informing him that a highlight of the Conference was the insight that housing should henceforth be considered as a "social service to be conducted centrally and taken out of the hands of the private speculator". He opined in his letter that Singapore should think along these lines in earnest, and he was conscious of the importance of social and psychological aspects of housing, something that hitherto had only been recognised by a limited number of people.

Lincoln Page (see Figure 14.1) had arrived in Singapore in 1939, first working for Langdon Williams at the SIT and then for Jim Fraser (see Figure

FIGURE 14.1. A dinner party: (left to right) Winifred Page, Denis Komlosy, Victor Doggett, Peggy Komlosy, Marjorie Doggett and Lincoln Page.

14.11) from 1941 upon the tragic death of Williams. After spending much of World War II as an internee, Lincoln and his wife returned to Singapore in 1946 to pick up where Williams left off. His efforts resulted in a number of critically important post-war social housing developments in Singapore.

To achieve the successful rollout of a social housing programme in Singapore, physical infrastructure was required, backed by enough financial support to enable it to happen on a scale that would make it effective. None of this infrastructure existed in 1949, but things were about to change. In December 1951, it was announced that a 20-person team, under the general supervision of Sir George Pepler (1882–1959), would arrive from London to carry out the 3-year "survey of Singapore". The Chief Planning Officer for the survey would be Denis Komlosy, who arrived by air from London on 5 January 1952. Komlosy's family and two members of his team (including Ken Watts, see Figure 14.2(a), arrived in Singapore on 7 January 1952 onboard the *H.M.S. Canton*, a P&O passenger ship.

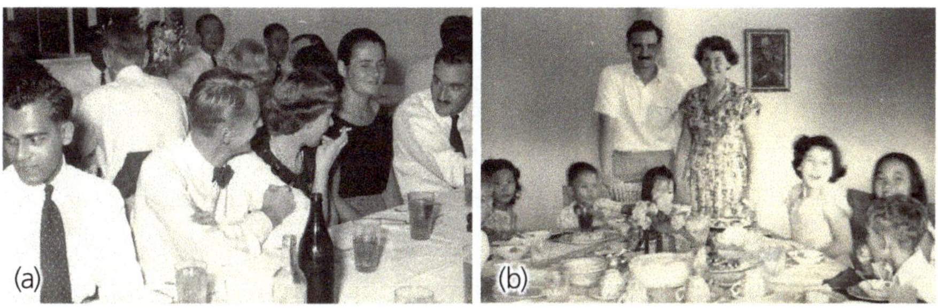

FIGURE 14.2. (a) Denis and Peggy Komlosy (right) at an SIT staff event with Ken and Joan Watts. (b) The Komlosys at home in Kay Siang Road.

Sir George confirmed that a preliminary plan would be ready after 6 months, and the team had already enlisted help from Singapore Government Departments, including the Department of Statistics.

On 5 April 1952, *The Straits Times* published an interview with Komlosy, who announced that he and his team had completed its trial survey of the City area (with a focus on congested and over-populated areas) and that their preliminary survey for the Master Plan of Singapore would be completed by the end of July 1952, and the actual Master Plan would be ready by August 1953. It would not just be a paper plan, but a plan capable of expression in "bricks and mortar", and the economic viability of the plan would be tested in the light of labour, materials and finance. Singapore's housing problem was a particular area of concern for him.

The work of Komlosy and his team continued, and in July 1952, a story appeared in *The Straits Times* about 30 geography students from the University of Malaya carrying out a survey of every single building and piece of land in the City. Komlosy informed the newspaper that 12,000 properties had been inspected, with only a 2% refusal rate. By the time the survey finished, he said that the students would have inspected 40,000 properties in the City area, a total of 70,000 separate housing units.

Komlosy was very active in keeping the media and business circles informed of progress in his team's activities. For example, he talked to the Singapore Rotarians about their work and said that if Sir George's eventual proposals were approved, this would be the blueprint and vision for Singapore in 1972: the creation of a limited central area, bounded by a green belt in which industrial, commercial and shopping centres would be judiciously distributed. There would be no more slums, and new housing would be built outside the green belt and linked to the city by efficient transport — these housing districts would also have their own industries to lessen the dependence on the city area. Agriculture for food production would also form part of the plan. But Komlosy highlighted the key challenges: the current SIT building programme would not be enough to provide the 12,000 housing units with the year needed to meet the projected doubling of the population. Also, the question of economic growth to support this population growth was a major issue.

However, Komlosy took pains to point out that it would also require public and private sector cooperation and investment for the plan to happen and work. Indeed, in June 1953, Komlosy was the Head of the newly formed Industrial Study Group that was set up to ascertain likely industrial development in Singapore over the next 20 years. The study group was to talk to 680 businesses employing more than 10 people in Singapore to understand their views on the amount of land required for industrial development in the future. This, in turn, would guide the Master Plan to make the proper zoning provision for industrial development across the island. Questionnaires were sent out, with Raffles Institution students following up with those who did not respond. The questionnaires covered matters on production, employment, fluctuations, power, water, raw materials, marketing, transport, floor space, capacity, equipment and future plans.

FIGURE 14.3. (a) Denis and Peggy Komlosy together at a social event in Singapore. (b) Komlosy making a speech outside his family home in Kay Siang Road.

Later in 1952, Komlosy was at the centre of discussions surrounding the future of the Kallang airfield site, an area of significant size and in a key location in Singapore that was being replaced by Paya Lebar as Singapore's international airfield. His vision was for the Kallang Basin area (including land within the flood plain and to the west of the river towards the end of Beach Road) to be transformed into a residential and industrial area, complemented with open spaces and playing fields. A committee chaired by the Singapore Commissioner for Lands soon held its first meeting to discuss the future of the entire area, including Tanjong Rhu. One issue that they needed to resolve as a committee was the future of the businesses that had flourished in the area, including shipbuilders and saw-millers. (A visit to the Kallang area today clearly demonstrates the rewards for all of the committee's efforts.)

The theme of slums and overcrowding continued into 1953 when Komlosy was quoted in the *Singapore Free Press* as saying (to an audience at the University of Malaya Geographical Society) that 160 people (including 33 families) lived in one single dilapidated two-storey bungalow in Queen Street. Living conditions were unbearable ("hell") for the occupiers, with the bungalow divided into cubicles with little room to breathe and access to water, toilets and bathing facilities often being the subject of arguments due to queueing times. For some residents, crying babies kept them awake until the early hours. To Komlosy, many of the residents looked sick and ill. Komlosy's comments came to the attention of J.M. Jumabhoy (1919–2012), then a City Councillor and later a Minister in the first elected Singapore Government, who urged the City Council to take action immediately to resolve the overcrowding problem. Jumabhoy was a strong advocate for affordable housing, and he was a campaigner for safer and more spacious housing for the poor long before the tragic Bukit Ho Swee fire of 1961.

The lack of suitable playing fields for children was also something that was of concern to Komlosy — his survey work was looking into the concerns expressed by the Singapore City Council's Playing Fields Committee about this — and he stated that this would form an important part of the survey work. However, that was in July 1953, and he was then indicating that the survey work would take another year to complete. This was longer than originally indicated but was a clear indication of the complexity of the work, coupled with the need to get the survey work done right.

In the same year, Komlosy was again quoted by the *Singapore Free Press*, this time urging clan associations and private concerns to allow City burial grounds (such as those then existing in Orchard Road), especially those that had not been used for some years, to be reclaimed for development. At that time, the City contained 2,000 acres of used and unused burial grounds, and it was expected that a large portion of this land would need to be reclaimed for development.

In May 1953, Komlosy confirmed that he and George Edgar Magnay (Senior Planning Officer) had recruited 300 of the 350 students needed (from certificate classes) to help carry out a full 3-week traffic survey of major roads on the island to help inform the wider survey. The Singapore Legislative Council approved the extra provision of $27,400 to cover the survey costs. By August 1953, the survey team reached the rural areas, and Komlosy asked the rural district committees to provide details of land and property use for all the land and property holdings within their respective districts.

Moving on to 1954, Komlosy restated his desire to ensure adequate playing fields in Singapore's built-up areas, especially given the criticism of the SIT's failure to provide this in their developments to that point in time, although this was not universally popular — landowners such as S.T.O. Alsagoff stated that for his Shenton Way development, he would not be willing to provide such facilities for his land, given that the general public would not be paying the expenses. There was another angle: some people had suggested through the Press that developers who leave their land vacant should allow their land to be used as playgrounds by the masses.

In June 1954, it was announced that Komlosy's 25-person team was almost finished, and that in three months, he hoped to present the Master Plan to the SIT Trustees and then to the Governor-In-Council. They were awaiting the arrival of Sir George Pepler on 13 July 1954 on a 6-week stay to review and put the final touches to it. The plan would consist of three maps: a small-scale plan of the entire island, a large-scale map of the city and suburban area, and an even larger map showing proposals for the redevelopment of old City Centre areas. All the survey work had been completed, and reports from the survey work had all been prepared, but a final meeting with the traffic advisory body was needed to discuss how the survey findings would be applied to meet the deficiencies in the present road system. Sir George arrived with his wife on a Qantas flight on the night of 13 July 1954 to be greeted by Komlosy and Jim Fraser. Once Sir George had reviewed the Master Plan, he stated his worry about the big issues of traffic and housing, and reiterated that unless the Land Acquisition Bill was enacted, the Master Plan (due to be issued by the end of that year) could not be fully implemented, as the Government would not have the powers to get hold of land needed for development.

The Land Acquisition Ordinance of 1955 certainly did help, but its scope was limited, and it was incapable of coping with the needs of Singapore for the next 15–20 years. This was recognised, and in 1966, a new Land Acquisition Act was passed. This time, this Act provided sufficient legislative powers for a rolling programme of comprehensive urban renewal — it enabled the Government to acquire private land suitable for housing, industrial development, commercial activity and education purposes.

By August 1954, Sir George and his wife returned to the UK, safe in the knowledge that Komlosy's work was coming to fruition. In the same month, Komlosy told *The Straits Times* that his report would include a recommendation that a suburban railway system encircling Singapore (and through the centre of Bukit Timah) be introduced. Some years later, Komlosy's recommendation came to fruition with the creation of Singapore's MRT system.

It was during this time that the phrases "Green Belt" and "Green lungs" were used for almost certainly the first time in the Singapore context and quite possibly in Asia too. *The Straits Times* reported that a Green Belt around the city would provide a set of Green lungs and limit unplanned development. Komlosy's vision for the Green Belts was a place where the "jaded townspeople can escape in their search for fresher air", but he acknowledged that Singapore needed ten times the green space than it currently had, and due to financing and the so-called "land freeze", it would take some years to come to fruition.

In particular, the following were mentioned in the Green Belt context — some are still applicable in Singapore today, while others are long gone, given the huge population and economic growth of Singapore since 1972 (the date when the Master Plan was due to reach full fruition):

- Areas to be set aside for playing fields in the city
- Agricultural zones to allow easy access to cheap vegetables and other farm produce
- Cemeteries in the city should the planned City Crematorium "not catch on"

- Cemeteries that have not been used for burials for many years to be converted into parks
- Parks at Mount Fraser and the Gap (the Gap is now the area around Kent Ridge and South Buona Vista Road, also nicknamed "99 turns")
- A Padang and playground area on Shenton Way
- 200 acres of land at Kallang Airport would, upon its closure, be allocated as playing fields. The newspaper article mentions this in the context of the new Stadium, i.e., it would most likely be sited there
- Sites would be reserved for Hospitals, Schools, Community Centres, Clinics and other community needs
- Nature reserves would be kept, as would buildings and sites of historic and scientific interest
- Historic buildings would be kept where possible, and a list of all such buildings was then in the stage of being collated by Komlosy, who added that whilst some would be lost due to more pressing development needs, all efforts would be taken to preserve them all.

Komlosy's plans for use allocation were carefully thought out and based on a number of key assumptions coming out of the detailed survey work that his teams undertook. *The Straits Times* article also rightly pointed out that whilst at one point it was questioned whether Singapore could afford the "luxury" of a plan, as Komlosy rightly said, the reality was that Singapore could not afford not to have one.

Another key feature of the plan, announced by Komlosy in December 1955, was the idea of having multi-storey car parks to alleviate congestion. At that time, Singapore was suffering major traffic congestion, but the concept of multi-storey car parks had not yet been brought across from the UK. The location suggested for the first such multi-storey car park was on "Cecil Street", which in reality was at the junction of Cross Street, Market Street and Cecil Street, and was opened in June 1964 by then-Minister for National Development Lim Kim San. This car park was known as the

"Golden Shoe"; it was demolished in 2017 for redevelopment purposes.

In August 1955, it was reported in *The Straits Times* that Denis Komlosy had just returned from four months of leave in London. It was called "leave" and it was chiefly a holiday but Komlosy carried on working through this period, such was his dedication to the task, and he even returned to Singapore early to continue his work. The newspaper article reported that during this 4-month period, he was not only busy preparing the Master Plan, but he was also visiting town planners in London and its satellite towns, no doubt catching up on current planning initiatives. He observed that the satellite "new towns" in England were striving to be self-sufficient and that this may be the blueprint for Singapore's satellite towns. He mentioned Basildon as being very similar to Singapore's proposed new towns because it had 20,000 squatters in the area (the word "squatters" is discussed further in Chapter 21).

He stated that the Master Plan would be ready for Government review in September 1955 once the SIT had approved it.

By December 1955, these approvals were in place, and so it was that in 1956, the Master Plan for Singapore was made public. Although Komlosy had stated that it would comprise three main maps, there were, in total, 71 maps. All of them are now held in the National Archives of Singapore and are available for public viewing on prior request.

As the Master Plan was being displayed for public feedback at the Victoria Memorial Hall, the Governor, Sir Robert Black, and the Minister for Housing, Lands and Local Government, Inche Abdul Hamid bin Haji, both visited in the first week of January 1956. They also encouraged all Singapore citizens and residents to visit and take a constructive interest in the Master Plan. They congratulated Komlosy and his team for their work and said that it was a very important piece of work despite it presenting a "challenge". Indeed, a large number of objections were received — some were from major landowners whose (prime) development land was due to be part of the planned "Green Belt". The general view was that if they were adequately compensated by the Government for their perceived loss of development value, they would be happy to hand over the land to the

Government. Whilst an Inquiry was set up to look at the objections, work continued, and in 1957, Komlosy announced plans to build a "New Town" at Woodlands for 80,000 people on a 3,000-acre site — its purpose was to house those working in the Bukit Timah industrial areas and would be self-sufficient with shops and all other amenities.

In August 1958, a week after the Governor-in-Council approved the Master Plan, Komlosy was interviewed on Radio Malaya. He explained that the draft plan was displayed two years before at the Victoria Hall, as a result of which a number of objections and observations were considered, and decisions were made, with some changes made to the plan as a result. The radio interview allowed Komlosy to discuss a few misconceptions about the plan. First, there was a misconception that the plan was designed to facilitate Government land acquisition. He explained that the plan was more than just about land acquisition — it was a master plan that would finally provide a broad, structured framework for the entire island, and it would enable the Government to control land use across the island regardless of whether or not the land is proposed for a purpose involving acquisition. The question of compensation for land acquired for the Government came up, and Komlosy explained that compensation (or lack of) for restrictions on use would be different from payment for acquisition of land — where the Government acquires land to enable development of land in accordance with the master plan, a fair price would be paid under the Land Acquisition Ordinance. There was some discussion about the definitions of rural and agricultural land. In conclusion, Komlosy stated that the success of the Master Plan would depend on how it moved with the times, i.e., it should not be static but ever evolving. As an example, a week after the plan was approved, his team was carrying out traffic studies to see if the plan needed to be altered in the light of changing traffic conditions.

As a follow-up to the approval of the Master Plan on 8 August 1958, two new Bills — Planning and the Singapore Housing and the Development Bills — were tabled before the Legislative Assembly for approval. Inche Abdul Hamid bin Haji stated that the Master Plan would provide for a population of 2 million, and whilst that would be achieved five years ahead of schedule, this would not invalidate the Plan. Rather, it accelerated the need for these Bills to enable the full implementation of the Plan, with the best use of all land being vital to Singapore, given the limited availability of land resources.

As the plan was adopted, Komlosy's work continued with a look at the rural areas, and his expertise also extended beyond the shores of the main Singapore island — in 1955, he looked at plans to develop one of the offshore islands as a Pleasure Resort, which later came to fruition in the form of Sentosa.

Conclusion on Komlosy

Denis Komlosy's contribution to Singapore should not be understated. He was a leader, planner and surveyor that possessed great energy, commitment and dedication to detail, and he was held in high regard by many people. Komlosy and his team's work from 1951 to the implementation of his Master Plan in 1958 created the blueprint for Singapore and laid the foundations for the successful implementation of later updates to the plan by Professors Erik Lorange and Otto

FIGURE 14.4. An undated photo of Denis Komlosy (centre) meeting H.M. Queen Elizabeth II and H.R.H. the Duke of Edinburgh.

Koenigsberger and, indeed, for the 1971 URA Concept Plan as well. His plans for decentralisation to Satellite Towns, for Green spaces, a Green Belt and the designation of retail sites on Orchard Road have proved successful to this day. After leaving Singapore to return to England, he worked in the Planning Inspectorate of the Ministry of Housing & Local Government in London but would return on occasions by invitation to see the many areas where Singapore was benefitting from the Master Plan implementation.

William Smart

William Henry Smart (1912–96), in addition to managing the team responsible for all of the new SIT estates around Singapore, was an active member of the R.I.C.S. Malaya Branch from the late 1940s (he qualified as a Chartered Surveyor in 1949) until the Malaya-Singapore split, after which he was an active member of the R.I.C.S. Singapore Branch until he retired in 1961. During the 1950s, as a senior SIT manager, he and his family lived in one of the Kay Siang Road houses.

He was an integral part of the team that enabled the SIT to take part in the International Exhibition of Low-Cost Housing: Village Centre, Delhi, from January to March 1954, along with his “boss” Jim Fraser, the SIT (General) Manager.

The 1954 Exhibition was the brainchild of Jaqueline Tyrwhitt, a globally respected town planner with many years of experience and close links to the UN. In May 1953, she was appointed by the Indian Government as (inter alia) Technical Adviser to the Ministry of Housing for the Low-Cost Housing Exhibition that was to take place in New Delhi in early 1954. The involvement of the SIT in the Exhibition was very much driven by Tyrwhitt, who knew SIT planner Ken Watts when he was one of her students at the London School of Planning and Research for Regional Development, which was where Lincoln Page had also studied.

On retirement from public service in Singapore in 1961, Smart returned to Auckland in his native New Zealand.

SIT Developments and the Role of Smart and His Team of Chartered Surveyors

The role of **William Smart** with the SIT (and his active role as an R.I.C.S. Branch Committee Chair and member) has been mentioned several times. As the Surveyor-in-Charge at the SIT, based at Fullerton Building and later at their Municipal offices in Upper Pickering Street, he was involved in all of the SIT's housing projects from 1950 until the Housing & Development Board (HDB) replaced the SIT in 1960. As mentioned earlier, aside from **Smart** (and **Denis Komlosy**), other Chartered Surveyors who worked for the SIT included **John Carter** (Deputy Lands Manager), **Max Parker** (Senior Surveyor) and **John Hill** (Estates Officer). They managed all of the early Queenstown developments (for example, the Duchess Estate) and, later on, in the 1950s, the Dakota Crescent flats. The majority of these housing developments have been demolished in recent years for land development, but some blocks still exist and hold fond memories for many Singaporeans — this is largely because of the generous unit sizes, open spaces and playground areas between the blocks, which had hitherto not been a prominent feature of Government housing in Singapore.

Princess Elizabeth Flats

The Princess Elizabeth Flats (see Figure 14.5) at Farrer Park at the junction of Racecourse Road and Hampshire Road junction (opposite the Foochew Methodist Church) were built in 1950, and William Smart's team was responsible for managing the flats until their handover to the HDB in 1960. These 3-storey "workmen's flats" were two-roomed flats designed to meet a particular social need.

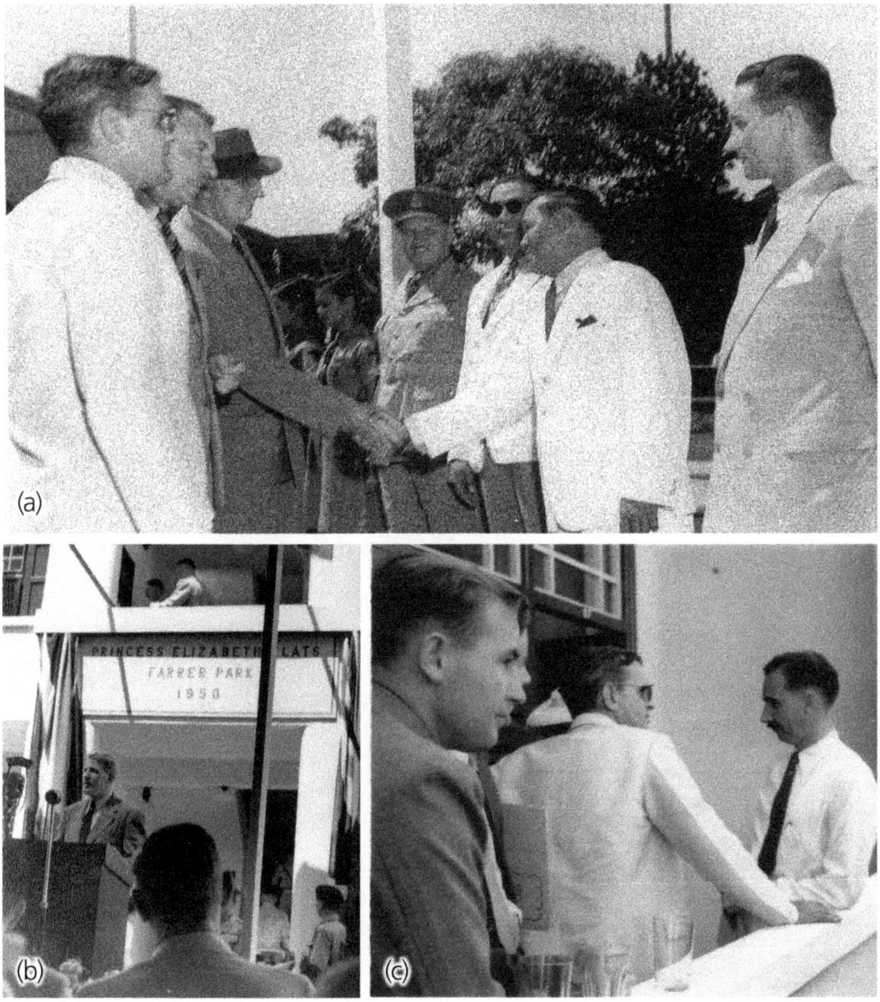

FIGURE 14.5. The 1950 opening of the Princess Elizabeth Flats in Farrer Park. (a) (from left to right) Lincoln Page, Jim Fraser, the Governor (Sir Franklin Gimson), along with senior SIT staff. (b) Rolf Koren, SIT architect. (c) Max Parker, Lincoln Page and Frank Howie from the SIT.

Contract II aka Neighbourhood II (Duchess Estate), Queenstown

The Neighbourhood II (Duchess Estate) Flats (see Figure 14.6) in Queenstown are examples of Lincoln Page's work, for which William Smart's team was responsible for estate management until their handover to the HDB in 1960.

FIGURE 14.6. Photographs of SIT flats on the Neighbourhood II on the Duchess Estate, Queenstown, circa 1952–53.

Kay Siang Road houses (see Figure 14.7)

Lincoln Page was also responsible for the design of the houses at Kay Siang Road that were built to house SIT senior employees, including the Komlosy, Page and Smart families. Many of these fine properties still exist today and are highly sought-after residences in the rental market. Some have been converted for commercial uses.

FIGURE 14.7. One of the Kay Siang Road houses that were designed by Lincoln Page for housing SIT staff in senior positions. Most of the Kay Siang Road houses constructed in the 1950s are still standing today.

Princess House

Princess House (see Figure 14.8) was designed by Lincoln Page and completed in 1957. It served initially as offices for the SIT, and later, the HDB chose it to be its Head Office on its formation in 1960.

Princess House was gazetted for conservation in 2007 after the Ministry of Environment vacated it in 1989. Given that most of the SIT

FIGURE 14.8. Princess House, Alexandra Road, Queenstown.

public housing in Queenstown is now gone, Princess House remains the final legacy from the SIT era of the 1950s.

Dakota Crescent SIT Flats and the Role of Smart's Team

William Smart and his team of Chartered Surveyors were also responsible for managing the Dakota Crescent SIT flats (see Figures 14.9 and 14.10) until their handover to the HDB in 1960.

However, time has since caught up with these SIT flats — all were fully vacated by 2018, with some facing demolition. The design of the Dakota Crescent SIT flats (finished in 1958) was largely modelled after UK towns and designed by a team of architects at the SIT, who introduced wide open spaces and other features more attuned to the Singapore climate.

FIGURE 14.9. (a) 12 Dakota Crescent: It used to house commercial shops, including the Tian Kee & Co. Provision Shop, at the bottom level; the upper level was for residential use. The flat is currently vacant but is on the Government's Preservation List. (b) One of the 3-storey walk-up flats. However, this flat has since been demolished.

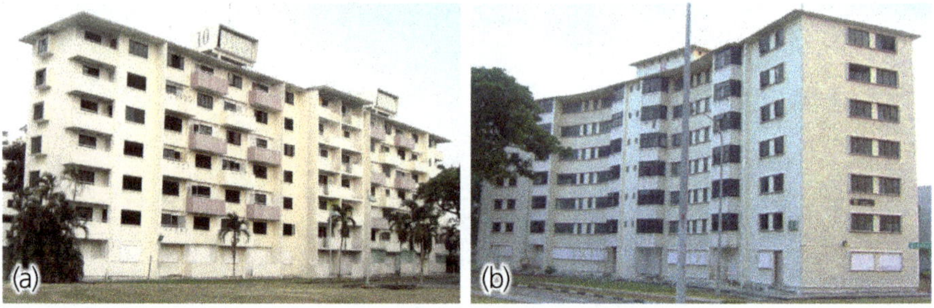

FIGURE 14.10. (a) The butterfly-shaped walk-up apartment block at 10 Dakota Crescent is one of the last few surviving blocks from the SIT era and faces an uncertain future although it is currently on the Government's Preservation List. (b) 6 Dakota Crescent, one of the original butterfly or "pi"-shaped blocks, was demolished in 2020. Along with Blocks 2, 4, 6, 14, 18, 22, 28 and 30, these 7-storey blocks were shaped like butterflies to allow as many blocks as possible in Dakota Crescent.

John Hill

John Walter Hill was a Lands Manager who worked under William Smart at the SIT from 1954 onwards. Like Smart, Hill stayed on with the HDB until 1961. Hill and his family returned to the UK in late 1961 and was the last Expatriate member of the SIT Estates team to leave Singapore.

During the 1950s until his return to the UK in late 1961, Hill invested much time into training up several Singaporeans who later became Chartered Surveyors with illustrious careers. Amongst those they trained at the SIT were **Loh Jwee Siam, Tan Tian Boon, Chan Wan Hang** and **Ng Boon Ong.** In later years he worked in local Government and then in private practice in Hertfordshire, England, where he passed away in October 2022, aged 94.

Max Parker

Aside from his work for the SIT, Max Parker (see Figure 14.5(c)) was a famous sporting figure in Singapore from 1951 to 1960, as full-back of the Singapore Cricket Club's Rugby team. He was the team captain and was instrumental in many famous team victories, leading to him representing Singapore and being an integral part of the All-Malayan team in games against Hong Kong, Fiji, and visiting Rugby teams to the region. For example, he was selected to play for the Malayan RU XV to play against the Combined Universities (Oxford/Cambridge) in a floodlit match at the Merdeka Stadium on 1 September 1959.

His legacy to the Singapore Cricket Club was the "Parker Plan" that saw Asian rugby players being able to play alongside the "Expat" rugby players for the first time. No doubt, the plan was unpopular amongst the diehards, but it proved extremely successful. It now seems incredible that it was only in 1960 that this change happened.

In late 1960, just after the transition from the SIT to the HDB, he moved back to his native New Zealand to join Waimairi County Council, near Christchurch, as a Planning Officer.

John Carter

John Murray Carter (1919–2014) was born in England and became a Chartered Valuation Surveyor in 1956, and later a Fellow. He remained a Chartered Surveyor until his death in Singapore in December 2014, aged 95 years.

He left England in 1951 to come to Singapore, where he joined the SIT as an Estates Officer. He was involved in a number of land acquisition projects that would lead to low-cost housing development (and later, housing developments by the HDB), and, for this purpose he was at one point seconded to the Government Land Office.

After leaving the SIT in the late 1950s, he worked as a rating assessor for the Singapore Inland Revenue Property Tax Division. He set up his own valuation practice in the 1960s, originally at 14 Laidlaw Building, followed by 12 Collyer Quay, and then at other offices in the Raffles Place area until his retirement in the late 1980s. He also set up an office in Malacca in the 1970s.

Carter (see Figure 14.11) was regarded as a valuer with a great deal of ability, but nonetheless, he had a reputation for being a difficult and

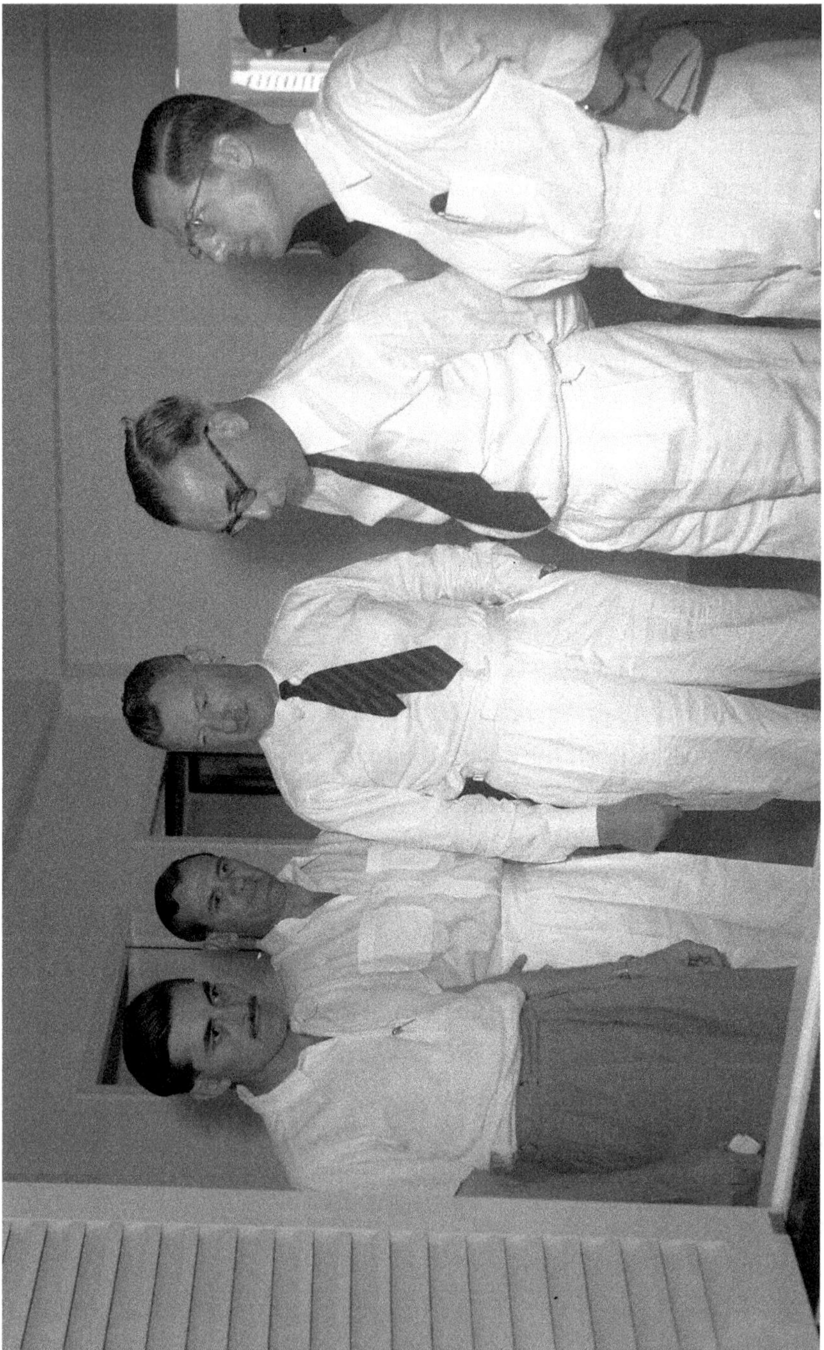

FIGURE 14.11. The acting President of the Municipal Commissioners, Norman Ward (second from right), at Tiong Bahru on 27 April 1951. With him are (from left to right) K.A. Brundle, Deputy Government Architect, Lincoln Page, SIT Senior Architect and Town Planner, Jim Fraser, SIT Manager, and John Carter, Acting Estates Officer.

confrontational person. In his negotiations with the Singapore (and Malaysia) Property Tax staff on assessment appeals, he was regarded as an extremely tough negotiator.

As someone who remained in Singapore after the Malayanisation programme in the late 1950s and became a Singapore citizen, he was an instrumental figure in the formation of the Real Estate Valuers Association Singapore (REVA(S)) in 1963. Indeed, he was the inaugural President. To this end, he was held in high regard as someone who upheld the highest standards of professionalism (and expected the same of other Chartered Surveyors), and this, in turn, influenced the young emerging Chartered Surveyors of the 1960s and 1970s who he came into contact with.

Part IV

Surveyors in the Pre and Post-Independence Eras

Chapter

15

1959–65: Singapore Independence and the Transition to Local Professional Bodies

In previous Chapters, the process of "Malayanisation" towards the end of the 1950s was beginning to show evidence of major changes. The number of expatriate Chartered Surveyors started to drop, particularly in the public sector, and at the same time, more local talent was starting to emerge.

In the period from 1961 to 1963, the number of R.I.C.S. qualified professionals dropped in Singapore from its 1958 levels. The Malayanisation process was the main reason for this, but some of those who departed were replaced by Singaporeans who had successfully attained their R.I.C.S. qualifications. Inevitability, numbers rose from 1963 until the early 1970s due to an influx of Chartered Surveyors working for the British Military service who continued to pass through Singapore. Some spent several years here before returning to the UK after the final British military withdrawal and handover of land and property assets to the Singapore Government. The withdrawal was largely completed by 1971, but the final handovers had all taken place by 1976.

At the same time, from 1961 onwards, Singapore had its own R.I.C.S. Branch for the first time, with the Branch office in the early 1960s being managed by correspondence through the Public Works Department. See Figure 15.1.

Whenever R.I.C.S. Singapore Branch meetings took place, people such as Gerald Chilton and Derrick Morris played an active role in conducting Branch activities and actively engaged the returning Singaporean Chartered

FIGURE 15.1. Former Public Works Department offices at High Street, Singapore: this was the Registered Office of the R.I.C.S. Singapore Branch in 1961.

Surveyors in these too. Apart from this and the annual Dinners, the other major happening in these years was the development of the Singapore Measurement Standards (which will be discussed later in this Chapter).

At this time, the roles of Chair, Vice-Chair and Honorary Secretary of the Branch rotated on an annual basis, no doubt to share the burden but also to ensure that as many members as possible get involved. Indeed, in 1964, Chilton and Morris had stepped aside, and Stuart Deans (Inland Revenue Assessor) became the Branch Chair and Terry Bonfield, a recently qualified Chartered Quantity Surveyor working for Langdon & Every (Far East), the Honorary Secretary. At that stage, membership of the Branch had risen to 78 members, but it should be noted that this number included members in Brunei, Sarawak and Sabah.

Chapters 16 and 17 will go further into the years from 1966 until the Branch was phased out and replaced by a Corresponding Secretary.

The following noteworthy Chartered Surveyors were in Singapore between 1961 and 1965:

Private Practice Firms (Quantity Surveying, Valuation Services)

Denis Crisp

Denis John Crisp (1924–2012), an R.I.C.S. Fellow, was born in Reading, England. Having served as a Pilot in the Royal Air Force during World War II, Denis migrated to Adelaide in September 1948 with his wife Phyllis. He qualified as a Chartered Quantity Surveyor later that year and worked for Qantas on construction projects in Australia and also around Asia. He took Australian citizenship in the early 1950s.

Having worked for Qantas, he formed Crisp, Kavanagh & Partners in Adelaide with Kevin Kavanagh in the mid-1950s. Their success led them to Singapore and Kuala Lumpur in 1961. Their first office was at Shaw House on Orchard Road, and in the 1970s (until Denis Crisp's retirement back to Adelaide circa 1982), their office was at 726 Colombo Court.*

Crisp was an active and passionate member of the R.I.C.S. Singapore Branch Committee, but at the same time, he was a strong advocate of the need for Singapore to create and grow its own professional bodies. Indeed, he was a founder member of the Singapore Institute of Surveyors (SIS).

In 1976, Ahmad Nizaruddin bin Mohd Hashim entered into a partnership with Crisp, Kavanagh & Partners, and after Crisp retired from business, the firm was renamed CKP Dan Nizaruddin Jurukur Bahan. Today, it is known as CKP Nizaruddin Jurukur Bahan Sdn. Bhd.

Crisp, Kavanagh & Partners was a successful business in Singapore from 1961 through to Crisp's retirement circa 1982, so much so that he is believed to have been the first person in Singapore to own a Rolls Royce.

One of the many projects for which Crisp, Kavanagh & Partners was appointed as the Quantity Surveyor was the construction of the Peninsula Plaza office and retail development on Coleman Street between 1976 and 1977.

---

* Colombo Court was a 1960s office building that lay opposite to what is now Funan Centre and next to the Adelphi Shopping Centre — it was demolished in the 1990s to make way for the new Supreme Court building.

On retiring from Quantity Surveying, Crisp and his family purchased and operated a farm in Wayville, Adelaide, South Australia.

Crisp passed away in Wayville in October 2012.

Frank & Vargeson (later Frank, Vargeson, Wicks & Partners)

Frank & Vargeson was created in 1958 with the opening of their first office in Kuala Lumpur, and their first major commission was as the appointed Quantity Surveyor on the Houses of Parliament construction project in Kuala Lumpur.

The original partners were **Henry Frank**, **Desmond Vargeson** and **Robert Rhodes**. All three had extensive and successful careers in the public service and took the plunge to work together in practice.

In the 1960s, the practice expanded and opened offices in Singapore, Brunei, Kuching, Kota Kinabalu and London.

Lily Ong (secretary to William ("Bill") George Wicks of Wicks & Partners, Kuala Lumpur) recalls that Frank & Vargeson opened their first office in Kuala Lumpur at the same time as them in 1958. Wicks had then been engaged in property management work by the Chartered Bank (and later, the Methodist Church), which necessitated him opening a Branch office in Singapore.

Wicks had discussions with Frank & Vargeson, resulting in a merger to form Frank, Vargeson, Wicks & Partners in 1966. As well as the Chartered Bank (now part of Standard Chartered Bank) account, they handled property management, sales and lettings. Ong was in charge of the sales and lettings side of the business.

As Frank, Vargeson, Wicks & Partners, from 1966 to 1969, they were based at 407 Shaw House, Orchard Road. Alongside Vargeson, Rhodes and Wicks by 1969 were Derrick Morris (featured elsewhere in this book) and Liong Ah Nai, who became a Chartered Quantity Surveyor in the early 1970s.

However, by 1969, Frank & Vargeson was once again a separate entity (having parted amicably from Wicks & Partners), and by 1969, Wicks &

Partners was run by Wicks (in Kuala Lumpur) and Goh Mia Hoon (in Singapore). There will be more on Wicks and Goh in Chapter 17.

From 1967 until Wicks' retirement in 1976, Lily Ong recalls that Wicks would come down from Kuala Lumpur once a month to spend a few days (sometimes a week) with Goh and his staff at their office in Shaw House. Whilst Wicks was very strict and expected his staff to be in the office well before 9 am every day, he was also appreciative of their efforts and had lunch with them every day when he was in Singapore.

By the late 1970s, Frank & Vargeson had also expanded to Hong Kong, by which time none of the original partners was involved in the business.

Henry Frank

Henry Julius ("H.J.") Frank was born in England in 1912 and first came to Singapore in 1938 to work for the Air Ministry Works Directorate. In 1946, after World War II, he returned to Singapore, and from 1948 to 1952, he was a Quantity Surveyor and Building Assessor with the War Damage Commission. In 1958, he joined forces with Vargeson and Rhodes to form Frank & Vargeson. He became a Chartered Quantity Surveyor in the early 1960s.

He announced his retirement from Frank & Vargeson in May 1964, after which Vargeson and Rhodes would continue as Frank & Vargeson. However, he came out of retirement in 1968 to replace Malcolm Burrells (see later in this Chapter) as General Manager of CDL (City Developments Limited) in Singapore. His final retirement was in the early 1970s.

Robert Rhodes (see Figures 15.2 and 15.3)

Robert Noel Seaton ("R.N.S.") Rhodes (1919–73) was born in England and was in South-East Asia during World War II serving as a Chindit. He was a Chartered Quantity Surveyor. He came to the region in 1953 with his wife and children to work for the Public Works Department in Kuala Lumpur.

After the formation of Frank & Vargeson and some time spent in Singapore running its office there, he moved back to England in 1964 to run Frank & Vargeson's office in London.

He finished his career in Bournemouth in the early 1970s with the Department of the Environment. He passed away in nearby Poole in 1973.

FIGURE 15.2. (a) Robert Rhodes. (b) Robert with his wife Marion and friends in Singapore circa 1962/64.

FIGURE 15.3. Public Works Department (now known as the Jabatan Kerja Raya Malaysia or JKR) in Kuala Lumpur circa 1956 — Robert Rhodes is seated fourth from right.

Desmond Vargeson (see Figures 15.4 and 15.5)

Desmond ("Des") George Vargeson (1927–91) was born in Norfolk, England, and qualified as a Chartered Quantity Surveyor in 1951, aged 24. He came to Kuala Lumpur around the same time as Robert Rhodes and, like him, worked in the Public Works Department in Kuala Lumpur. There, they formed a strong working relationship with PWD architects W. Ivor Shipley (later Sir Ivor Shipley) and Peter Merer, a relationship which would later stand them both in good stead when Shipley was involved in the design and construction of the Houses of Parliament building (still standing today) in Kuala Lumpur as well as other Government buildings in the early 1960s.

Vargeson was highly regarded as a man of good temperament and honesty and someone who recognised that being a Chartered Surveyor meant upholding the highest level of professional standards.

He passed away in Norfolk, England, in 1991.

FIGURE 15.4. A photograph taken in Kuala Lumpur circa 1959: (second from left to right) Desmond Vargeson, Ivor Shipley, unknown, Stella Shipley and Peter Merer.

FIGURE 15.5. Photographs of Desmond Vargeson in the 1970s.

Langdon & Every (Far East)

Langdon & Every (Far East) remained at 14 Laidlaw Building until 1964 when they moved to the 4th floor of the Malaya Building, 41 Robinson Road.

In the mid-1960s, its partners were Cecil Tom Every and Giles Every (London), Mike McLeod (Kuala Lumpur), Peter Sanderson (Hong Kong) and Seah Mong Hee (Singapore). By 1969, Nicholas Chong (another Chartered Surveyor) had joined to run their Kuala Lumpur office. There will be more about the practice's transformation to Davis Langdon & Seah and then to Arcadis in future editions of this book, which will cover the story of Chartered Surveyors through to 2018, when the R.I.C.S. celebrated its 150th anniversary. In the early 1960s, Francis "Frank" Angell rejoined the Singapore office. From 1 January 1967, the practice decided to drop the words "Far East" from its title, and by 1969, it had become Langdon Every & Seah to acknowledge the contribution of Seah Mong Hee to the growth and success of the business, which was now primarily focused on Asia.

During the 1960s, the number of Quantity Surveying firms started to increase in Singapore as a result of an increasing number of Government-led construction and infrastructure projects, which created an opportunity for new businesses despite the dominance of Langdon Every & Seah.

This Book will not go into detail about the practice's successes from the 1970s onwards — a future expanded edition of this book through to the present day will attempt to do that. Suffice to say, its Singapore office, headed by Seah Mong Hee, remained at 41 Robinson Road during the 1970s, and then by Malcolm Smithson (see Figure 15.6) into the 1970s when Seah retired, although the latter remained a consultant to the business.

FIGURE 15.6. Malcolm Smithson, a Chartered Quantity Surveyor who joined Langdon & Every (Far East) in 1965 in Singapore. He then moved to its Brunei office (1965–69) but was back at its Singapore office during the 1970s. He passed away in Singapore in 1995, aged 55.

Seah Mong Hee retired in 1975 after 40 years of service. (He started with Waters & Watson in 1935, as mentioned earlier in this book.) In his time with the practice, he had gained a reputation of being the leading Quantity Surveyor in Asia and was highly respected by all those who did business with him (indeed by anyone who came into contact with him). The practice was built on a platform of the highest professional and technical standards, and he oversaw its growth to a dominant industry position in Asia by the mid-1970s. He was replaced by Kuala Lumpur-based Nicholas Chong as Senior Partner and then by his son Seah Choo Meng (see Chapter 21), who became International Partner. *Quantifying Asia: The History of Langdon & Seah* (compiled by John Peacock) provides more in-depth insight into Langdon Every & Seah during the 1960s (and 1970s).

Lau Teo & Yong

Lau Fook Sing and Lionel Teo Cheng Ann

Lau Fook Sing (1923–2008) (photographed in Figure 16.2, Chapter 16, with the R.I.C.S. President Guy Biscoe) was born in Malaysia and qualified as a Chartered Quantity Surveyor in 1963 whilst working for Langdon & Every (Far East).

That same year, Lau was persuaded by Lionel Teo to leave Langdon & Every (Far East) to set up their own partnership together. Initially, they called themselves Lau & Partners, and this changed to Lau, Teo & Partners when Teo himself qualified as a Chartered Surveyor in 1965.

Lau is, of course, mentioned in Chapter 11 as having passed the First R.I.C.S. Examination in Quantity Surveying in 1958 whilst working for Langdon & Every (Far East).

In 1968, as the success of the practice grew, Lau and Teo joined forces with another Chartered Quantity Surveyor, **M.S. Yong**, who would be based in Kuala Lumpur. At this point, the practice became Lau Teo & Yong.

In the 1980s, Lau retired from Singapore to New South Wales, Australia, where he passed away in 2008. The name Lau Teo & Yong still exists as a Quantity Surveying practice, but ownership has long since changed from the original partners.

Lionel Teo Cheng Ann (1931–2007) — see Figure 15.7 — was born in Singapore, and after senior schooling in Australia, he moved to England in 1955 to pursue his career ambition of being a property professional. He attended the College of Estate Management in Kensington and worked in a West End firm for a year before sitting for his Finals.

Despite having married in London in 1956 and with a family of three children by 1962, he returned to Singapore. At that stage, Teo had not yet passed the Final R.I.C.S. Examinations, but despite that, he managed to persuade Lau to leave Langdon & Every (Far East) to set up their own partnership in 1963. Initially, they called themselves Lau & Partners, but this changed to Lau, Teo & Partners when Teo qualified as a Chartered Surveyor in 1965.

Their first office was located at 173b Cecil Street.

Together, they worked on their first projects, such as the Balmoral Road flats for the Chan Wing family, as well as Government projects, such as the Singapore Telephone Board (now Singtel) buildings in Siglap and Thomson Road.

When Teo was studying in London, his student friends were sent by Singapore and the Kuala Lumpur PWD to study in England at places such as the Brighton Technical College.

By 1968, the firm's name had changed to Lau Teo & Yong with the addition of Yong as a partner based in Kuala Lumpur. Yong had been one of Teo's Kuala Lumpur PWD friends from his days in London who had studied at the Brighton Technical College. As jobs came up in Kuala Lumpur, such as the development of new buildings at the University of Malaya complex in the late 1960s and public works projects, Yong proved to be a valuable partner.

They thrived together and were renowned for their high standard of professional ethics. The NUS at Kent Ridge project in the 1970s was possibly their biggest project. They continued until the early to mid-1980s, at which point Teo had decided to retire from the practice. Lau and a junior partner (Jeffrey Low) carried on, and indeed, Low continues to run the company alone to this day.

Teo remained a Chartered Surveyor after his retirement until his death in 2007.

FIGURE 15.7. (a) Lionel Teo Cheng Ann. (b) A photograph of Lionel with his wife Vera.

W.J.F. Tillyard & Partners

Walter Tillyard and Alan Davis

Born in Surrey, England, in 1909, **Walter James F. Tillyard** — see Figure 15.8(a) — was a Quantity Surveyor based in London. He founded the Tillyard Quantity Surveying firm shortly after the end of World War II. That firm, and other Quantity Surveying businesses around the world bearing the Tillyard name, still exists today, albeit the Tillyard family has not been associated with the firm since 1966. Tillyard opened a Singapore branch of his London practice around 1957, originally from an office on the 8th floor of the Asia Insurance Building, Finlayson Green (the building is now known as the Ascott Raffles Place). He also went into partnership with another Chartered Quantity Surveyor, Albert Sydney Zoller, who ran their Melbourne office during the 1960s.

His original Singapore partner was Tony Peacock, who returned to the UK to set up their Cheltenham office. From 1963 onwards, another Chartered Quantity Surveyor, I.K. McArthur, ran the Singapore office. McArthur continued to run the Singapore office for several years after Tillyard retired from the Tillyard firms in 1966. After retiring in 1966, Tillyard took up permanent residence in Victoria, Australia, in 1967–68, where, until a few months before his passing, was a farmer and grazier.

By 1970, another Chartered Surveyor — **Alan John ("A.J.") Mytton Davis** of Contract Services Group — see Figure 15.8(b) — had entered into a partnership with W.J.F. Tillyard & Partners from an office at Shaw House on Orchard Road.

Davis, born in Surrey, England, in 1936, and qualified as a Chartered Surveyor in 1970, remained in Singapore until the late 1980s and for many years was General Manager of Singapore Land Limited (now known as Singland Properties Limited and part of UIC [United Industrial Corporation Limited]). During the 1970s and 1980s, Davis' name was synonymous with Singapore Land's prized real estate assets, such as Clifford Centre and Singapore Land Tower. Davis returned to England in the late 1980s, where he was a consultant to Thorstone Land & Property Limited for some years. He is now retired but still a Fellow of the Institution.

FIGURE 15.8. (a) Walter Tillyard, photographed in 1967 at his daughter's wedding. (b) Alan Davis.

Suckling & McDonald

Keith Suckling and Johan McDonald

FIGURE 15.9. Johan Mcdonald.

Keith Suckling and **Johan ("John") Hector McDonald** — both Chartered Surveyors and Fellows of the Institution — formed Suckling & McDonald in 1966. Their office was on the 4th floor of the Union Building, Collyer Quay.

Keith Suckling (he is shown in the photograph at Figure 16.3) has already been featured in this book — he was a strong proponent of the R.I.C.S. in Singapore in the 1950s and 1960s, and upon Singapore's independence in 1965, he was instrumental in the R.I.C.S. stepping back to encourage and enable the formation of local professional bodies.

By 1969, the firm had become Suckling, McDonald & Mohd Isahak, and from 1969 to 1970 only, their practice was in partnership in Singapore with John Rawlinson & Partners.

It was later in the 1970s called Pakatan International, based at 37B Tanglin Road.

There will be more about Johan McDonald (see Figure 15.9) in future editions of this book covering the 1970s and 1980s. McDonald qualified as a Chartered Surveyor in Glasgow, Scotland, in 1957, and later in his career, he was President of the Institution of Surveyors Malaysia (now the Royal Institution of Surveyors Malaysia) from 1973 to 1974.

C.H. Williams, Valuer & Estate Agent

Colin Williams (see Figure 15.10)

Colin Harold ("C.H.") Williams (1927–2019) arrived in Singapore as a qualified valuer and Chartered Surveyor in 1952 to take up a position as an Assistant Assessor in the Assessment & Estates Department of the

Singapore City Council. His job was to conduct property tax assessments on landed property, and his first assignment was in the Geylang and Kallang areas, where large areas of land were swamps covered along tracks with illegal *attap* huts. The common language in the area was a mixture of Hokkien and Teochew, and Williams sought help from a teacher to learn enough of these dialects to be able to do his job effectively.

FIGURE 15.10. Colin Williams.

By 1956, his assignment was in the business district where the common language was another dialect, Cantonese, which presented him with a new challenge.

He left the City Council in late 1957 and set up his own business in Singapore in January 1958 under the title "C.H. Williams, Valuer & Estate Agent". The new office (in Hong Kong Bank Chambers, Collyer Quay) comprised Colin, his wife, Peggy, and an office boy/driver.

By 1960, there was a shortage of work in Singapore due to a drop in real estate values, which, in turn, led to a sharp drop in valuation work and little else in the way of work apart from a couple of property tax objections by Banks, where the tax was higher than the restricted rents that they could legally charge their tenants under the Control of Rent Ordinance.

So, Williams took up an opportunity to open an office of C.H. Williams & Co. (Sdn. Bhd.) Chartered Surveyors Valuers and Estate Agents in Kuala Lumpur, where there was much property revaluation (objection) work following a recent rates (property tax) revaluation exercise by the Kuala Lumpur City Council.

After a couple of years, he moved to Kuala Lumpur full-time and put Peter Newman in charge of the Singapore office. He would often return to Singapore for business despite making Kuala Lumpur his home.

Williams was held in the highest regard, a fatherly figure that helped to develop and train his staff, many of who became Chartered Surveyors. He was an R.I.C.S. Fellow as well as an active member of the Institution of Surveyors (Federation of Malaya), now, of course, the Royal Institution of Surveyors Malaysia.

Francis Lai Pang Fee worked for Williams as a trainee from 1961 to 1963 alongside John L. Chen, another trainee who, like Lai, learned much from Williams and was mentored by him.**

Lai regards Williams as his mentor who had a big impact on his life and professional career. Lai opines that Williams treated all his employees with care and respect and, in turn, expected surveyors in his firm to care for each other and ensure the well being of their peers and junior staff.

In 1974, the company merged with Talhar & Co., a chartered surveyors and valuers firm set up by Mohd Talhar Abdul Rahman in Johor Bahru, and having Datuk Wong Choon Kee from C.H. Williams & Co. onboard too, WTW was thus founded with Williams as Chair — he was succeeded by Mohd Talhar, who also remembers Williams with a great deal of fondness.

In 1977, Idris Pearce (later Sir Idris Pearce, and a global President of the R.I.C.S. in 1990) persuaded Johnny Loh, Tony Sallmann and Justin Seward to merge with C.H. Williams & Co. in Singapore to form Richard Ellis C.H. Williams & Loh, with Loh (for Richard Ellis) and Peter Newman (for C.H. Williams) as Joint Managing Directors. They were based on the 24th floor of the Shaw Centre, Scotts Road.

Williams passed away in England in 2019 at the age of 92. There will be more on him in a future book on Chartered Surveyors in Malaysia (and in further editions of this Book covering the 1970s and 1980s). There will also be more on Peter Newman in Chapter 17.

C.S. Toh and Partners

Toh Chee Seng (see Figure 15.11)

Toh Chee Seng (1920–92) was the proprietor of C.S. Toh and Partners. He was a Chartered Quantity Surveyor who attained his R.I.C.S. membership in 1963 by a correspondence course via the College of Estate Management

**John L. Chen qualified as a Chartered Surveyor in 1967, and after a period with American Assurance Co. at 1 Robinson Road, formed J. L. Chen & Associates, Chartered Surveyors and Valuers in Singapore in the early 1970s. This practice continued for over 40 years but has now ceased trading since Chen retired.

and then successfully taking the Final R.I.C.S. Examinations. At that time, he was working for Langdon & Every (Far East). He later became an R.I.C.S. Fellow and was known professionally as **Dick Toh** or **Dick Toh Chee Seng**. He is earlier mentioned in Chapter 8.

FIGURE 15.11. Toh Chee Seng.

After a period with the HDB from 1962 to 1964 as a Contracts Officer, he set up his own Quantity Surveying practice in 1965 in Singapore at 21 Selegie House, Short Street. He moved to Hong Kong in 1965 to set up an office with his eldest sons in Hang Chong Building, 5 Queen's Road Central, and then he decided to settle with his family in Hong Kong.

Meantime, for the Singapore office, he brought in W.C.C. Press to the partnership, and they were then known as C.S. Toh Sons & Press, based on the 17th floor of International Plaza.

Gradually he built up the firm in Hong Kong, with its client base eventually covering China and the region, and although the Singapore office has long since closed, the Hong Kong office still continues to this day as C.S. Toh & Sons & Associates.

His sons, Tony, Eddie and Philip, have continued to practice as Quantity Surveyors and are still professionally qualified with the R.I.C.S. and the Hong Kong Institute of Surveyors (HKIS).

Toh was proud of his R.I.C.S. qualifications, and after his move to Hong Kong, he helped to build up the Hong Kong Branch of R.I.C.S., of which he was Chair from 1972 to 1973.

He was a big supporter of the HKIS, which he helped to establish in 1984, and he was also a founding member of the Hong Kong Branch of the Chartered Institute of Arbitrators, being its Chair in 1973. He had a reputation for hard work, both professionally and in his life outside of work, and was well regarded as a friendly, courteous man.

His death on 9 September 1992 saddened many. His human remains rest in a niche in Bright Hill Buddhist Temple in Singapore, but his legacy lives on.

Seah Choo Meng recalls him as a close friend of his father, Seah Mong Hee, and as someone held in high regard as an industry pioneer in both Singapore and Hong Kong.

Wolferstan Trower & Partners

In 1962, Wolferstan Trower & Partners established their first office in Singapore on the 5th floor of the 6-storey Bank of Tokyo Building at Nos. 9 and 10 Phillip Street. Also known as the Hong Building and built in 1957, the Bank occupied two floors and the rest of the building was leased out as commercial offices. The building was demolished in the late 1970s — No. 3 Phillip Street, Royal Group Building is now on its site. They moved to the Ngee Ann Building on Orchard Road in the 1970s, with Chartered Quantity Surveyor **Hans Bufe** at the helm. They were involved in some of the early developments in Jurong in the 1960s, and at the point of their Singapore office's opening, they were the appointed Quantity Surveyors for the construction project for the new Subang International Airport in Kuala Lumpur.

Bufe was later joined by **Eric Cromby**, another Chartered Quantity Surveyor, in running the Singapore office and then later in the 1960s (and into the 1970s) by **Herbert J. Maddocks**, alongside **Peter Summerhayes,** both of whom were also Chartered Quantity Surveyors.

Wolferstan Trower & Partners is, of course, today better known as WT Partnership, and as of 2022, they are celebrating their 60th year in Singapore.

Chartered Surveyors in the Public Sector

The creation of the Housing & Development Board (HDB) under the Housing and Development Ordinance paved the way for a massive public housing programme, and the Land Acquisition Act of 1966 enabled the HDB to undertake massive slum clearance through financial compensation for private land needed for public use, which was something that the SIT had found very difficult to achieve. This section of the Chapter celebrates the pioneer Chartered Surveyors who worked for the HDB during those early years.

Chan Wan Hang (see Figure 15.12)

FIGURE 15.12. Chan Wan Hang.

Chan Wan Hang (1923–2007) was with the SIT in the mid to late 1950s alongside Ng Boon Ong. He qualified as a Chartered Surveyor in 1960 whilst with the HDB through the College of Estate Management correspondence course, coupled with the R.I.C.S. Examinations in Singapore hosted in the 1950s and early 1960s by the British Council in Singapore at their Stamford Road offices.

He was the Contracts Officer in the Contracts Department, which was in the 1960s under the management of Teh Cheang Wan, who was the Chief Architect and Head of the Building Department. He managed the team of Quantity Surveyors during this time, based at the HDB offices at Princess House, Alexandra Road, Queenstown.

After a spell in the 1970s with Borneo Housing Development Limited (now known as Borneo Housing Mortgage Finance Berhad), Chan migrated to Australia, where he continued to work as a surveyor in New South Wales until his retirement.

He passed away in Sutherland Shire, New South Wales, Australia, in 2007.

In the early to mid-1960s, Chan's team included future or newly qualified Chartered Surveyors who he mentored, such as Francis Lai Pang Fee, Toh Chee Seng and Michael Wee Sek Kay, who are, of course, mentioned elsewhere in this book.

Loh Kum Seng

Loh Kum Seng (whose photograph is in Figure 16.2, Chapter 16) qualified as a Chartered Surveyor in 1963.

He was with the PWD at its High Street offices throughout the 1960s under the supervision of Derrick Morris, James Khoo and Chelliah Tiruchelvarayan. He worked for Francis Lai in the 1970s, and then in the early 1980s, he set up his own practice, **KS Associates.**

Rae Littlejohn

Rae James Garrett Littlejohn (see Figure 15.13) was a Chartered Surveyor who worked as a Planner for the Singapore City Council in 1958. He later became a Planning Inspector at the Ministry of National Development (MND), Upper Pickering Street, and from the end of 1968, he was based at the MND Building on Maxwell Road after the Planning function came under the MND.

He later took employment with Crooks Mitchell Peacock Stewart in Singapore and then moved with the same firm to New South Wales, Australia, in the early 1970s. During his time with the City Council, Littlejohn and his family lived in Seton Close, Tanglin.

His role with the City Council and MND in the days leading up to Singapore's independence and immediately after independence was extremely important — he was responsible for handling many applications for new developments across the island in these years.

FIGURE 15.13. Rae Littlejohn during his time in Singapore.

In his spare time, he was a car rally enthusiast and won several car rallies in Malaysia during the late 1960s.

After returning to Australia in the 1970s, he continued his planning work, including a role as Strategic Planning Officer for Brisbane City Council, where he was involved with the long-term development of the Brisbane area.

Rae Littlejohn passed away in Australia in 2012, aged 82.

Chartered Surveyors Working for the British Military Service

From 1959 to 1965, a large proportion of Chartered Surveyors working in Singapore were employed by the British Military service. They were primarily involved in managing the lands and buildings of the British Army, Naval and Air Force across the island, as well as the region, e.g., Hong Kong, Malaya, Maldives and Brunei, from a Singapore base.

After the British Military Withdrawal from Singapore between 1967/68 and the end of 1971, the Singapore Government successfully negotiated the free transfer of land held by the British, which enabled the Government to oversee the shift towards commercial uses for the land and buildings that helped Singapore's shipbuilding and aviation industries grow and prosper.

Indeed, from 1958 to the late 1960s (and indeed up until the final asset handovers in the early 1970s), the work of Chartered Surveyors shifted away from land surveying towards asset management, asset maintenance and the collation of land and building asset data, with a view to handing over all British Military assets to the newly formed nation of Singapore.

In the period from 1959 to 1974/5, more than 30 Chartered Surveyors spent time in Singapore working for the various parts of the British Military service that held land and buildings in Singapore. The roles they played ultimately ensured the smooth transition from British control to Singapore ownership. This part of Chapter 15 features some of these individuals.

Director of Lands Ministry of Defence (War Department) Far East Land Forces

Edmund Rowlstone Speak, FRICS (see Figures 15.14(a) and 15.15, was the Director of the Lands Ministry of Defence (War Department), FARELF. His job was to head up the Land Directorate, and he had a team of Chartered Surveyors working for him in the Lands Office, including **Ken Massey**, who was heavily involved in R.I.C.S. Branch activity in the 1960s in Singapore.

Speak had two separate postings in Singapore, from 1961 to 1963 and from 1969 to 1972.

The FARELF's Lands Office was at Tyersall Park from 1961 to 1963 but later moved to Loewen Road, where it remained until the handover to the Singapore Government in the early 1970s (see Figures 15.14(b) and 15.15(b)). Today, the Loewen Road Lands Office building still stands, and as of 2021, is the home of the Museum of Ice Cream Singapore.

The job of Speak and his team from 1969 to 1972 was to manage the effective handover of land and building assets to the Singapore Government prior to the British withdrawal from Singapore and facilitate the retention of the ANZUK Office upon its withdrawal.

A significant amount of work was involved in documenting land and property assets for the handover, which was why his team comprised at least four Chartered Surveyors at any given time. He also managed the

FIGURE 15.14. (a) Photograph of Edmund Speak, taken in the UK prior to his first posting in Singapore. (b) The FARELF Shield, which was affixed to the wall of the Directorate of Lands Office in Loewen Road.

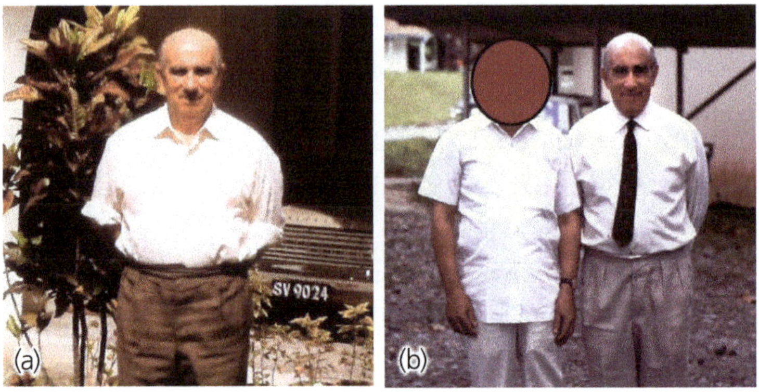

FIGURE 15.15. (a) Edmund Speak at his home in Alexandra Park, FARELF Quarters, and (b) outside the FARELF Lands Office in Loewen Road, which is now the home of the Museum of Ice Cream Singapore. The identity of the person next to Speak is hidden for privacy reasons.

District Works office (responsible for ongoing building and infrastructure maintenance pending transfer), where **Peter Eldridge** was amongst the Chartered Surveyors who worked there. At the District Lands Office, there was **Alec Cowan**, another Chartered Surveyor who was a friend and colleague of **Eric Jubb** (more on him below) from Defence Lands Scotland.

There was another small team of Chartered Quantity Surveyors based at the Works Office on Adam Drive, with various Chartered Surveyors staying for periods of 3 to 4 years from the late 1950s until 1971 — they were responsible for continued building maintenance pending the handover.

The Works Office became part of the Ministry of Public Building and Works in 1965.

Finally, the Ministry also had an Area Office at Keene Road, Tanglin, where several Chartered Quantity Surveyors were also based until the late 1960s.

After Speak left Singapore in 1970, more Chartered Surveyors came to Singapore from the UK to work for the British Government. They will feature in future editions of this book. When Speak left, his role was taken by Tom Steers, OBE, a Chartered Land Surveyor who was responsible for the final handover of some of the FARELF's key assets to the Singapore Government.

Speak passed away in England in 2003 at the age of 88.

Tom Steers

Tom Steers, a Chartered Land Agent and Surveyor, was the man responsible for the final handover of FARELF assets in 1971, including the transfer of Gillman Barracks for a dollar. A photograph of him appeared in *The Straits Times* on 21 August 1971, accepting a one-dollar note from Leong Foke Meng (on behalf of the Singapore Government) for Gillman Barracks in the presence of Lieutenant-Colonel Larry Lamble (Commander of the Royal Engineers) and Major Gurcharan Singh of the Singapore Armed Forces.

Ministry of Defence (Army), Far East Land Forces

Eric Jubb (see Figure 15.16)

Eric S. Jubb (1912–93) qualified as a Land Agent and joined the Naval Works Department in the mid to late 1930s. He served in the Royal Navy during World War II and returned to the Department after the war. He had been responsible for the Royal Navy's Scottish Land Estate from 1961 to 1966 when he moved to Singapore under the banner of the then newly formed Ministry of Defence to take responsibility of the Defence Land Forces, in what was then referred to as the FARELF.

His role was similar to that of Edmund Speak but covered a different group of Military assets. Also, his remit was not just Singapore but the entire Asia region.

Jubb played a key role in the transfer of the H.M. Naval Base to the Singapore Government, after which he returned to Scotland in 1970 to take charge of the Ministry of Defence's land assets in Scotland, where he retired in 1972. He passed away in Scotland in 1993.

During his time in Singapore, Jubb lived in 2 Canterbury Road in Alexandra Park, a property that is still there today. The FARELF's Defence Lands office was then at the Tanglin Garrison compound. Two other Chartered Surveyors were part of his team at this time, Derek Rice and B.L. Porter, who were based at RAF Tengah. Jubb and Rice were also friends outside of work and used to sail at the Royal Singapore Yacht Club at Jurong, where they took part in races together.

At the same time as Jubb and his team were preparing for the handover, a small team of Chartered Surveyors worked at the Admiralty's Navy Works Department at the Naval Base in Sembawang, continuing to maintain the buildings until the handover.

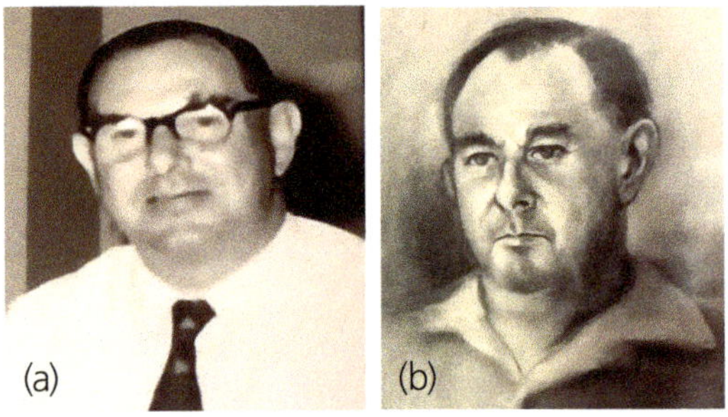

FIGURE 15.16. (a) Eric Jubb during his time in Singapore. (b) A sketch of Jubb drawn by his Singapore golf caddie Khalil.

Chief Engineers' Department, Far East Land Forces, Sime Road

After Frederick Croft left Singapore in 1962, a small team of Chartered Quantity Surveyors continued the task of building and infrastructure maintenance until 1966.

Corporate Real Estate

Hongkong & Shanghai Banking Corporation

Stanley Godfrey (see Figures 15.17 and 15.18)

Stanley ("Stan") Montague Godfrey (1930–2008) was born in London. He qualified in 1959 and arrived in Singapore several years later. He is, of course, mentioned elsewhere in this Book as someone who was both active as an R.I.C.S. Committee member as well as being instrumental in the R.I.C.S. stepping back to allow local professional bodies to be created and, in effect, take the place of the R.I.C.S.

He moved to Sydney circa 1971 to work for Jones Lang Wootton and continued his career with them in Melbourne. He passed away in Melbourne in 2008.

Hong Kong & Shanghai Banking Corporation and City Developments Limited

Malcolm Burrells

Malcolm Sidney Burrells (1929–2020) was born in England and passed away in Australia. He was at Dulwich College during World War II and was in the school swimming team in his last two years. After leaving Dulwich, he deferred his National Service while training to become a Quantity Surveyor. After successfully passing the R.I.C.S. examinations, he was then called up.

He attended a Selection Board course at the War Office and was commissioned into the Royal Engineers, serving three years in Singapore. After demobilisation, he returned to Singapore for a further 12 years, first as an Assistant Property Manager for Hong Kong & Shanghai Banking

FIGURE 15.17. Hong Kong & Shanghai Banking Corporation (HSBC) event in Singapore in the mid-1960s, featuring Malcolm and Jean Burrells (second row, sixth and seventh from the left) and Stanley Godfrey (back row, centre).

FIGURE 15.18. Hong Kong & Shanghai Banking Corporation (HSBC) event in Singapore in the mid-1960s, featuring Malcolm Burrells (second from left), Jean Burrells (second from right) and Stanley Godfrey (far right).

Corporation (HSBC) (see Figures 15.17 and 15.18) and then as the first Director of City Developments Limited (CDL) — see Figures 15.19 and 15.20 for some of the development projects led by Burrells. Whilst in Singapore, he shared a passion of sailing in a Snipe, a small dinghy.

FIGURE 15.19. City Towers, Bukit Timah, which was Singapore's first high-rise residential development and a CDL project led by Malcolm Burrells.

FIGURE 15.20. Clementi Park, Singapore, another CDL project that was led by Malcolm Burrells. This view is believed to be what is now Sunset Vale.

In January 1968, Burrells and his family emigrated to Australia, where he initially joined Jones Lang Wootton in Sydney, and, having become Senior Partner there, he later opened his own property consultancy practice in Sydney. He was "enticed" to join Hendersons the Property People, which later became part of the Chesterton Group.

In 1951, Burrells married Jean, and they had two sons. In Australia, Burrells was an enthusiastic sailor, successfully racing yachts on Sydney Harbour, and he became an active member of the Old Alleynians in Sydney, taking a great interest in the history, progress and development of Dulwich College.

In 1992, Burrells retired from the Chesterton Group, and he and Jean moved to Bowral in the Southern Highlands of New South Wales, where they restored a neglected old property to create an award-winning "English style" garden. He was also involved in the local opera society that he helped establish.

Malaya-Borneo Building Society

Gerald Chilton

Gerald Randolph Chilton (see Figures 15.21 and 13.4.) was an active and engaged R.I.C.S. Branch member from 1961 until 1966. During these years, he was the Vice-Chair (for one year) and Chair (for another year). He was employed by the Malaya-Borneo Building Society (MBBS). Like Derrick Morris, he is remembered as a kind man who freely shared his professional knowledge with Singaporeans and encouraged them to attain an R.I.C.S. qualification.

Within his organisation, he employed Singaporean talent with a view to succession planning. To this end, he employed Michael Oh Chong Guan as a school graduate and encouraged him to take his R.I.C.S. examinations, which he did.

The work they did for the Building Society included buildings surveys, inspections and valuations for mortgage purposes, providing valuation advice in relation to mortgage applications for property development in Malaya, Borneo and Singapore (commercial, industrial and residential) as well as building inspections for loan disbursements and progress payments for mortgaged land/property being developed with the benefit of an MBBS mortgage.

The Building Society was particularly successful in the late 1950s and early 1960s in granting loans for the construction of houses in Singapore, Malaya and Borneo. Indeed, at one point in 1959, the Society had to stop granting further housing loans for a period of time because

FIGURE 15.21. (a) Late 1950s photograph of the Chiltons (Barbara, far left, and Gerald, far right) at a Malaya-Borneo Building Society (MBBS) staff event. John Burgess (the General Manager) is believed to be in the centre. (b) Gerald Chilton on a building inspection in the late 1950s.

of a lack of funds to satisfy growing demand and the uncertainty due to the risk of an increased bank rate — at that time, it was accepting 250 new applications per month with loan releases of $3 million each month, which, in turn, would mean having to raise another $20 million of new capital each year.

It is also worth noting that the Society did not just fund the construction of residential property. At that time, 75% of all new private housing purchases in Singapore were funded by mortgage loans from the Society, with the maximum loan being $15,000 and a house price range of $9,000 to $25,000. Aside from this, the Society acted as an agent to the Federal Government (Federation of Malaya) for the construction of low-cost housing, in particular, in Petaling Jaya.

Chilton and his team played an important assurance role to MBBS around its mortgage arrangements with its clients (members). The Society still exists today, and is now known as Malaysia Building Society Berhad (MBSB).

In 1966, he moved to Africa, and his place at the Malaya-Borneo Building Society was taken by Oh, who, in turn, was replaced by another of Chilton's young graduates, Peter Lin Teck Hsiong, who joined MBBS after passing his first R.I.C.S examination. The offices of MBBS were at MacDonald House, Orchard Road.

Chilton later moved to Western Australia, where he retired and spent his last years.

Oh Chong Guan

Michael Oh Chong Guan was born in 1942 and qualified as a Chartered Surveyor in 1965. After his time with the Building Society, he joined Malayan Credit Properties Pte. Ltd. at their office at No. 300 Orchard Road in the mid-1970s.

In 1979, he joined Lai Pang Fee & Simon Lim Chartered Surveyors. He left the property industry some years ago.

Standard Method of Measurement of Building Works — for Use in the Federation of Malaya, Singapore and Borneo

In late 1959 and the early 1960s, the now-separated R.I.C.S. Singapore and R.I.C.S. (Federation of) Malaya Branches collaborated on what was a significant piece of work — a Standard Method of Measurement of Building Works — for use in the Federation of Malaya, Singapore and Borneo.

From the days when Waters and Watson first introduced Bills of Quantities to the industry in Singapore in the mid-1930s, the R.I.C.S. Standard Method of Measurement of Building Works has been used and adopted as a broad base for measurement of building construction works, with modifications taking place to suit local custom and practices and to meet the needs of the various Government Procurement and Contracting Departments.

The original problem was that each firm and Government Department, as well as British Military Quantity Surveyors working for the FARELF and the War Office, were all using their own slightly different versions of the R.I.C.S. Standard Method of Measurement of Building Works to suit their own needs, which were also adapted to what each Chartered Quantity Surveyor thought was the best way to measure building works. Indeed, the War Office introduced their Rules of Measurement in 1956 to standardize Bills of Quantities for work done by the War Department, Air Ministry and Admiralty in the region, but this was for internal use only.

An attempt to land on a local Standard prior to World War II did not happen, but in 1957–58, the R.I.C.S. Malaya Branch formed a Sub-Committee to revise and alter the "British" Standard Method of Measurement to meet local needs and practice. The result in November 1959 was the Standard Method of Measurement of Building Works, and updated in later years, it was replaced (in Singapore) by the Standard Method of Measurement of Building Works (Second Edition), as published by the Singapore Institute of Surveyors and Valuers (SISV) in 1986.

The November 1959 publication was largely based on the R.I.C.S. Method of Measurement of Building Works (Fourth Edition), but the

Sub-Committee deleted non-applicable sections such as those involving labour only and incorporated them into other major basic items while adding new items to suit local labour requirements and trade practices.

Of significance is that the new Standard amounted to an amalgamation, harmonisation and compromise of the local firms' and Government Departments' practices, along with practices of the War Ministry, Admiralty and Air Ministry.

The final document was circulated to all practitioners, who made suggestions for change and adaptation that were taken onboard in the final version that was published.

The credit for this pioneering piece of work goes to the Sub-Committee, which was chaired by **Seah Mong Hee,** who is, of course, featured throughout this book. The members of his Sub-Committee have all already been featured in this book: Robert Rhodes, Stuart Duncan, Derrick Morris (who succeeded Duncan when he returned to the UK) and Keith Suckling. However, there were many other Chartered Quantity Surveyors in the Federation of Malaya, Singapore and Borneo who contributed to the Sub-Committee's work.

The Standard was divided into the following sections:

- General Principles
- Preliminaries (including demolitions and works on site)
- Excavation
- Concreting
- Bricklaying
- Drainage
- Asphalting
- Paving
- Masonry
- Roofing
- Carpentry
- Joinery
- Steelwork and ironwork

- Plastering works
- Plumbing works
- Heating and ventilation engineering
- Electrical works
- Glazing works and painting works

The development and implementation of this Standard was the culmination of many years of effort by Seah and his colleagues to support the construction industry and improve professional standards in the industry in Malaya and Singapore.

R.I.C.S. Singapore Branch Dinner at the Tanglin Club

In June 1965, the R.I.C.S. Singapore Branch was honoured with the presence of Prime Minister Lee Kuan Yew and Deputy Prime Minister Toh Chin Chye.

The dinner took place at the Tanglin Club on Monday, 14 June 1965. See Figure 15.22.

Photographs of the event were taken, and one is published here with the kind permission from the Singapore Government and with the kind help and facilitation of the National Archives of Singapore.

This was one of the last such gatherings of the R.I.C.S. Singapore Branch, as it slowly but surely stepped back to allow the formation and advancement of local professional bodies.

FIGURE 15.22. R.I.C.S. Singapore Branch dinner at the Tanglin Club, 14 June 1965. Here, Prime Minister Lee Kuan Yew is addressing guests. Deputy Prime Minister Toh Chin Chye is seated to his right. The R.I.C.S. Singapore Branch Chair (Seah Mong Hee) and Vice-Chair (Gerald Chilton) can both be seen in the photograph, but Chelliah Tiruchelvarayan (Branch Hon. Treasurer), who played a prominent role in organising the dinner, is seated (middle, right) with his back to the camera. Rae Littlejohn is sitting diagonally opposite Tiruchelvarayan, with his face partially obscured. Permission has been granted by the relevant approving authority (via the National Archives of Singapore) to use this image of the late Mr. Lee Kuan Yew in this book.

Chapter

16

R.I.C.S. President's Visit to Singapore in 1966

In 1966, the R.I.C.S.'s President Guy Biscoe undertook the very first Presidential World Tour, accompanied by his wife Sheila and the R.I.C.S. Secretary, Francis Bennion. The Singapore leg of the tour started on 16 October 1966 and concluded on 19 October. Fortunately, Bennion kept a diary of certain parts of the World Tour, including the Singapore leg, and his family estate has kindly given permission for extracts from his diary to be included in this Book.

The President, his wife Sheila and Francis Bennion arrived in Singapore from Perth via Qantas at 8 pm on **15 October 1966**. They were met by various members of the local Branch Committee, who had prepared an extremely heavy programme for his three days in Singapore. The programme was perhaps too heavy, due in part to the fact that a letter written by Bennion from Melbourne to Stan Godfrey of the R.I.C.S. Singapore Branch, explaining the President's wishes regarding the programme, had not been distributed to the Branch Committee because Godfrey had gone on leave prior to the letter arriving in Singapore.

To quote directly from Bennion's diary: "The Goodwood Park Hotel where we are staying is an enormous, rambling place with the longest hotel corridors I have ever seen. Nevertheless, it seems a comfortable enough place. The sticky heat of Singapore was a great contrast to anything we had experienced previously on the tour, and I was relieved to find that the bedrooms were air-conditioned."

Monday, 17 October 1966

The Presidential team was taken by Seah Mong Hee to the Langdon & Every (Far East) offices in the Malaya Building on Robinson Road. There, Seah introduced the Presidential team to senior staff members. Seah told them that the "main question of concern to the Singapore members at present was the formation of the proposed Singapore Institute of Surveyors", which he described as a breakaway body from the Institution of Surveyors Malaysia (now, of course, known as RISM) — the earlier background to this is given in Chapter 13. He gave them a letter addressed to the R.I.C.S. about this. The main point seems to be whether the Real Estate Valuers Association (Singapore) (REVA(S)) should be incorporated into the new body. Although some Chartered Surveyors belong to REVA(S) — headed by a Chartered Surveyor (John Carter) — the concern of Seah and others was that some REVA(S) members had not passed any examinations.

The team was informed that both the existing and proposed surveyors' institutions were umbrella bodies, comparable in range to the R.I.C.S.

Seah then took them to the (adjacent) conference hall and trade union house **Singapore Conference Hall and Trade Union House on Shenton Way, now home to the Singapore Chinese Orchestra.** The Presidential team found the building to be very architecturally interesting — it had been designed by a firm of local architects and erected at only £4 per sq. ft. (a total of £500,000) — this included fitting out a 1,000-seat auditorium. The President commented on how much scope the tropical conditions had given to the architects in producing an interesting and striking design. The next stop on the itinerary was the Toa Payoh HDB Housing Estate, which Bennion recorded in his diary as having been built for "low income occupiers on reclaimed land, hitherto consisting of bush and swamp, occupied by squatters' huts".

They were then shown around an HDB development by Wee Sek Kay (otherwise known as Michael Wee — there is more on him in Chapter 21) of the HDB Lands Department, and they observed that unlike England, the car parks adjoining the Housing Blocks were occupied by hawkers (the crowds of people at the hawker area added an air of interest for the Presidential team). They also observed that each flat had a small balcony from which clothes-drying poles were protruding (four to each flat) with colourful clothes, giving the blocks the appearance of a "battleship".

Their hectic morning concluded with a garlic steak lunch at the Adelphi Hotel, including a glass of beer. The Presidential team learned that Singapore's best meat came from Australia and that a pint of beer cost five shillings (now 25 pence in English money, 40 cents in Singapore currency) — this shook them in comparison to England prices, even more so when they were told that beer was more expensive than wine even though the former was locally produced, and that taxation was not the reason for this.

In the afternoon, the Presidential team visited the new Singapore Polytechnic — this is covered in Chapter 18. Later in the evening, they attended a Members' Presidential cocktail party, where they met with 35 R.I.C.S. members who expressed their pleasure about the visit. Topics of conversation included the questions of subscriptions, the rebuilding of R.I.C.S. Headquarters, the formation of the proposed new Singapore Institute of Surveyors and the general conditions in the area. Members were very supportive and full of praise. After dinner, "the three of us had a quiet dinner at the Goodwood Park, and watched one or two items of cabaret before retiring to bed".

Tuesday, 18 October 1966

The day started with a visit to the offices of the Economic Development Board (EDB) to meet Lim Ho Hup (1929–2021), the EDB's Managing Director, H.A.J. (Jack) Fryer, the Chief Surveyor and Planner, and Jack Phillips, the Senior Surveyor. The two Jacks were both Australia-trained Land Surveyors who had embarked upon a training programme for their own staff in Singapore. Lim became the first Managing Director of the EDB when it was created back in 1961, while Phillips played an important role in the development of the profession in the 1960s and 1970s — there will be more on him in the next Chapter.

During the visit, Lim explained the rationale and planning for the entire Jurong Industrial Estate and "New Town", and he fully expected the Presidential team to visit Jurong as planned that afternoon, but because the programme for the Presidential team was too heavy, the team declined the visit. This caused a great deal of disappointment to Lim due to the immense importance of Jurong to the EDB — it was regarded as the show place that all visitors had to see.

Bennion's diary then turns to the next meeting with the Chair of the HDB, which he described as "very useful". The Chair was Howe Yoon Chong (1923–2007), who was the first Chief Executive of the HDB and also its Chair from 1966 to 1969. Bennion described him as being very dynamic, as Howe was both HDB Chair and Permanent Secretary in one of the Ministries. Howe explained the reasons for the remarkable success of HDB in delivering low-cost housing for lower-income groups were its ability to: operate on a larger scale than private developers, using a set design that was not varied during construction, and borrowing at very low interest rates that could be repaid very quickly, which in turn led to contractors securing credit for materials quickly too.

Howe said that it was his policy to employ Chartered Surveyors and he had sent a number of students to England for training. He expressed his disappointment that the failure rate was high, reinforcing his apparent suspicion that R.I.C.S. operated a quota system for passing Asian candidates. Although the Presidential team reassured him that there was no quota system limiting the pass rate, Bennion's diary records the need for R.I.C.S. to do more to help students from Asia to secure placements in the UK and that senior UK members had a duty to help. A further suggestion by Howe was that R.I.C.S. should have a basic system of common papers with global applicability, plus local papers with particular local relevance, and he agreed to help with the exchange of information around this. Howe was particularly keen on urban renewal, dealing with such matters as cost-benefit analysis.

Lunch was at the Cockpit Hotel — a colonial-style mansion that used to stand on Penang Road — followed by a meeting with the Acting Director of Public Works Hiew Siew Nam* and senior PWD staff.

On the evening of 18 October 1966, the Branch hosted a Cocktail Party at the Tanglin Club for the President — see Figure 16.1. One of the guests, Mr. Ramaswaram, the Chief Surveyor, confirmed his intention to set up a Land Surveyors Board on lines similar to those in Australia, and if such a

Royal Institute of Surveyors President on visit to S'pore

SINGAPORE, Wed. — The President of the Royal Institute of Surveyors, Mr. Guy Biscoe, arrived here with Mrs. Biscoe from London yesterday as part of their tour of Commonwealth nations to visit officials and members of the Institute's branches.

Mr. F.A.R. Bennion, the Institute's Secretary, is also accompanying Mr. and Mrs. Biscoe on the tour.

While in Singapore, they met surveyors in Government and private practice and discussed matters of mutual interest. They also had private talks with local engineers, contractors, architects and officials of the S'ngapore branch of the Institute.

The visiting officials of the Institute were entertained to cocktails at the Tanglin Club last night.

From left: Mr. A.E. Lewis (Secretary of the S'ngapore Branch), Mr. Burrells (Vice-Chairman of the Singapore Branch), Mr. Biscoe, Mr. W.M. Stevenson (Chairman of the Singapore Branch), Mr. Bennion and Mr. M.H. Seah (past Chairman of the Singapore Branch).

FIGURE 16.1. Photograph (from an unknown newspaper) of the R.I.C.S. Singapore Branch's Cocktail Party at the Tanglin Club on 18 October 1966.

* Dr. Hiew Siew Nam, who passed away in 2019 aged 103, was the Founding Honorary Secretary and 2nd President of the Institution of Engineers, Singapore (IES).

Board was set up, he would expect it to be a reciprocating Board on equal terms with the Boards of Australia and New Zealand. After cocktails, the Presidential team was hosted at a dinner at the Tanglin Club by Branch Committee members Keith Sucking, Anthony Lewis and their wives.

Wednesday, 19 October 1966

The Presidential team reached the sad conclusion that their programme would not enable them to do any sightseeing at all in Singapore or visit any places of interest other than those where surveyors were working, so the sightseeing experience was limited to tantalising glimpses seen through car windows. They viewed Singapore as a crowded, bustling city, with the City Hall and Supreme Court buildings standing out for them.

Another building that they glimpsed through the car window but frustratingly could not visit was the National Theatre, described by Bennion as "an extraordinary structure resembling a letter 'T' standing on one end of the cross piece. The sides of the auditorium are completely open, and the roof is a cantilever structure of striking appearance".

Their morning session was led by William Murdoch Stevenson, Chief Quantity Surveyor at the British Ministry of Public Building and Works, which was in charge of all British installations in the Far East. The first stop was the HQ of the Military Properties, Works and Buildings (MPWB) in Adam Road to meet the Regional Director, A.F.W. Grant, a Civil Engineer, as well as several Quantity Surveyors employed there. He told them of the enormous range of his responsibilities and that he had suffered heavily financially through the bankruptcy of a firm of estate agents in Harrow, England. His experience led him to welcome the prospect of regulating estate agents — Bennion agreed to get details from him later.

A visit followed to the Ministry offices of the Southern Area of Singapore Island, which were responsible for looking after all British Government construction services, including the Changi Airfield, where they met quantity surveyors, quantity surveying students and technicians. The following is quoted directly from the diary: "One of the senior men, an

Englishman, who is proposing to take the Direct Final examination shortly was firmly convinced that the R.I.C.S. operate a quota policy under which the level of the pass mark is adjusted according to how many candidates the Institution desire should qualify. The President assured him that there was no truth in this story, but he said it was widely believed and appeared unconvinced".

A visit followed to the offices of the Ministry in the HQ Far East Command to meet quantity surveyors and students. Lunch followed at the Raffles Hotel. Again to quote directly from Bennion's diary: "[The Raffles Hotel] ... fully justified expectations. It retains a great deal of the atmosphere of the Empire Days with much bamboo and rattan furniture, whirling punkahs and a general air of Somerset Maughan. The hotel is built round a small central garden called the Palm Court, where tropical flowers and tall palms contrast with the white walls of the old building. The tropical rainfall gives such gardens as this an intensely green look. We had lunch in the Elizabethan Grill, which, although famous and of long standing, seems completely inappropriate since it is furnished and decorated in an English Tudor style. As seemed fitting, I had a curry, which was very good."

After lunch, they visited the British High Commissioner, John Vernon Rob CMG (1915–71), who showed great interest in the tour. The conversation was mainly on the subject of education problems. Rob was very anxious to assist in any way he could, and he asked Stevenson (the Branch Chair) several times to keep in touch with him about developments. Rob had thought that the President's idea of a first-year course at the Polytechnic mainly concerned with improving academic standards was a very good one. He undertook to talk to the Principal of the Polytechnic, Mr. Aldis, about this.

In the afternoon, there was a meeting of the Committee of the Singapore Branch of the R.I.C.S. The main item on the Agenda was the proposed Singapore Institution of Surveyors. The meeting considered a draft letter to the R.I.C.S., asking for guidance on whether affiliation with the Institution would be possible. They also sought the R.I.C.S.'s view on the proposal to include certain members of the Real Estate Valuers Association (Singapore)

(REVA(S)), who had not qualified by examination, in the new body. It was explained that it would be necessary for the new Institution to obtain registration under the Society's Ordinance, and if members of the REVA(S) were not included, the Government might oppose the registration. The President and Secretary gave their comments on the draft letter, and the Branch proposed to alter its wording in the light of these comments.

It is worth mentioning at this point that the REVA(S) was created in 1963 by local and expatriate valuers and estate agents, with John Carter (mentioned in Chapter 12) as its first President. As well as Chartered Surveyors (primarily valuers), it included members of the Chartered Institute of Auctioneers and Estate Agents (CIAEA), whose members had a different qualification route to becoming R.I.C.S. members, which was a major concern expressed by members at the R.I.C.S. Singapore Branch meeting.

As footnotes to this discussion:

(i) The CIAEA eventually merged with R.I.C.S. in 1970.

(ii) The R.I.C.S. encouraged the phasing out of the R.I.C.S. Singapore Branch to enable the creation and growth of the Singapore Institute of Surveyors (SIS) and REVA(S). This is discussed further in Chapter 17.

(iii) In 1977, REVA(S) changed its name to the Singapore Institute of Valuers (SIV). In 1982, SIV merged with the SIS to form the Singapore Institute of Surveyors and Valuers (SISV).

The discussion at the R.I.C.S. Singapore Branch meeting continued, and the following points emerged, which are, all to some extent, pertinent in today's context:

- The Committee felt that there was insufficient information put out by the R.I.C.S. about members moving into the area. They asked for information to be sent out by HQ whenever it was known that a member had moved into or left. They also asked for someone to represent the Branch in London, such as the 1962/3 Branch Chair Kenneth ("Ken") Massey, who was by then back in London. They also complained that they made attempts without success to get further copies of the booklet explaining various aspects of the profession. They would like to have 200 copies of this sent out.

- They thought it would be useful for an article in the R.I.C.S. journal to state why candidates failed examinations. They welcomed some indication from the examiners of where candidates from Singapore were proving ineffective, e.g., failure in the English language, lack of technical knowledge, faulty examination technique, etc. The Presidential team explained the difficulties in giving the reasons why a candidate had failed. The Branch agreed that there should be no lowering of the standard of examinations.
- The Committee supported the idea of introducing a greater local element into examination papers. They discounted the argument that all Chartered Surveyors should be examined in British techniques and laws because they might go and practice in England and pointed out that the reverse was also true. In addition, the intensive training required for the R.I.C.S. examinations fitted someone with the necessary flexibility to adapt readily to conditions in another country. They thought the revised Quantity Surveying syllabus, which concentrated more on setting problems than testing memory, was a great improvement.

On the evening of 19 October 1966, the Presidential team was given a farewell dinner by members of the Branch Committee and their wives at the "Pekin" Restaurant — this was quite possibly the "Peking" Restaurant, which was then a popular restaurant on Orchard Road.

The ceremonial Cantonese dinner began with "thousand-year-old" eggs and continued with shark fins, roast suckling pig, fried crabs' claws, and an entire roasted duckling. Bennion's diary records that the Presidential team used chopsticks and were pleased that they managed to do so with success!

The event was captured on camera, and the sons of the late Walter Reeve and Malcolm Burrells have kindly permitted these photographs to be used in this Book — see Figures 16.2 and 16.3.

FIGURE 16.2. The R.I.C.S. Singapore dinner on 19 October 1966. Back row, left to right: Loh Kum Seng, Walter Reeve, James Khoo, Francis Bennion, unknown (possibly Terry Bonfield), William Stevenson (Branch Chair) and Lau Fook Sing. Front row, left to right: Unknown, June Reeve, Guy Biscoe, R.I.C.S. President, unknown and Valerie Biscoe.

FIGURE 16.3. The R.I.C.S. Singapore dinner on 19 October 1966. Back row, left to right: William Stevenson, Seah Mong Hee, Keith Suckling, Francis Bennion, Anthony Lewis, (Michael) Wee Sek Kay, Chelliah Tiruchelvarayan, Malcolm Burrells. Front row, left to right: Mrs. Stevenson, unknown, Valerie Biscoe, unknown, Guy Biscoe, R.I.C.S. President, Jean Burrells and unknown.

Chapter

17

1966–68: The Immediate Years After Singapore's Independence

From 1966 onwards, the R.I.C.S. gradually wound down its Branch activity during the time of Stan Godfrey as its Singapore Branch Honorary Secretary and Malcolm Burrells and Keith Suckling as Chair in 1967–68 and 1968–70, respectively.

With the full support and encouragement of the R.I.C.S., Seah Mong Hee, Malcolm Burrells, Goh Mia Hoon and **Stan Godfrey** — with help from others — enabled local surveyors to form the Singapore Institute of Surveyors (SIS), registered on 23 July 1968.

REVA(S) was reconstituted as the Singapore Institute of Valuers (SIV) in September 1977. Both the SIS and SIV merged in 1982 to become the Singapore Institute of Surveyors and Valuers (SISV). Stan Godfrey moved to Sydney in 1971 to work for Jones Lang Wootton, but he had left his mark in Singapore with the start of winding down of R.I.C.S. activity in Singapore as well as encouraging the growth of the SIS and REVA(S).

Godfrey, Seah, Burrells, Goh and several others were important figures in the transition from R.I.C.S. to the newly formed Singapore professional bodies, because they ensured that the full support and encouragement of the R.I.C.S. was in place to enable a smooth transition. They achieved this by enabling the R.I.C.S. Singapore Branch to be formally closed and replaced with a Corresponding Secretary role, which would provide a degree of ongoing communication and support to R.I.C.S. members in Singapore.

Goh took on the R.I.C.S. Singapore Corresponding Secretary role in 1971, and so it was at this time the Branch and Committee was replaced by the role of Corresponding Secretary.

Lily Ong (who managed the office operations for Collier Goh & Tan in these years) supported Goh on any R.I.C.S. related correspondence, and in particular, with regard to the R.I.C.S. Examinations from 1971 until Goh's untimely death in 1982. During this period, Ong's role was to ensure that her team parcelled up the completed R.I.C.S. Examination papers that came to the Collier Goh & Tan offices (once a year during the Examination time) and despatched them to the UK for assessment and marking.

Goh himself took a keen interest in the process. When he died in 1982, Lily continued the process for his successor, Tan Kim Choon, who was another Partner in the firm. She did this until she retired in 1990. Tan showed less interest in the R.I.C.S. — the general feeling at that time was that the R.I.C.S. was no longer relevant in the Singapore context, and more importantly, the general sentiment was that priority should be given to the growth of Singapore professional bodies. Indeed, another factor was that students seeking to become Quantity Surveyors or Valuers started to do the Colombo Plan scholarship at the University of Auckland, with only a minority doing their scholarship in London.

More details on the R.I.C.S. in the 1970s under the Corresponding Secretary role will be featured in a future updated edition of this book. Suffice to say that in the late 1960s and early to mid-1970s, under the leadership of Seah Mong Hee, the SIS held monthly Council meetings, yearly General Meetings and regular Educational Training sessions.

The appointed R.I.C.S. Corresponding Secretary (Goh) was always present at these meetings (and those of the REVA(S), later known as SIV), and so it was that information sent by the R.I.C.S. to Goh was disseminated to local professionals.

However, these were times when the Singapore Government was rightly keen to "localise" professional bodies, and so it was that R.I.C.S. "correspondence" gradually became lower on the agenda for discussion. At this time, SIS and REVA(S) meetings took place at the Singapore Professional Centre, Outram Park.

The importance of the Singapore Professional Centre in Outram Park (see Figure 17.1) cannot be understated. It opened in 1971 and was an important centre and focal point for new and emerging Singapore professional bodies as they became established.

FIGURE 17.1. Members of the Singapore Institution of Surveyors giving a presentation about their profession to Singaporean schoolgirls at the Singapore Professional Centre, Outram Park Shopping Centre. Left to right: Francis Lai Pang Fee, Chelliah Tiruchelvarayan, Lee Eng Kwang, Simon Lim Ban Hoong and Seah Choo Meng.

As a final note, Chartered Surveyors in Singapore between 1969 and 1978 were mostly Quantity Surveyors, but a growing number of General Practice Surveyors and Valuers were also establishing themselves.

In the late 1960s and early 1970s, a number of Quantity Surveyors who studied in England or New Zealand were coming to the end of their 5-year "bonded" period in public service and were seeking employment in the private sector or, alternatively, setting up their own firms. Similarly, those who studied in Malaysia at the Technical College were also looking to set up independent firms if they could not find decent positions with firms in Singapore.

Ironically, R.I.C.S. membership increased from 80 in 1967 to 120 in 1970. This occurred for the following reasons:

- Sarawak, Sabah and Brunei were brought under the purview of the Singapore Branch from 1968 to 1970. This added to the number of members.
- A number of Singaporeans continued to attain their R.I.C.S. qualification as well as seeking membership in the emerging local bodies, and
- At this time (indeed, until 1976) R.I.C.S. professionals continued to come to Singapore from the UK to assist in the transition of ownership of the many land and property assets held by the UK Navy, Army and Air Force to the Singapore Government.

In the period from 1966 to 1968, the following noteworthy Chartered Surveyors came to prominence in Singapore:

Antony Lewis and Alan Runnicles

Antony ("Tony") Edward Lewis (1927–2004) was a Chartered Quantity Surveyor, Fellow of the R.I.C.S. and founding member of the Singapore Institute of Surveyors who arrived in Singapore around 1961 — he has already been featured in Chapter 16 as the Secretary of the R.I.C.S. Singapore Branch Committee. From 1968 until the early 1970s, he and **Alan Runnicles (1944–2010)** were important figures in the $70 million development of the new **Ocean Financial Centre.**

Due to his extensive expertise in contract administration for large developments, **Lewis** was appointed as the Project Controller by Ocean Properties for the development. **Runnicles** (see Figure 17.2) who qualified as a Chartered Surveyor in 1963 specialising in Valuation and Town

Planning, was there to represent the interests of Capital & Counties.

Construction was completed in 1974. It was demolished in 2008 to make way for the current Ocean Financial Centre, which was completed in 2011.

FIGURE 17.2. Photographs of Alan Runnicles.

Francis Lai Pang Fee (see Figures 17.3 and 17.4)

Born in Singapore in 1940, **Dr. Francis Lai Pang Fee** worked for C.H. Williams as well as Langdon & Every (Far East). He then moved to London in the early 1960s to study at the College of Estate Management (CEM) on a Colombo Plan scholarship, as well as serve as a Property Valuer to the London Borough of Southwark — he graduated in both Valuation and Quantity Surveying. Lai lived at 25 Harrington Gardens, Kensington, when he studied at the CEM.

He was unique in this era because he qualified as both a Chartered Quantity Surveyor (1969) and a Chartered Property Valuer (1966). He holds a Master's Degree in Business Administration from Deakin University, Australia, and a Doctorate Degree in inter-religious philosophy from Assumption University in Thailand.

After returning to Singapore and spending time as a bonded scholar from mid-1966 with the Housing & Development Board (HDB) as a Contracts Officer reporting to Teh Cheang Wan, who was the Chief Architect and Head of the Building Department. In 1970, he set up Lai

FIGURE 17.3. (a) High Street Centre, Singapore. (b) Dr. Francis Lai Pang Fee in Singapore prior to his time in London.

Pang Fee & Co. in an office at 49a New Bridge Road. In the mid-1970s, he moved to 279F Selegie Complex, Selegie Road, and in 1977, he joined forces with another Chartered Surveyor, Simon Lim Ban Hoong (of Mansfield Property), who qualified as a Chartered Surveyor in 1971. They operated a joint practice on the 15th Floor of the CPF Building on Robinson Road (this building was demolished in 2016). Their practice was called Lai Pang Fee & Simon Lim Chartered Surveyors. The Company soon became known as Simon Lim Oh and Partners with the addition of Michael Oh Chong Guan as a partner.

Dr. Lai was also, in the 1970s, a founder member of the Singapore Industrial Training Board.

Lai Pang Fee & Simon Lim Chartered Surveyors was one of the first — if not the first — Singapore practice to offer UK commercial property for sale to Singaporean investors. In 1977, they marketed the former Hertford Hotel on Bayswater Road in Kensington, London, for sale as a freehold investment built. This Hotel had been built in 1961/62 and was subject to a long lease to Trust House Forte, which also included the lease of a Petrol Filling Station situated in front of the Hotel. The asking price for the investment was $2.5 million. The Hotel and Petrol Filling Station still stand today — the former is now known as the Thistle Kensington Gardens Hotel, and Shell continues to operate the latter.

Dr. Lai was a strong proponent for the idea of a Brain's Trust or Brain Centre comprising Singaporean surveyors to advise the Government on matters pertaining to the built environment, and from the 1960s onwards, he was an early proponent of Singapore talent and the need to set up local training and examination facilities to ensure the sustained and continued growth of local talent. His view was that local developers should be encouraged to award contracts to Singaporeans to "cut out the middle man", given that when foreigners were awarded contracts, they tended to simply sub-contract to Singaporeans. He also believed that local examinations should be developed to supplement the R.I.C.S. Examinations because those set by the R.I.C.S. were not locally relevant in terms of building construction, law and practice.

He later joined Fletcher Construction in New Zealand in June 1979, followed by Rawlinson & Co in Auckland in June 1982, and Belton's (also in Auckland) in 1986. After returning to Singapore to take up a position with the National University of Singapore (NUS) as a Senior Lecturer from 1992 to 1995, he returned to New Zealand, where he still resides.

Amongst his achievements in his Singapore career are the following:

- Peoples' Park Centre — advisor to Ho Kok Cheong while still in the Civil Service
- People's Centre — consultant to the Contractor
- High Street Centre — project consultant Quantity Surveyor (see Figure 17.3(a))
- Admore Park Residential development — project Quantity Surveyor
- Outram Park Development (now demolished as part of General Hospital) while in Government Service

Dr. Lai was also an advisor to Alan Choe on the planning and valuations of Sentosa Development. He met Choe in 1966 when Choe was head of urban renewal at the HDB. In 1966–69, they worked together on working groups such as Urban Renewal tenders for sites, starting up the Singapore Tourism Board, the future of Sentosa, plus many other projects, including the establishment of the Building Industry Training Board.

He was involved in the real estate industry for all his active working life, in real estate projects in Auckland, Singapore, Thailand, Hong Kong and New Zealand. Amongst his other achievements were the Discovery Bay Holiday Resort (Hong Kong), Duta Merlin Hotel (Jakarta), and the Shangrila Gardens (Auckland).

In Thailand, Dr. Lai assisted property valuation firms like Simon Lim & Partners, N&A Valuers and Pasupat Realty Ltd and assisted the Thai Government in the setting up of Property Valuation Standards. Dr. Lai has been involved in the training of many successful property developers, real estate agents, property valuers, and quantity surveyors.

As well as NUS, Dr. Lai also taught at the Assumption University of Thailand and the Unitec Auckland, assisting many people into the many property-related fields.

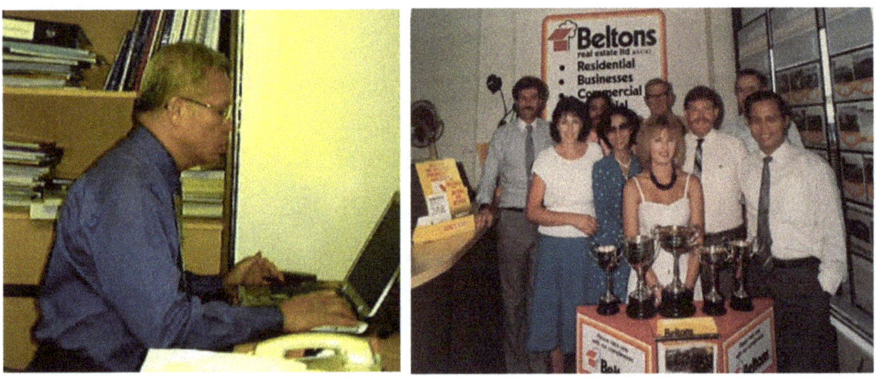

FIGURE 17.4. Dr. Francis Lai during his time with (a) Belton's in Auckland and (b) Assumption University of Thailand.

Dr. Lai's strength is in his ability to relate to people and assist many of his friends in achieving their property-related dreams. His multi-lingual skills include Thai, Malay and several Chinese dialects, on top of his excellent grasp of the English Language, which developed during his years in London. Dr. Lai's interests extend beyond properties to the logic of Sudoku, where he has developed and published books on the "Francis' Rules" for solving all Sudoku challenges. He is now retired but is available for consultancy assignments.

Johnny Loh Jwee Siam (see Figure 17.5)

FIGURE 17.5. Johnny Loh Jwee Siam.

Johnny Loh Jwee Siam (1929–2016) was born in Malaysia in May 1929 and passed away in Australia in February 2016, aged 86.

He was encouraged to become a Chartered Surveyor by John W. Hill, who was his manager in his first position in the industry as a Lands Inspector for the SIT in 1953. He then became a Lands Officer with its successor, the HDB, in 1960.

Loh worked during the day and studied at night for his surveying examinations, which he sat for in Singapore as a distance student. Study material from the College of Estate Management in England would come by post; his daughter, Sallyann, recollects, and it came in 3 parts over 4 years. It was so tightly bound that he would place a heavy book to straighten the papers. He had them bound later in the hope that Sallyann would also take the same examinations.

He qualified first as a Chartered Auctioneer and Chartered Real Estate Agent (using the designation "AAI", which its Associate members were entitled to do), and then he secured his R.I.C.S. qualification in 1969 — another year, and he would have automatically become "ARICS" through the 1970 merger of the R.I.C.S. with the Chartered Auctioneers' and Estate Agents' Institute.

In 1969, he joined DBS Bank, and then in 1970, he did a three-year stint at the Port of Singapore Authority (PSA) as its property manager and then as Deputy Director.

PSA had existed in some form or other for over a century, but the need for the skills of Chartered Surveyors such as Loh increased in 1966 when the Singapore Government made a strategic decision to develop a Container Port, thus allowing the Port to expand from Keppel Harbour/Telok Ayer Basin into Tanjong Pagar and Telok Blangah for shipping berths and also warehouse/logistics facilities. Prior to Loh joining the PSA, another Chartered Surveyor was at the helm there in the mid-1960s: Lieutenant Commander H.L.F. Vaughan, a Hydrographic Surveyor.

In 1970, he successfully applied for an auctioneer's licence whilst at the PSA to auction chattels — he was possibly the first R.I.C.S. member to do so since Henry Lake Coghlan 60 years earlier, as many auctioneers were not then qualified as Chartered Surveyors in Singapore.

Three years later, in 1973, Loh set up his own firm, which became Richard Ellis Sallmann Seward and Loh (RESSLoh). In 1977, Idris Pearce (later Sir Idris Pearce, who became the R.I.C.S. President in 1990) persuaded Loh, Tony Sallmann and Justin Seward to merge with C.H. Williams to form Richard Ellis CH Williams & Loh, with Loh (for RESSLoh) and Peter Newman (for C.H. Williams) as Joint Managing Directors. At this time, their offices were at 20 Ngee Ann Building, Orchard Road.

Loh left the practice in 1980 to operate on his own once again. In 1982, he brought in his daughter Sallyann as an Associate. He and his family later migrated to Australia. Loh was a member of the Tenants Compensation Board in the 1970s until his migration.

Loh is remembered with affection by all those who worked with him. He was also a respected adviser and mentor to his staff at the HDB, PSA and RESS — this was something he was proud of then and in his retirement years in Australia.

William Jansen

William Raymond ("Allan") Jansen (1924–2016) was born in Singapore in October 1924 and passed away in Singapore in July 2016, aged 91 years.

He qualified first as a Chartered Auctioneer and Chartered Real Estate Agent and then a Fellow of the R.I.C.S. in 1970 in the merger of the R.I.C.S. with the Chartered Auctioneers' and Estate Agents' Institute.

Jansen worked for the SIT, and then from 1960, with the HDB in its Estates Department, which in 1966 merged with the Lands Department and Lands Section of the Urban Renewal team to form a single, coordinated

Estates Department. He was based at Princess House, Alexandra Road, from the early 1960s onwards as well as seconded for a short time into the State & City Planning Department. He, along with Ho Lum Khuan (see Chapter 21), managed individual Estates around the City under the leadership of Tan Tian Boon, who was the Head of the newly consolidated Estates Department in 1966.

Ng Boon Ong (see Figures 17.6 and 17.7)

Ng Boon Ong (1923–86) was born in Perak, Malaysia. In terms of formal education, he had only the Senior Cambridge Certificate (the equivalent of the "O" levels) because his mother — a widow — could not afford to send him for further education. He and his family moved to Singapore after the end of World War II when he secured a position with the SIT as a Housing Officer (Property Management). In the mid-1950s, he was encouraged by his SIT managers to get professionally qualified, and so he did just that, studying remotely by correspondence courses via the Chartered Auctioneers' and Estate Agents' Institute. This Institute merged with the R.I.C.S. in 1970, at which point he became a Fellow of the R.I.C.S.

FIGURE 17.6. Ng Boon Ong in: (a) the late 1950s and (b) 1958 or 1959 in London, during his time working there.

Having qualified, he was given the opportunity for a one-year secondment to London, England, as a Housing Manager in 1958. This was a very difficult decision for him to make, as it meant leaving his mother, siblings (and their children), wife and daughter behind. Travelling by British Overseas Airways Corporation, now known as British Airways (BA), he proudly returned to Singapore in 1959 and was promoted upon his return. At that time, his family was living in an SIT home on Rawang Road.

In 1960, with the formation of the HDB and dissolution of the SIT, he took on a Property Management role with the HDB and was based at the HDB office on Maxwell Road. The role of a Housing (Property) Manager was tough back in the 1950s and 1960s. For example, Ng had the task of rehousing the 16,000 residents who lost everything in the Bukit Ho Swee fire of 1961 — and the Resettlement Plan understandably had a very tight timeline. He was also closely involved in the removal of squatters from land being acquired by the Government for development by the HDB, and he often had difficult negotiations with the squatters (and people purportedly representing them), which frequently led to him and his colleagues receiving threats of violence.

In 1971, he was promoted to Senior Principal Estates and Lands Officer, and in addition to his managerial duties, he was involved in major land acquisitions for the Government to support the HDB's housing programmes. He was Deputy Head of Department for some time, and then circa 1977, he was promoted to Head of Estates Management when Tan Tian Boon left to join Jurong Town Corporation (JTC). When the Estates Management and Lands Departments merged, he became the head of the combined Departments.

In recognition of his public services, he was awarded the Public Administration (Silver) Medal. A devoted family man, he was also a violin player — as a boy, he longed to play the violin but could not afford to purchase one. When he got his SIT job, he saved up and initially bought a cheap violin. A Filipino violinist (a graduate of the Philippines Conservatoire) who played at the Raffles Hotel taught him.

Going back to the 1960s, he and a small group of others were persuaded by Francis Lai Pang Fee and Tan Tian Boon to teach valuers in the evenings and weekends at the Singapore Polytechnic and later on at NUS in the 1970s.

During this time, this group, including Ng, ran two training part-time schools within the HDB with about 30 trainees in Valuation (for Trainee Estate Officers) and 10 in Quantity Surveying (for Building Inspectors). Their HDB in-house training programme produced Singapore's next generation of surveyors.

He passed away suddenly in Singapore in 1986, aged 63, but left a strong legacy behind for future generations. He is a wonderful example of someone who rose from the very bottom of the SIT/HDB ladder to become a Department Head through hard work and endeavour and, of course, with the benefit of his professional qualification.

FIGURE 17.7. Ng Boon Ong in (a) Trafalgar Square London in 1958/59 during his time working there and (b) Singapore in his later years.

Tan Tian Boon (see Figure 17.8)

Tan Tian Boon (1927–2017), also known as Paul Tan, qualified in the late 1950s as a member of the Chartered Auctioneers' and Estate Agents' Institute. He later became a Fellow of this Institute. Upon its merger with the R.I.C.S. in 1970, he became a Fellow of the R.I.C.S.

He played a very important role, not only as one of the first — if not the first — Singaporean Chartered Surveyor to reach a senior position in Government service as well as playing a key role as a mentor and catalyst for other Singaporeans to later become Chartered Surveyors.

Tan started with the SIT in the 1950s, and in 1960, he joined the HDB and became its Estates Manager in 1961. On 1 April 1966, he became the Head of the newly formed Estates Department — on its merger with the Lands Department — and he remained in this role until the late 1970s.

He was largely responsible for the creation of the highly efficient decentralised team, which had an efficient central direction but with local direction from nine area offices (Alexandra, Bukit Merah, Tiong Bahru, Bukit Ho Swee, Toa Payoh, Kampong Java, Macpherson, Kallang and Queensway). This decentralisation allowed for more direct, personal contact with tenants. It also allowed Estates Officers to quickly be on hand to meet tenants, supervise repairs, manage the activities of hawkers (many of whom at that time continued to operate from estate car parks rather than availing themselves of one of the newly-built units inside the hawker centres), and get a much deeper understanding of the areas under their supervision.

He worked closely during this time with Teh Chiang Wan, Chief Architect and Head of the Building Department.

After leaving the HDB, Tan was succeeded by Chong Kim Chang, with Tham Yew Fang as Chong's Deputy, but the discipline set by Tan was continued by those who followed him. Indeed, Dr. Liu Thai Ker recalls that they were renowned for their focus on continuous improvement.

During his time with the HDB, Tan would ensure that their area Estates Managers continued to provide feedback to enable the Chief Architect and his staff to continually fine-tune HDB apartments layouts to achieve optimal layouts and improve the management of HDB estates.

Tan also led the way in monitoring supply and demand to ensure the right balance. After he moved on, this was perfected by Tham, who was very meticulous in ensuring only a 0.5% gap between supply and demand

— supply always had to be higher. This approach led to a significant improvement in the management of HDB estates.

Tan worked closely with Teh Cheang Wan to find ways for the HDB to succeed where the SIT had not been successful from a social housing perspective. Tan and Teh did the following to effect change:

- Upskilled contractors and increased capacity
- Improved supply chain of materials

Tan then joined JTC. Initially, he was seconded from the HDB to JTC as its General Manager by Teh after Teh had been appointed as the (acting) Chairman (and later, Chairman) of JTC in September 1976. Upon taking early retirement from JTC in 1981 to spend time in London with his wife and family, Tan's JTC role as General Manager was taken by Francis Mak, who had previously been a Deputy Director in the PWD.

FIGURE 17.8. Photograph of Tan Tian Boon, whilst General Manager of Jurong Town Corporation (JTC), giving a briefing to Chinese Foreign Trade Minister Li Qiang at Jurong Town Hall.

At this time, another Chartered Surveyor, Swee Kee Siong, worked under Tan (and then Mak) as Head of JTC's Lands Division. Swee became a Chartered Surveyor in 1970 upon the merger of the Chartered Auctioneers' and Estate Agents' Institute with the R.I.C.S. He later became JTC's Deputy CEO and, subsequently, CEO of Ascendas' Singapore operations and the Singapore Science Park.

Tan Tian Boon was regarded as a great leader and someone who inspired others to perform to the best of their abilities.

Lim Poh Guan

Lim Poh Guan was one of those who decided to work full-time in Singapore and do his studies in his own time. He took his surveying examinations in Singapore (at the British Council on Stamford Road) rather than heading to London. He qualified in 1968 as a member of the Chartered Auctioneers' and Estate Agents' Institute, and upon the Institute's merger with the R.I.C.S. in 1970, he became a Chartered Surveyor.

He was an Estates Officer with the HDB from the mid-1960s onwards and was mostly based in its Toa Payoh area office.

In 1968, he left the HDB to join DBS in Shenton Way. With DBS, along with other Chartered Surveyors, Yeo Soo Pio and Johnny Loh, the duties were wide-ranging. They included title verifications and property inspections for mortgage purposes, providing valuation advice in relation to mortgage applications for development projects in Singapore (commercial, industrial and residential), as well as loan disbursements and progress payments for mortgaged land/property.

After leaving DBS in the mid-1970s, he joined Hong Fok Corporation as their General Manager.

A highlight of his career was the development of The Concourse for Hong Fok Corporation.

Andrew Tan Cheng Yoon (see Figure 17.9)

Andrew Tan Cheng Yoon is a Malaysian whose career in Singapore, albeit short, was entwined with some of the pioneers mentioned in this book.

He migrated to Canada in the 1970s and is now an R.I.C.S. Fellow living in Vancouver, British Columbia, where he worked for (amongst others) Austin Real Estate.

FIGURE 17.9. Photograph of Andrew Tan Cheng Yoon taken from his 1970 Singapore Visit Pass for his employment with the Port of Singapore Authority.

Tan started as a trainee valuer with the Malaysian Government's Treasury Valuation Division in Kuala Lumpur. He enrolled in the College of Estate Management (CEM) correspondence course and passed the Intermediate examination. He then secured a Government scholarship to take his Finals in London. He continued the CEM course whilst working for Gerald Eve & Co. in Savile Row, London, where he secured his R.I.C.S. membership in 1967.

Aside from a spell at the HDB, Tan worked at the PSA from 1971 to 1972, where he reported to Johnny Loh Jwee Siam (who features elsewhere in this chapter). During that time, he had many tough negotiations with Ng Tee Geok, a Government Valuer and member of ANZIV, on a wide range of valuation and associated property matters.

Ng Thin Wah

Ng Thin Wah (see Figure 17.10) worked and studied in London, and after his qualification as a Chartered Surveyor during his time in London, where he got married in 1966, he returned to Singapore to serve his bond.

He worked in JTC's Land Division from 1971 until 1978.

FIGURE 17.10. Ng Thin Wah.

However, he decided to take a law degree and change professions. He then practised as a lawyer through his own law firm.

Ng passed away in Singapore in 2018.

Tan Kim Choon (see Figure 17.11)

Tan Kim Choon (also known as John Tan) started his professional career with the Kuala Lumpur Municipal Valuation Department in 1959. He trained in England and qualified as a Chartered Surveyor in 1963. He left Malaysia in 1966 to come to Singapore, where he joined JTC as Director, Land & Estates Division. In 1973, he joined Ocean & Capital Properties Pte. Ltd, where he was associated with its Ocean Building and Bukit Timah Plaza developments before joining M.H. Goh, Wicks & Partners in 1976. During the 1970s, and then from 1980 onwards, when the firm joined forces with Colliers International to become Collier Goh & Tan, it became the largest real estate consultancy in Singapore.

FIGURE 17.11. (a) Tan Kim Choon. (b) An advertisement confirming the appointment of Tan as a Partner of M.H. Goh, Wicks & Partners on 5 January 1976. This resulted in the firm changing its name to M.H. Goh, Tan & Partners.

Tan remained with the firm — now part of the Colliers International firm — until his retirement, after which he remained a consultant to the firm until the 2010s.

During his time with M.H. Goh, Tan & Partners (later, Collier Goh & Tan), Tan was also the R.I.C.S. Corresponding Secretary from 1988 to

1994. At this time, the R.I.C.S.'s focus was on growing its base in Europe via Brussels with the creation of European Association of Chartered Surveyors (EACS), and then in 1993, its Europe strategy was completed with opening of an office in Brussels, and so its focus on places like Singapore diminished.

The local professional bodies could no longer rely upon the R.I.C.S. for support, which made Tan's role as Corresponding Secretary difficult. Despite this, from the late 1990s until 2007/08, he continued to be a supporter of the R.I.C.S., and indeed, he was an APC assessor throughout this time, encouraging others to do likewise. He passed away in 2018, aged 90.

Yap Eng Bock

Yap Eng Bok (see Figure 17.12) was a Malaysian who qualified as a Chartered Surveyor in 1967.

During his time in Singapore as a newly qualified Chartered Surveyor, he worked for JTC as an assistant to Tan Kim Choon. In the late 1970s, he returned to Kuala Lumpur to work for Jones Lang Wootton, where he stayed for many years. He was, from 1992 to 1993, the President of the Institution of Surveyors Malaysia (now known as RISM).

FIGURE 17.12. Yap Eng Bock.

Yew Fan Tatt

Yew Fan Tatt (see Figure 17.13) was a Malaysian Chartered Quantity Surveyor who became Chartered in 1964. He studied for his R.I.C.S. Examinations whilst working in London. He also worked for Desmond Vargeson and R.N.S. Rhodes at the PWD in Kuala Lumpur in the late 1950s prior to qualifying and later on again for them post-qualification in Kuala Lumpur and Singapore. During his time in London, he became friends with the artist Ibrahim Hussein (later to become Datuk Ibrahim Hussein), who was also studying in London. They, along with other students from Malaysia, would spend time with the Rhodes family in Poole, England, during the holidays.

FIGURE 17.13. Photograph of Yew Fan Tatt taken whilst visiting the Rhodes family in Poole, England, circa mid-1962.

Yew remained in Singapore until the early 1970s, by which time he was a Senior Quantity Surveyor for Mertz & McLellan at Jurong Power Station. After his time in Singapore, he returned to Malaysia, where he worked for **Universiti Teknologi MARA (UiTM)** on Real Estate and Construction research projects.

Justin Ong Huck Jin

FIGURE 17.14. Justin Ong Huck Jin.

Justin Ong Huck Jin (see Figure 17.14) was originally from Penang but settled in Singapore during the 1940s. His father-in-law was Cheong Koon Seng (1880–1934) of Cheong Koon Seng & Co., a well-known and well-respected pioneer local auctioneers and estate agents' firm in Singapore.

He qualified in the late 1950s as a member of the Chartered Auctioneers' and Estate Agents' Institute. He later became a Fellow of this Institute, and upon its merger with the R.I.C.S. in 1970, he became a Fellow of the R.I.C.S. Later, he was a Fellow of the SISV.

He worked for the SIT and then for the HDB from 1960 onwards, where he took over from John W. Hill as the Lands Manager in 1961. On the merger of the Lands and Estates Departments in 1966, he became an Estates Officer, where he was later seconded to the State & City Department, where he undertook various roles, including a role as Collector of Land Revenues in Singapore.

He later ran his own real estate firm called Benchmark Real Estate Pte. Ltd. He passed away in 1998.

Lim Peng Song

Lim Peng Song was a Chartered Surveyor who qualified in 1968 and worked for the Public Utilities Board from 1969 until the 1980s in their Architectural and Maintenance Department.

Peter Newman

Peter Howard Newman (see Figures 17.15 and 17.16) was a Chartered Valuation Surveyor who was hired to manage the Singapore office of C.H. Williams from the mid-1960s onwards when Colin Williams moved to Kuala Lumpur. As with Johnny Loh, he played a prominent role in the growth of C.H. Williams & Co. Pte. Ltd. Valuers and Estate Agents, and as General Manager-cum-Managing Director, he was a pivotal figure in the 1977 C.H. Williams merger with Richard Ellis Sallmann Seward and Loh (RESSLoh), which is now CBRE Singapore.

In addition, he also managed their Hong Kong and Jakarta offices in the 1970s until his retirement in 1984/85.

He was an active member of the Tanglin Club during his time in Singapore and a past President. He died in London in 2008, aged 71.

FIGURE 17.15. Peter Newman (fourth from right) with the CB Richard Ellis Singapore team, circa 1981. Seated next to him is Willy Shee, who later became Managing Director of CB Richard Ellis from 1991 to 2005 and then its Chair.

FIGURE 17.16. Peter Newman (second from left), circa 1981.

Goh Mia Hoon (see Figure 17.17)

Willie Goh Mia Hoon did his R.I.C.S. Examinations in England from 1955 to 1960 whilst working for Edwards Bigwood & Bewlay, a firm that joined forces with Colliers International in March 1981 to create the UK entity known as Colliers Bigwood & Bewlay. He became a Chartered Surveyor in 1960 (see Figure 17.18), at which point he returned to Singapore, and as a bonded scholar, he worked as an assessor in the Property Tax Division.

From 1961 to 1967, he was the Regional Properties Manager (later Director) of American International Assurance Co. Ltd, overseeing their Malaysian building and mortgage development programme throughout Malaysia and latterly initiating its mortgage development.

In 1967, he became a Partner of Wicks & Partners. The firm was renamed M.H. Goh Tan & Partners in 1976 when Bill Wicks made the decision to retire and sold his stake in the business. At the same time, Wicks also sold his Kuala Lumpur business to Jones, Lang Wootton, St Helier, Jersey, Channel Islands (hereinafter referred to as "the Jersey Partners") and Jones, Lang Wootton, Sydney, Australia (hereinafter referred to as "the

Australian Partners"), as well as entered into a partnership agreement in 1977. Wicks then retired to live in Benidorm, Spain.

In 1980, Goh was instrumental in the firm becoming a member of the Colliers International Property Consultants group, at which point the firm became Collier Goh & Tan. Although the firm had strong links in the 1970s and 1980s to the SISV, it retained strong links to the R.I.C.S. through Goh, who was the R.I.C.S. Corresponding Secretary.

FIGURE 17.17. (a) An advertisement from 1980 confirming the company's name change to Collier Goh & Tan. (b) Goh Mia Hoon in 1955, taken immediately before his flight from Singapore to England, where he went to work and study. (c) Goh in the late 1960s.

Collier Goh & Tan offered professional services in accordance with the strict code of conduct laid down by the R.I.C.S., and their promotional material from that time confirms that they adopted the R.I.C.S.'s high standards of qualification and competence and the observance of strict rules of integrity and independence in the transaction of their clients' businesses.

Sadly, at the height of his career, and very unexpectedly, Goh died suddenly in 1982. He left a strong legacy and a pathway for other international consulting firms to come to Singapore. He was also a strong ambassador for the R.I.C.S. in Singapore. At the time of his death, he was also on the Board of Directors of OCBC Bank.

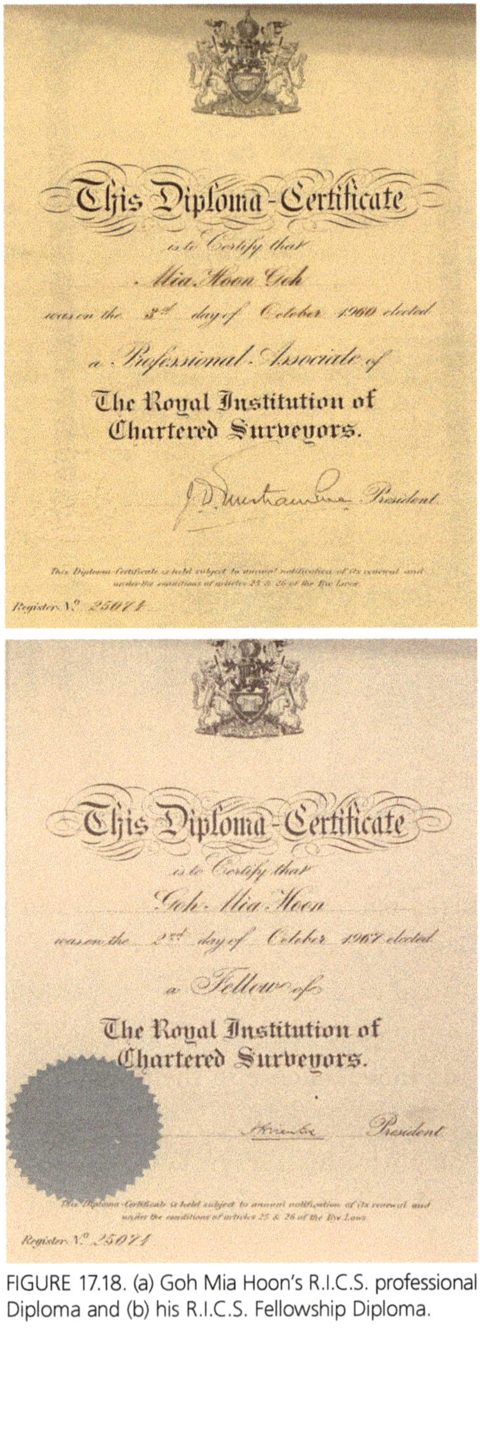

This Diploma-Certificate
is to Certify that
Mia Hoon Goh
was on the 3rd day of October 1960 elected
a Professional Associate of
The Royal Institution of
Chartered Surveyors.
President.
This Diploma-Certificate is held subject to annual notification of its renewal and under the conditions of articles 25 & 26 of the Bye Laws.
Register No. 25071

This Diploma-Certificate
is to Certify that
Goh Mia Hoon
was on the 2nd day of October 1967 elected
a Fellow of
The Royal Institution of
Chartered Surveyors.
President.
This Diploma-Certificate is held subject to annual notification of its renewal and under the conditions of articles 25 & 26 of the Bye Laws.
Register No. 25071

FIGURE 17.18. (a) Goh Mia Hoon's R.I.C.S. professional Diploma and (b) his R.I.C.S. Fellowship Diploma.

Final Notes

[1]

Both **CB Richard Ellis** and **Colliers** have received mentions in this book.

However, during the late 1960s and into the 1970s, **Jones Lang Wootton** was engaged by the Singapore Government's City Council to do comprehensive planning of land uses in Singapore. **Richard Luscombe**, a Chartered Surveyor from Jones Lang Wootton's Sydney office, was seconded to Singapore for that task, and he opened a permanent office here in 1974 on the 3rd floor of the Ramayana Building, Robinson Road.

By 1977, Luscombe had moved the office to the 3rd floor of the OCBC Centre on Chulia Street, at which point he was working alongside another Chartered Surveyor **(S.C.K. "Dick" Bird).**

Singaporean Chartered Surveyors such as Wan Fook Kong (who qualified in 1971 whilst working for the HDB as an Estates Officer) came to prominence in Jones Lang Wootton in the 1970s.

There will be more on Jones Lang Wootton and Jones Lang LaSalle in future editions of this book covering the 1970s and 1980s.

[2]

At the end of the 1970s, there was another R.I.C.S. Presidential Visit to Singapore — June 1979 saw the R.I.C.S. President Peter W. Grafton and his wife Margaret visit Singapore via a Concorde plane.

Whilst this is outside the 1868–1968 "timeline" for this Book, it is noteworthy, as many Singaporeans may be unaware that a Concorde plane with a dual BA (then BOAC) and Singapore Airlines livery used to operate on the London-Singapore-London route, with a stopover in Bahrain.

The visit was hosted by Goh Mia Hoon (in his capacity as R.I.C.S. Singapore Corresponding Secretary) and his wife, Celia. Grafton also attended the SIS's annual dinner, and almost immediately after the Presidential Visit had concluded, Margaret wrote to Celia on 15 June

1979 to thank her for the hospitality the Graftons had received during their stay in Singapore (see Figure 17.19).

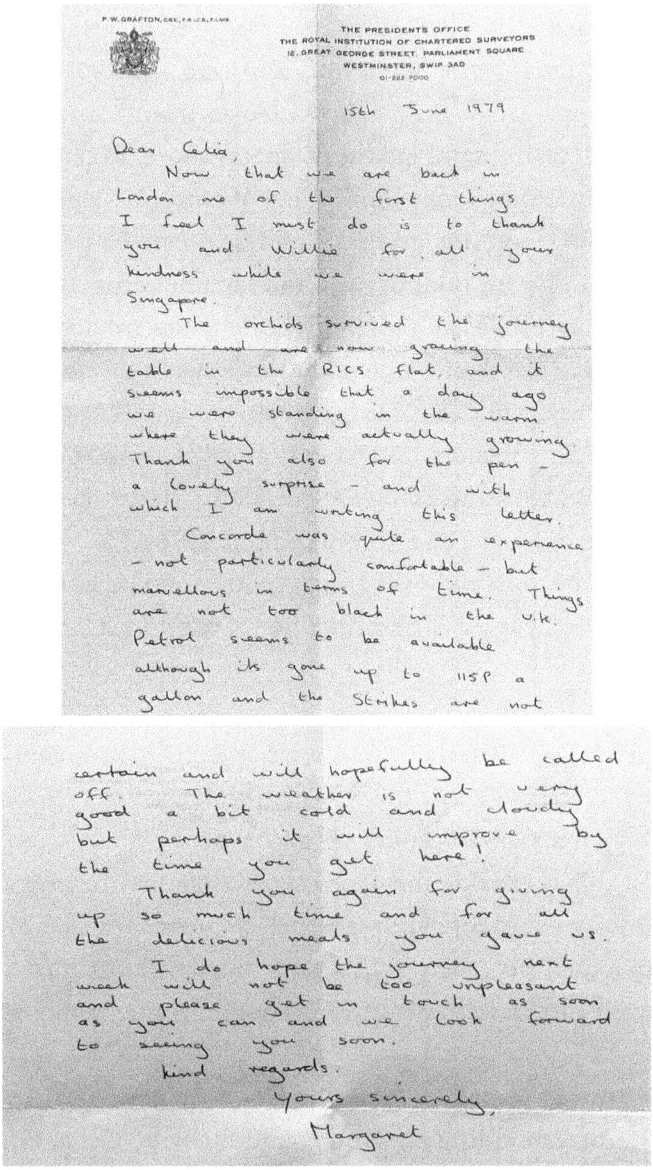

P. W. GRAFTON

THE PRESIDENTS OFFICE
THE ROYAL INSTITUTION OF CHARTERED SURVEYORS
12, GREAT GEORGE STREET, PARLIAMENT SQUARE
WESTMINSTER, SW1P 3AD
01-222 7000

15th June 1979

Dear Celia,

Now that we are back in London one of the first things I feel I must do is to thank you and Willie for all your kindness while we were in Singapore.

The orchids survived the journey well and are now gracing the table in the RICS flat, and it seems impossible that a day ago we were standing in the warm where they were actually growing. Thank you also for the pen – a lovely surprise – and with which I am writing this letter.

Concorde was quite an experience – not particularly comfortable – but marvellous in terms of time. Things are not too black in the U.K. Petrol seems to be available although its gone up to 115p a gallon and the strikes are not certain and will hopefully be called off. The weather is not very good a bit cold and cloudy but perhaps it will improve by the time you get here!

Thank you again for giving up so much time and for all the delicious meals you gave us.

I do hope the journey next week will not be too unpleasant and please get in touch as soon as you can and we look forward to seeing you soon.

Kind regards.

Yours sincerely,

Margaret

FIGURE 17.19. Letter dated 15 June 1979 from Margaret Grafton to Celia Goh to thank her for the hospitality the Graftons had received during their R.I.C.S. Presidential Visit to Singapore.

[3]

Jack Phillips was mentioned in Chapter 16 as meeting the President of the R.I.C.S. on 18 October 1966 during his time in Singapore. However, Phillips was not a Chartered Surveyor but is mentioned in this Book as he worked closely with many Chartered Surveyors in Singapore from 1950 until his untimely death in 1982, and he played a pivotal role in the advancement and development of the surveying profession in Singapore, leading to the transition to (and formation of) local professional bodies.

FIGURE 17.20. Jack Phillips.

Jack Arthur Phillips (1932–1982) — see Figure 17.20 — was born in Singapore on 28 January. His family was Eurasian — his mother was a de Souza and his father (Arthur James Phillips — a teacher) was of mixed English and Eurasian heritage. During the Japanese occupation of Singapore in World War II, Phillips and his father were interned at Changi Prison whilst his mother and three younger brothers managed to avoid internment. He joined the Singapore Survey Department in 1950, working under Harry Ward, Chief Surveyor. After successfully obtained his Land Surveying Diploma at the Kuala Lumpur Technical College, he was appointed a Technical Cadet in 1955 and then an Assistant by 1956.

Having served a two-year Scholarship in Brisbane, he passed the Australian and New Zealand Survey Board examinations and returned to the Singapore Survey Department and was appointed as Surveyor in 1960, where he was involved in surveying field duties. In 1965, he moved to the Economic Development Board (EDB) as Senior Surveyor to support the effort to open up land for industrial development.

His EDB role led to a role with JTC, where he was appointed the Director of Survey and Planning.

During his time in private practice in Singapore from 1970 until his death, he was a leading industry figure and held appointments such as being on the Singapore Government's Construction Trade Advisory

Committee as well as a member of the Land Surveyors Board. He was also President of the SIS from 1974 to 1976. One of the projects he was heavily involved in during his private practice days was the survey work for the land reclamation for and then construction of Changi Airport.

He died tragically in a car accident in Brunei in 1982. At that point in time, he had already opened a branch of his surveying practice there.

Chapter

18

Singapore's First Female Chartered Surveyor

On 5 June 1970, the R.I.C.S. merged with the Chartered Institute of Auctioneers and Estate Agents (CIAEA) in a unification of three bodies: R.I.C.S., CIAEA and the Chartered Land Agents' Society. As a result of that merger, a number of Singaporeans (particularly in the Public Sector) who were members of CIAEA became Chartered Surveyors.

Some of these individuals feature in this book as they played a prominent industry role prior to 1970. This Chapter features one of them — Dr. Anamah Tan.

FIGURE 18.1. Dr. Anamah Tan.

Dr. Anamah ("Ann") Tan — see Figure 18.1 — was the first woman to qualify as a Chartered Surveyor in Singapore. She was born in Singapore in 1940 and started working as an Estates Officer with the Housing & Development Board (HDB) in 1963. During that time, she met her future husband, Tan Tian Boon, who was one of the first Singaporean Chartered Surveyors (see Chapter 17) and worked with the Singapore Improvement Trust (SIT) in the 1950s prior to its transformation into the HDB in 1960. He had served the SIT from the Colonial days and through the period of Singapore's independence and spent the early part of his career in the HDB, which then housed 85% of the population of Singapore. He later moved to Jurong Town Corporation to head up the industrial arm of Singapore. He remained there until his retirement.

Whilst with the HDB as an Estates Officer over a period of 7 years, Dr. Tan learned the trade from other professionals, with her main inspiration coming from a Senior Maintenance Inspector who helped to develop her all-rounded skills as a surveyor, including the so-called "dirty" jobs that involved climbing up steep ladders to inspect roof cavities.

She studied in whatever spare time she had and paid for herself to take the Chartered Institute of Auctioneers and Estate Agents examinations via a correspondence course. Her papers were also sent to London for marking. She passed her examinations to become a member of this Institution in 1965. At this point, Dr. Tan became a Chartered Surveyor, and she was the first woman to do so in Singapore. She has remained a Chartered Surveyor ever since, albeit her focus for many years has been her law practice.

Her move from property to law (her new role title was "Estates Officer (Legal)") came through the HDB practice of buying land for development and the evenings that she spent visiting HDB flat dwellers to persuade them to take up flat ownership (i.e., paying a bit extra each month to enable them to own their own flat after 10–15 years). Through this work, she set up the Legal Department, and then in the 1970s, she entered private practice to become a Partner in Laycock & Ong, one of Singapore's oldest legal firms. She later went to the UK to get her English Solicitors' qualification and gain experience in international financial instruments. She then set up her own law practice, which she still operates to this day.

Dr. Tan was a founding member of the Singapore Association of Women Lawyers (SAWL) in the 1970s and later the Singapore Council of Women's Organisations (SCWO), and her work on women's rights went regional and international. In 1980, Dr. Tan played an influential role in galvanising the leading women's associations in Singapore into a unified body under the umbrella of the SCWO. The now 500,000-member-strong Council has since worked on many women's projects and causes, including developing a shelter for victims of domestic violence.

She has been active for many years in the ASEAN Confederation of Women's Organisations (ACWO) and was the President of the International Council of Women (ICW-CIF) from 2003 to 2009. Furthermore, she served as an elected member of the United Nations' Committee on the Elimination of Discrimination against Women (CEDAW) from 2004 to 2008 with the support of then-MCDS Minister Abdullah Tarmugi and former Senior Minister of State Yu-Foo Yee Shoon. Her many other achievements include a period from 1998 to 2000 when she was an Advisor on Human Rights in the Standing Committee on the Status of Women at the International Council of Women (ICW-CIF) (Paris), one of the earliest women's rights organisations founded in 1888 in Seneca Falls in the United States.

In Singapore, Dr. Tan was one of the key persons who made representations to the Select Committee, on behalf of the Singapore Council of Women's Organizations and the Council of the Law Society, on the amendments to the Women's Charter in 1996. The Charter is Singapore's primary family law statute that protects the rights of women, especially in marriage and family matters. Dr. Tan worked for the establishment of a Family Court, which has since been set up in Singapore, to help, among other things, dysfunctional families and victims of abuse, who are usually women, to access and obtain legal protection.

Dr. Tan's contributions to the public service have been highly commended both by the Singapore Government and in the private sector. She was awarded a National Medal for Community Work in 1993 (for helping the poor, underprivileged and women) and a National Medal for Public Service in 1998. She was also granted the SCWO Award in 2002 for her contributions to the women groups in Singapore and the Leader Mentor Award in 2003 by Research Communication International (a private organisation) for her contribution to grooming leaders in the community. Throughout her career, Dr. Tan has combined her significant achievements in the women's movement with a successful family and matrimonial law practice in Singapore.

Chapter

19

1950–68: Local Education and the Influence of Chartered Surveyors

From the mid-1950s onwards, education opportunities for aspiring Chartered Surveyors in Singapore increased significantly, for it was during this time that scholarships opened the door for aspiring Singaporeans to study overseas without having to encounter the obstacles mentioned in earlier Chapters.

Colombo scholarships came to prominence in 1951, followed a few years later by Government scholarships. As a result, aspiring Chartered Surveyors in the region were able to go to New Zealand under the Colombo Plan scholarships or to London to work and study through the Colombo Plan or a Government scholarship, taking their examinations in England set by the College of Estate Management (now known as UCEM).

For those unable to get a Colombo or Government scholarship in the 1950s and 1960s, their journey started with a Diploma in Estate Management or Quantity Surveying at either Singapore Polytechnic or Kuala Lumpur Technical College, coupled with on-the-job training and private study to enable them to take the R.I.C.S. Examinations.

During this time, the College of Estate Management was based at 1 St. Alban's Grove Kensington, London, until moving to Reading University's Whiteknights campus in 1972.

As explained by Yu Shi Ming in *Singapore's Real Estate: 50 Years of Transformation*, early Singapore tertiary education in the land, construction and the built environment was offered primarily by the College of Estate Management via the Singapore Polytechnic and through the Diploma in valuation from the New Zealand Universities under the Colombo Plan scholarship scheme, sponsored by the New Zealand Government to help Singapore as a young nation develop its professional expertise in valuation. Returning scholars had to serve their time for a fixed period as "bonded scholars" in the Inland Revenue Authority (or another Government body), after which they could stay if a position was available or move on to private practice or even overseas.

This scholarship scheme did not just apply to surveyors — aspiring engineers would journey to the Brighton Technical College in England in the 1950s and 1960s and then head back to Singapore to serve their time as bonded scholars.

(Background: The Colombo Plan for Cooperative Economic and Social Development in Asia and the Pacific was conceived at the Commonwealth Conference on Foreign Affairs in Colombo, Ceylon (now Sri Lanka), in January 1950. It was launched on 1 July 1951 as a cooperative venture for the economic and social advancement of the peoples of South and South-East Asia (later, Asia and the Pacific). It has grown from a group of seven Commonwealth nations — Australia, Great Britain, Canada, Ceylon, India, New Zealand and Pakistan — into an international organisation of 26.

The story of the Kuala Lumpur Technical College is mentioned elsewhere in this book and will be explored further in the future book on the R.I.C.S. and Chartered Surveyors in Malaysia. However, in the Singapore context, from 1957 onwards, the creation of the Polytechnic Diploma courses was the start of a journey over the next 66 years to the present day, where Singapore offers a wide range of opportunities for Built Environment professionals to attain a professional qualification in land, property and the built environment.

It was announced in August 1957 that Singapore Polytechnic courses had started (albeit the new Polytechnic building was still under construction), and the new intake included 21 students taking a surveying course to prepare them for the R.I.C.S. Examinations. Earlier that year, the City Council's Finance and General Purposes Committee had agreed to allow suitably qualified staff in the Assessment & Estates Departments to undertake part-time teaching at the Polytechnic (for the courses developed to take students forward towards their R.I.C.S. examinations), and it was agreed that they could keep their training fees. In May 1958, an advertisement was placed to source for full-time Lecturers in Quantity Surveying and Building Construction (also surveying, general building and architecture), with a view to starting in time for the October 1958 intake. The salary band was $1,600 to $1,750 per month, including allowances. The advertisement made it clear that being a qualified R.I.C.S. or IOB (now CIOB) professional was a job requirement.

By August 1961, the relationship with the R.I.C.S. was further cemented, as the first batch of Polytechnic students graduated with their Diplomas in a ceremony at the Victoria Theatre, where the R.I.C.S. donated a prize along with prize donations from other professional bodies, the Shaw Foundation and local businesses.

The ceremony was attended by then-Minister for Education, Mr. Yong Nyuk Lin, and the Polytechnic Chair and Deputy Prime Minister, Dr. Toh Chin Chye, who called on all 192 students (191 men, 1 woman) to be bold and adventurous but not to let their initiative be stifled by patronage or the "smug satisfaction of possessing a Diploma".

In 1966, the R.I.C.S. President Guy Biscoe undertook a World Presidential Tour, accompanied by his wife and Francis Bennion, then-R.I.C.S. Secretary-General (equivalent to what is now the Chief Executive role within the Institution). During their short time in Singapore, they visited the Singapore Polytechnic (see Figure 19.1). Thanks to the diaries of Bennion, we know what happened on the afternoon of 17 October. The Presidential team was met by (Alfred) Robert Edis (who was the Polytechnic Principal from 1963 to 1973) and Mr. Notley (who was then the Head of the School of Architecture and Building — later in his career, he became Professor Notley of the National University of Singapore's School of Architecture). Also present were Victor Torrance and Eric Lim (more on them later in this Chapter).

FIGURE 19.1. An aerial view taken in 2019 of the original Singapore Polytechnic building at Prince Edward Road. This was later known as the Bestway Building. It is currently situated in the centre of the construction site for the Prince Edward MRT Station and Bus Interchange.

The external course examiner (not present) was Professor Norman Sidwell (1910–93) of Heriot Watt University.

The Presidential team was told that the Quantity Surveying course had just started (no students at that point in time), and it was a 4-year building course, with a further year focusing on quantity surveying. The lack of students taking the course was thought to be due to the syllabus not being at the level that would take students through to the R.I.C.S. Examinations, but the President was assured that the Polytechnic was anxious to provide suitable training for the R.I.C.S. Quantity Surveying Examinations and have the guidance and assistance of the R.I.C.S. to get their building course established. It was explained to the President that the Polytechnic urgently needed a Quantity Surveying lecturer, possibly seconded from a University as a visiting professor. The Polytechnic would finance this through the Ford Foundation (albeit their focus in South-East Asia was Malaysia and the Philippines at that time) if they could convince the Foundation that the incoming lecturer came from a reputable source and was of good standing. This, in turn, could generate a better salary to attract the right person for the role.

Despite the enthusiasm that was evident on the visit and despite the desire to cooperate with the R.I.C.S., the response from the President was not too hopeful. Indeed, the President and his Secretary-General advised that they could not, in the near future, agree to the request for an R.I.C.S. representative to assess the course because the course was not yet ready for inspection by the R.I.C.S., even though it had been recognised by the Institution of Electrical Engineers in the UK, with recognition expected soon from the Mechanical and Civil Engineers.

From the late 1950s to the late 1960s, a number of Chartered Surveyors were actively involved in teaching at the Polytechnic:

David Vickery and Roland Sheath

At the very start of the journey, **David J. Vickery (1922–2013)** and **Roland Harold Sheath** were important figures. Vickery was the Head of the Department of Building and Architecture from 1959 onwards, and Sheath (a Chartered Surveyor) was a Lecturer in Building at the Polytechnic from 1961 to 1964. Both Vickery and Sheath went on to have long and successful careers with UNESCO, with Sheath based in their offices in Bangkok until his retirement in the 1990s. They co-wrote many research papers on education-building requirements, including the 1997 key work "A Guide to facilities for early childhood care and education".

Eric Lim Ang Beng Hock

Eric Lim Ang Beng Hock was one of the very first "bonded scholars" to leave Singapore and study in England. It was in England, and in particular at the College of Estate Management in Kensington, London, where he met and became friends with Michael Wee Sek Kay and Francis Lai Pang Fee. He qualified as a Chartered Surveyor in 1964 and worked for the Public Works Department (PWD). Along with Derrick Morris, he taught part-time at the Polytechnic throughout the 1960s and eventually took on the role on a full-time basis. He was a strong advocate of the R.I.C.S during these years. In the late 1970s and until his retirement in the 1980s, he worked as a sole practitioner from an office in Selegie House Complex. Lim passed away in February 2008, aged 84.

Victor Torrance

FIGURE 19.2. Victor Torrance. Source: The Straits Times © SPH Media Limited

In the 1960s, **Victor ("Vic") Brownlie Torrance** (see Figure 19.2) was a Senior Lecturer in Quantity Surveying and the Senior Lecturer-in-charge of the Building Division of the Singapore Polytechnic in his first-ever education job — a Chartered Surveyor. He later became Professor Torrance, CBE, and was a past President of the Chartered Institute of Building (CIOB) with a long and distinguished career in education around the world.

Originally from Scotland, Torrance was at the Polytechnic from around 1965 until he was appointed as Head of the Building Section at the University of Singapore in 1970/71.

In his long career, he was highly respected in academic communities around the world. He was held in high regard due to his dedication to the task of sharing his knowledge with students. He was also recognised as someone who worked closely with the property and construction industry in Singapore in ensuring the delivery of a curriculum that fully met industry needs, which paved the way for graduates to seek professional qualifications.

He became a Professor in the Department of Building at Heriot-Watt University in 1972, and then later in his career, in the late 1990s, he returned to South-East Asia to Malaysia at the Universiti Teknologi MARA (UiTM) at Shah Alam, where, as Prof. Dr. Johan Victor Torrance Abdullah, he remained until his death on 18 January 2008.

Aside from Eric Lim and Victor Torrance, **Michael Wee Sek Kay** was the Lecturer in charge of the Polytechnic's Building Division at the School of Industrial Technology.

The Polytechnic teaching ranks were further strengthened in 1965 with the addition of another Chartered Surveyor, **M.J.P. ("Philip") Motha**, who was appointed as a Lecturer in Valuation.

However, things changed in 1969 with the establishment of the very first University of Singapore Architecture and Building School. Indeed, the very first Campus for the School resided (on a temporary basis) in the old Singapore Polytechnic building on Prince Edward Road. It was part of the Faculty of Architecture, comprising the School of Architecture and the Department of Building and Estate Management.

By 1970, it had moved to Ladyhill Campus, Lady Hill Road, and the School of Architecture (later called the Faculty of Architecture, Building and Real Estate) took over Kinloss House in the Ladyhill Campus — Kinloss House had previously been British Army barracks. Construction of Kent Ridge started in 1973, and the Faculty stayed at Kinloss House for a total of 6 years until Kent Ridge Campus was completed and ready for occupation. The University at this time also introduced a part-time professional diploma-to-degree conversion course; no doubt focused on the 1960s Polytechnic Diploma students.

At that time, the National University of Singapore (NUS) (as it was by then called upon the merger of the University of Singapore with Nanyang University) set up a B.Sc. programme with the objective of producing graduates capable of entering managerial and executive positions in the building industry and the allied professions of quantity surveying, valuation and estate management. Originally, the teaching staff comprised Expatriates, but later the number of local teaching staff grew.

The first 12 B.Sc. (Building) graduates and 15 B.Sc. (Estate Management) graduates were in 1972. This was a significant shift and was the first time that Estate Management degree courses had been delivered in Singapore.

By this stage, Motha was an Associate Professor at NUS (and Head of Department by 1977) and an extremely important figure in the establishment of the NUS B.Sc. courses. He left Singapore Polytechnic in 1971 to work in private practice in Sri Lanka but soon returned to Singapore to teach when the attraction of a senior role at the NUS brought him back.

Philip Motha

M.J.P. ("Philip") Motha (1926–2007) was one of the first (if not the first) individuals to go to England to study at the College of Estate Management under the Colombo Plan's Technical Cooperation Scheme. He headed to England in December 1951 from Colombo, Sri Lanka, where he was already working as a trainee Government valuer. Motha later returned to Colombo in 1954 to begin his "bonded scholar" work (10 years) as a Government valuer. At the end of that period, he applied for valuation roles in the UK and also for a role as a Lecturer at the Singapore Polytechnic. Having been offered the Singapore role, he decided not to go to the UK, and he arrived in January 1965 at the Polytechnic.

Other Noteworthy Educational Activities

Whilst this book does not cover the period from 1968 to the present day, three activities during the early 1970s arising through the Polytechnic are worthy of mention:

- **Francis Lai Pang Fee**, along with **Tan Tian Boon, Ng Boon Ong** and others (mostly R.I.C.S. and NZIV qualified people), would teach trainee valuers at the Singapore Polytechnic in the evenings and during weekends. Amongst those taught at this time at the Polytechnic was Edmund Tie. The same group also ran two training part-time schools within the HDB, with about 30 trainees in Valuation (for

Trainee Estate Officers) and 10 in Quantity Surveying (for Building Inspectors). Lai and Tan agreed that the training ought to be focused on the R.I.C.S. external examinations, in preference to the effort and complications involved in setting up an Examination Unit. One of the Building lecturers from the Polytechnic assisted with evening classes in Construction at the Polytechnic. This training would go on to produce Singapore's next generation of surveyors.

- Lai also ran a private Building Studies Group (BSG) in the 1970s to train technicians for the City and Guilds examinations to fulfil the need for competent Technicians. He recruited part-time teachers such Ng Chee Yoong, Teo Pin, Benedict Tan, Simon Lim and Michael Oh Chong Guan to assist. The total number of students at the peak of the BSG's existence was more than 300, with classes at different levels every evening and weekend.
- A private School opened in Singapore in the early 1970s. It was the Ellis School and it ran correspondence courses that helped people to progress towards their R.I.C.S. and Institute of Quantity Surveyors qualifications, as well City and Guilds examinations. They had acquired an office in International Plaza, Anson Road, by the early 1980s. The Ellis School was founded in England and was the first private school to run correspondence courses for Singapore-based surveying students.

In the years since the first NUS graduations in 1972, NUS and others have produced professionals who have contributed to the development of Singapore as a nation, and many of these individuals have become Chartered Surveyors.

Chapter

20

John Aspinall and his Role in Singapore's Post-War Property Tax System

Aspinall's Time as Municipal Rates Assessor from 1946 to 1958

After Charles Frederick ("Fred") Snell and his family returned to the UK after World War II, the role of Singapore Municipal Assessor went to **John George Aspinall (1906–66)** from 1946 to 1958.

He left Southampton on 10 November 1946 to come to Singapore to take up his new duties. The assessor role was an unpopular one. In the post-war years, it was more unpopular because, despite the urgent need for public spending coupled with the increase in development, the rate-paying community in Singapore was more aware than before of the major inconsistencies in the system.

Aspinall was a Professional Associate (and later, a Fellow) of the R.I.C.S. He was Singapore's Municipal (later City) Assessor in the Assessment & Estates Department and also valuer to the Singapore Improvement Trust (SIT). He was also an R.I.C.S. Malaya Branch Committee member from 1948 to 1950.

By 1947, the rates (property tax) system that post-war Singapore inherited was inefficient and disparate, which was resolved through the 1947 enactment of the Income Tax Ordinance and the creation of the Inland Revenue Department (later, Inland Revenue Authority of Singapore (IRAS)). However, during this transition period, the assessor's role proved to be a difficult one to fulfil.

Indeed, on 1 April 1948, Aspinall got off to a controversial start by announcing that assessments would henceforth be levied on all unauthorised buildings in the Singapore Municipal area that were constructed during the Japanese Occupation (1942–45) or the period of British Military Administration (1945–46). Aspinall's logic was that the occupiers of these properties were enjoying the benefits of Municipal services that Singapore ratepayers also enjoyed, so it was only fair that the owners of these buildings should pay rates. Some of these buildings were owner-occupied while others were rented out to tenants, and Aspinall added that if owners were getting rent from tenants, then it was only reasonable that they should consequently pay rates. Examples quoted were the Indian Military Hospital in Siglap and the former Dutch camp in Katong, now occupied by tenants. Aspinall said that the owners had to pay rates even if a future action was to be taken by the authorities to either regularize the existence of these buildings or to have them demolished. Newspapers of the time do not report whether any building owners later challenged these rulings, using the argument that the payment of rates was equivalent to the Municipality legitimising the existence of such properties.

From 1950 onwards, to improve communications around the role of the assessor and how the rates system worked, Aspinall often spoke to the newspapers to provide explanations about his work. Indeed, in 1950, he gave a Rotary Club talk (reported in *The Straits Times*) to inform people what the rates revenues were used for — public health, highways, bridges, house refuse, trade waste and disposal, recreation grounds, parks, swimming baths, public lighting, public sewers, fire brigade and the police. He said that in England, the rate levied was 42½ pence for every pound of assessed value, but in Singapore, it was only 29 cents levied for every dollar of assessed value. He argued that for Singapore to become self-governed, it needed to have adequate services for the city, and this had to be paid for somehow.

In 1952, Aspinall successfully brought into assessment all SIT properties that were "used for profit" (previously, they had been exempt), and he stated in the press that this would bring in a good amount of additional revenue to the Municipality. In return, the Municipality would pay a 2% Improvement Rate to the Assessment of SIT-owned properties as at 1952 were based on 25% of the annual value of the property.

Later in the same year, he was quoted in the Press as saying that the City Council was losing potential revenue on assessment rates because of the lack of professional and technical staff to support him and the challenge in effecting satisfactory amendments to existing legislation. This would not have done anything to endear him to ratepayers, many of whom did not have anything like an Expatriate salary and their associated benefits, and yet here was Aspinall wanting to bring in more highly paid Expatriate professional staff to ensure that even more revenue was collected from the ratepayers!

In December 1952, the City Council debate on this saw objections from P.R. Williams (Labour, Rochor), who stated that the $8.5 million in extra revenue was simply a gift to Britain and that the extravagant expense was an outrage. However, the proposal was broadly supported, although many Council members said that whilst they regretted the additional burden upon ratepayers, they saw no other option given the financial position of the City Council.

In January 1953, the Singapore Rate-Payers Association decided that enough was enough, and they pulled together a Petition to the City Council against the "evil effects" of the 25–30% increase in rates that people faced due to the decision to increase the rate levy. The Association stated that property owners would now be deterred from buying or would pass on the costs to the public in the form of higher prices, and for the rented sector, landlords would pass on additional rates to tenants, which would lead to higher prices being passed on to the public in the form of higher prices for goods.

Against the backdrop of protests from the Singapore Rate-Payers Association, Aspinall decided to set the record straight and explain the "mystery of rates". In February 1953, he explained that the rates that landlords and property owners were responsible for paying (at half-yearly intervals) were based on the annual value of the property. This, in turn, was derived from rent passing or (in the case of properties occupied by its owners) the estimated market rent. For properties covered by the Control of Rent Ordinance, the annual value of the property would be based on a rental amount the Rent Conciliation Board had pre-determined for it. This, of course, led to inequality because there were situations where two identical shops (both operating the same type of business) could be paying very different levels of rates — one at a market rent and the other at a much lower rent determined by the Board. Some landlords had been arguing that the proper notice of the rates assessment (and rates bills) had not been served on them, so Aspinall explained the process and opined that the City Council had "incontrovertible proof" that this was not correct.

Aspinall also explained that the personal circumstances of individuals could not be taken into consideration in assessing properties. However, he did not mention if the City Council would take personal circumstances into consideration in the collection of rates, though he mentioned that if a property was empty for one month or more, then the landlord could claim back the rates, provided that a report of the vacancy was provided within seven days of the start of the vacant period, and the actual claim had to be made one month after the half year in which the property was vacant.

He also explained that the City Council had no power to agree to claims for a refund where the property did <u>not</u> meet the following three conditions: (i) in full repair, (ii) every effort made to find a tenant, and (iii) the rent was reasonable but the building being vacant for the entire period of the claim. However, if the building was empty whilst being repaired, then the three criteria did not need to be met. In conclusion, if a property was demolished or destroyed by fire, then it was the duty of the owner to tell the City Council, because if the notice was not served on them of this within the relevant half year, no refund would be made during that period.

Although what he said was clear and technically accurate, it did not improve the situation for the average ratepayer. By 1956, the City's total rates assessment had risen from $21 to $61 million, and the number of assessed properties had risen from 38,000 in 1941 to 60,000 by 1956.

Aspinall was clearly determined that no landlord should evade their rates liabilities, and as *The Straits Times* reported on 18 July 1956, he "almost sings" this out in his annual report. His only disappointment was that of rent control, which in his opinion, was not only restricting landlords but was also restricting the City Council from assessing properties at the true market level and bringing in extra rates revenue. He also pointed out that the British model, whereby the occupier pays the assessed rates — Singapore's model of charging only the landlord was, in his view, unfair. For example, *attap*-dwellers paid more in rates than a "businessman" owning a commercial property simply because of rent control on the property owned by the "businessman".

In 1957, he indicated his desire to retire in April 1958 and return to England, given that he had exceeded the retirement age of 50 for Expatriates, and although the City Council asked him to stay on, he declined due to ill health. At the same time, as Chair of the Executive Committee of the Senior Officers Association, he was heavily involved in the compensation debate for Expatriates who were leaving in large numbers in the late 1950s under the Malayanisation programme.

On his retirement in 1958 after 11.5 years in charge of the assessment team, Aspinall told *The Straits Times* that greater revenues could be achieved for the City Council if the Rent Control Ordinance were amended, relaxed or removed altogether. He stated that there was an imbalance between the assessed value of controlled and non-controlled properties, which had led to an angry landlord writing to *The Straits Times* about the subject to support Aspinall. The angry landlord stated that in October 1954, David Marshall (as a member of the Singapore Rate-Payers Association) recommended the abolition of rent control, but nothing had happened since. He quoted the case of his own shophouses, where the rent was controlled at $400 per month by the Board (he had asked for $500), and he paid $1,683 in rates (based on the rent he received from his tenant), but the tenant was actually subletting his space at $1,200 per month without having to pay a cent to the City Council! The angry landlord explained that if Rent Control was abolished, then he, as the landlord, could charge the full market rent of $1,200 per month but happily pay a higher assessment.

Aspinall left Singapore in 1958. In 1960, he worked as an assessor for the Inland Revenue in the City of Westminster, London. He died in London in 1965.

Within a short time of Aspinall retiring from the City Assessment and Estates Department in 1958, his entire team also left: Thomas F. Stevenson, his Deputy and who stepped up to take his place on an interim basis, and two other British assessors — Gordon M. Lowden and Selwyn Lewis (see Chapters 10 and 12 for more on Lowden and Lewis, respectively). The only assessor that remained was Heng Chong How, an Assistant who was appointed as interim City Assessor in early 1959.

(Note: Rent Control in Singapore started in the period just after World War I and again in earnest after World War II due to unscrupulous landlords demanding exorbitant rents from tenants (and evicting them if they could not afford to pay). A whole series of Control of Rent Ordinances were enacted in Singapore from 1939 for many years, including the Control of Rent Ordinance 1947, at which point all premises built on or before 7 September 1947 were subject to rent control and security of tenure, and a Rent Conciliation Board played a strong role in determining rents. This Board continued (and the Act was further amended, e.g., the 1968–69 Special Provisions that allowed building owners to terminate controlled tenancies for redevelopment and for the possession of gazette properties/areas to allow for restoration, and again in 1990) until rent control was finally abolished in 2001 — the abolition was mainly due to the fast-decreasing number of pre-1947 premises under its purview.)

By early 1959, after the departure of Aspinall and his team, the vacant assessor roles were a problem. The Minister for Local Government, Lands and Housing reversed a decision proposal to appoint Karoly (known as Stuart) Berkovits to Aspinall's role. Berkovits was a Hungarian-born naturalised Singaporean who was a Registered architect and appraiser. He had lived in Singapore since 1937 and transitioned from his own private practice to a Consultant Assessor role with the City Council in the early

1950s. Instead, they appointed Heng Chong How on an interim basis whilst three temporary assessors (Messrs. Freeston, Doherty and Buck) were brought in from the UK on a temporary basis. This, in turn, would enable the team to give attention to the backlog.

The roles remained within the City Council offices on St. Andrews Road (now of course part of the National Gallery Singapore) and the team — once the vacancies were filled — was renamed the Property Tax Division, Inland Revenue Department. By 1961, this Division was migrated from the City Council to the Inland Revenue, and in 1960, following the Property Tax Ordinance 1960, the naming convention changed from Rates to Property Tax.

As a result of the Property Tax Ordinance 1960 (following on from the 1947 formation of the Singapore Income Tax Department), Property Tax became a fiscal levy, no longer specifically tied to the provision of "named" services and utilities. The term "Property Tax" was also being used for the first time, instead of "rates", although the revenue from the City Council and Rural Board was still the primary source of public finance, alongside stamp duty and estate duty. It would be some years before Income Tax and the GST would provide greater public finance sources. Indeed, the system introduced and run by the municipal authorities from 1819 to 1947 had become unworkable, and through the leadership and efforts of people such as Lim Soo Chin, Singapore is fortunate to have the benefit of the current property tax system, with the benefit of technology and mass appraisal systems.

Chartered Surveyors Responsible for Rates Assessment from 1958 to 1968

As a final footnote, from 1959 and through the 1960s, during the transition of responsibility to the Inland Revenue Department, a number of Chartered Surveyors were involved in assessments for Property Tax — amongst them were the following who worked as assessors during the 1960s:

Peter Forbes-Harriss

When **Peter Forbes-Harriss** arrived in Singapore in 1959, he was the only assessor employed by the City Council as all the interim staff had moved on. He quickly rose the ranks to Acting City Assessor and then Chief Assessor by the time of his departure at the end of 1961 after the preparation of the Valuation List 1962. He received a great deal of help during this time from Stuart Berkovits, who was a big help to Forbes-Harriss and his wife settling in Singapore. This was especially generous when one considers that Berkovits had been offered the role on an interim basis, only to find the offer withdrawn by the Minister for Local Government, Lands and Housing. However, Berkovits (see Figure 20.1) eventually did become Chief Assessor in 1964.

FIGURE 20.1. Stuart Berkovits (right) taken in 1961.

Within a short space of time, Forbes-Harriss built a new team around him — Stuart Deans, Ron Mathews and Goh Mia Hoon — which enabled him to prepare the 1962 Valuation List. He was also able to put into place a much-improved system for Rates/Property Tax appeals, as the previous system under Aspinall was perceived by many to be unfair. His task as Chief Assessor was helped by the appointment of highly capable senior leaders in the Department, such as Kwa Soon Chuan as (acting) Collector of Land Revenue and Goh Kee Song (who later formed Goh Tan & Co, a firm that is now part of Deloitte in Singapore).

Stuart Deans

A founder member of REVA, **John Stuart Deans (1906–86)** was already a vastly experienced Chartered Surveyor of almost 30 years' standing when he was appointed as an assessor for the Inland Revenue in the early 1960s. He was the R.I.C.S. Singapore Branch Chair in 1964. After leaving Singapore in the late 1960s, he retired to Malta, where he passed away in 1986.

Goh Mia Hoon

On his return to Singapore in 1960, upon qualifying as a Chartered Surveyor in England, **Willie Goh Mia Hoon** served his Government bond with the City Council — a role that saw him moved to the Inland Revenue Service as an Assistant Assessor. His story was, of course, discussed in more detail in Chapter 17.

Yoong Wah Pin

Yoong Wah Pin (see Figure 20.2) was a Malaysian who took his professional studies in London in the 1960s at the same time as Francis Lai Pang Fee (whose story is told in Chapter 16), and they became firm friends. After graduation, he worked for the Property Tax Department during his compulsory 5-year bond for the Singapore Government in return for his scholarship to London and qualifications as a Chartered Surveyor in 1970.

FIGURE 20.2. Yoong Wah Pin, taken in 1955 when he was with Second Kuala Lumpur Scouts.

On the first day after his bonding ended, he set up his own practice, Yoong & Co., which was originally based in Cecil Street but moved to the 9th floor of the International Plaza, Anson Road, in the 1980s. When he first set up his practice, he engaged in property management and to this end, worked closely with Lai, who advised him on property management and a number of property development projects.

He retired from this practice in the early 1990s, and he passed away in 2002, aged 71. His property portfolio was inherited by his two sons. He was well respected and highly regarded as a professional in the world of property in Singapore.

Footnote on Post-1992 Property Tax Developments

After the formation of IRAS in 1992, the system of property tax assessment and revenue collection was further improved with the introduction of strong processes and a pragmatic approach adopted under the leadership and guidance of then-Chief Valuer Lim Soo Chin, a Valuation Surveyor who qualified as a Chartered Surveyor in 1973. In a paper produced for Nanyang Technological University by AsiaCase.com, Lim opined: "We must shift from spending too much time on noting fine property differences, which would not materially enhance the accuracy of valuations. Instead, we should follow how people think in the marketplace — making fast decisions and getting more done through confident broad-based approximations."

Chapter

21

Individual Stories from the 1960s

The final Chapter of this book contains stories of Singaporean Chartered Surveyors who came to prominence from the 1970s onwards but whose origins are rooted in the pioneer days of the nation's independence in the mid-1960s.

Seah Choo Meng

Seah Choo Meng (see Figure 21.1) is the son of Seah Mong Hee (Bill Seah), who was the first Asian Chartered Quantity Surveyor. He became a Student member of the R.I.C.S. in 1969/70 and qualified as a Chartered Surveyor in 1974. He oversaw the transition of the firm from Langdon Every & Seah to Davis Langdon & Seah and eventually Langdon & Seah. He stepped down as Senior Partner in 2008. Subsequently, the firm was merged with European Engineering firm Arcadis NV in 2012 by the Continuing Partners.

Seah initially studied at Singapore Polytechnic in 1968 whilst working for his father, and in 1972, he went to work in Langdon & Every in the UK under the patronage (articleship) of Giles Every and studied at the South Bank Polytechnic whilst there. He returned to Singapore in 1974 as a fully qualified Chartered Quantity Surveyor and rejoined Langdon Every & Seah to start the expansion of its scope of services.

From 1974 onwards, the Urban Redevelopment Authority (URA) started to offer sites to the market, and this created many opportunities for

FIGURE 21.1. (a) Seah Choo Meng. (b) Three generations of Quantity Surveyors: Seah Choo Meng, Seah Mong Hee and Eugene Seah.

Langdon Every & Seah in the development of condominiums, commercial developments, hotels and industrial projects. Seah played a pivotal role in this as well as a key role in the expansion of its business across Asia.

Upon his return to Singapore, Seah was elected an Associate in 1975, a Partner in 1976, Managing Director in 1995 and finally, Executive Chairman in 1999. He retired in 2008 and became a Director of the firm from 2009 to 2016.

Due to the influence of his father on the profession, he decided to follow the same path, and his entire working life has been dedicated to the built environment through his affinity as a second-generation pioneer of the Quantity Surveying profession in Singapore.

He is now with Surbana Jurong, an Asian-based urban and infrastructure multidisciplinary consultancy group with diverse specialism across the entire spectrum of build assets and global infrastructure construction works.

As the Senior Advisor to the GCEO's office, he is directly involved in a multitude of tasks and the provision of advice covering the conduct of business of the various business units/companies under the Group's global portfolio.

He is concurrently Chairman of Threesixty Cost Management and Threesixty Contract Advisory under the Group. The Cost Management unit provides quantity surveying and cost management service, while the Contract Advisory provides mainly dispute management services such arbitration and adjudication, while at the same time providing internal advice to the various business units in contractual claims and disputes.

As well as his long service to the profession and the industry in Singapore and his long service to the Singapore Institute of Surveyors and Valuers, he has been an active and vocal supporter of the R.I.C.S., including spending many years on its Asia Pacific Board, and is a regular contributor to R.I.C.S. activity across the Asia Pacific region. Under his supervision, many Quantity Surveyors have progressed from being a trainee to full professional qualifications.

Wee Sek Kay (see Figure 16.3)

Wee Sek Kay (who is also known as Michael or Mike Wee) is a Singaporean who went to London to work and study. Initially, he shared a flat in Holland Park Avenue, but by the time that he got a job with the Middlesex County Council (MCC), he moved to a flat in Kensington Palace Gardens close to where the College of Estate Management (CEM) was then located.

He attended a full-time R.I.C.S. (Valuations) course at the CEM and did his practical attachment with the MCC. He then continued with a part-time evening course with the CEM for another two years before taking the Final R.I.C.S. Examination. He remained with the MCC for five years until his return to Singapore in 1964, at which point MCC merged with the London County Council to form the Greater London Council (GLC). He also qualified as a Chartered Auctioneer and Estate Agent in 1963.

Whilst in London, he met and formed a social acquaintance with Eric Lim Ang Beng Hock, who qualified as a Chartered Surveyor in Quantity Surveying, and who is featured in Chapter 19.

The Two-Year Prequisite — Practical Attachment

Finding an attachment to a firm after passing the Intermediate R.I.C.S. Examination turned out to be a frustrating and traumatic experience for Wee. It took him several months before he could find such an attachment. Even Mr. Brown, the Employment Officer at the R.I.C.S. at that time, was unable to find a placement with a firm. Mercifully, in the end, John Hearsum of the MCC was sympathetic and offered him a post. One of those who worked alongside Wee at the MCC was Mr. Dick Luff, who later visited Singapore when he was the President of the R.I.C.S. (1982–83). Wee was happy and honoured to have the opportunity to host him at his home for dinner during his visit to Singapore.

Housing & Development Board

On his return to Singapore in 1964, Wee joined the Housing & Development Board (HDB) as a Valuation Officer and was tasked mainly with the compulsory acquisition of land and properties in the Central and Northern areas of Singapore. His experience of doing similar work when working in the MCC served him well to handle this difficult and complicated task.

The role of his wider team included the acquisition of land pending development by the HDB, primarily by compulsory acquisition, the sale or leasing out of land for non-housing development such as Petrol Filling Stations and Light Industrial Factories, as well as wider non-HDB support to the Government, including land acquisition for the Paya Lebar Airport project.

Wee hosted the HDB leg of the visit of R.I.C.S. President Guy Biscoe when he visited Singapore on his 1966 World Tour, and a photograph of him with the President on that trip is shown in Chapter 16.

Wee left the HDB in 1967 to take up teaching at the Singapore Polytechnic, which was then located at Prince Edward Road. He was attached to the School of Architecture and Building as a Lecturer in the Building Department. At that time, students could pursue a Diploma Course in

Valuation, and the main subjects he taught related to Valuation and Building Construction. The Head of the Building Department was Victor Torrance, and Eric Lim was also a Senior Lecturer there at that time. The Valuation Diploma course was then upgraded in 1969 to a B.Sc. Degree course under the School of Architecture and Building when it was relocated to the National University of Singapore campus (see Chapter 19 for more on this).

Wee was also keen to move to the NUS but was asked to remain in the Singapore Polytechnic as the Acting Head of the Building, Architectural Draughtsmanship and Structural Engineering Department. However, he simultaneously taught Valuation as a part-time lecturer at the Building Department of NUS in the B.Sc. Valuation course.

Wee left the Singapore Polytechnic in 1970 to set up his own private practice under the name of Sek Kay & Associates, based on the 9th floor of Supreme House, Penang Road. There were three other established Chartered Surveying firms set up by Expatriates at that time, so he was initially apprehensive about the prospects of success against those firms. However, his fears were unfounded as he established himself as a reputable valuer with Banks and other Corporations. He continued with his practice for 25 years until his retirement in 1995.

He continues to live in Singapore, enjoying his retirement years.

Charles Ho Fook Weng

Charles Ho Fook Weng started his career in Malaysia, where he started a Diploma in Quantity Surveying in 1966, awarded in June 1969 from the Technical College in Kuala Lumpur.* Diploma holders were exempted from the R.I.C.S. Intermediate Examinations, so he sat for the Final Part 1 and Part 2 Examinations and became a Chartered Surveyor. He then practised in London (1972–78), and his career in Singapore spanned the period from 1978 to 1982, after which he worked for Landon & Seah in Brunei and

* In 1971, the Technical College was upgraded to a university status — Universiti Teknologi Malaysia.

again in Singapore from 1994 to 2006 on projects at Temasek Polytechnic. His career saw him manage a multitude of complex construction projects around the world. He became a Fellow in 1989 but is now retired.

Lim Chan Hoe

Lim Chan Hoe (see Figure 21.2) hails from Malaysia, where in 1960, he graduated with 1st Class Honours in Quantity Surveying from the Technical College, Kuala Lumpur. He was awarded a Federal scholarship in 1961 to leave Malaya to study at the College of Estate Management, England, in 1961, and in 1965, he passed the Final R.I.C.S. Examination, becoming a Chartered Surveyor in 1966. After moving from Malaysia to Singapore in the early 1970s and working in the construction industry for Low Keng Huat Construction Co. Pte. Ltd. as their General Manager and Director, he set up Lim Chan Hoe & Partners in 1981 in Singapore, as well as Lim Chan Hoe Bersekutu (Malaysia), a firm of quantity surveyors and construction cost managers.

FIGURE 21.2. Lim Chan Hoe receiving his R.I.C.S. Pioneer Appreciation Award in December 2015.

In 1990, the above became incorporated as Northcroft Lim Consultants Pte Ltd (Singapore) and Northcroft Lim Perunding (Malaysia). His firm has played a key role in many construction projects over the years, and he has been a champion for the development of local talent. Aside from attaining Fellowship of the Institution, he is also a Fellow of the SISV and RISM.

Since 1981, trainees in his firm have since gone on to become Chartered Surveyors as well as becoming qualified members of SISV and RISM. He has been a champion for the quantity surveying profession in Singapore and Malaysia for over 50 years.

One of the most recent projects that his firm was involved in is the Apple Store that opened in Orchard Road, Singapore, in 2018 — his firm was the appointed Quantity Surveyor.

Adrian Harding (see Figure 21.3.)

Adrian Isidore Gore Harding (1913–96) — known as A.I.G. Harding — was involved in the transition from the R.I.C.S. Singapore Branch to local professional bodies and played an active role in the late 1960s and early 1970s. Indeed, Harding was the President of the Singapore Institute of Valuers (then known as the Real Estate Valuers Association (Singapore)) from 1970 to 1971.

FIGURE 21.3. Adrian Harding during his time in Singapore.

Born in 1913 in Guildford, England, Harding spent more than 20 years living in Malaya and Singapore from the 1950s until the 1970s. Adrian met his second wife, Evelyn, in Kuala Lumpur in the mid-1950s, where she was working as a primary school headteacher for the Far East Land Forces (FARELF) schools in Malaya. He had been posted to Malaya as a Land Agent for the British Government, overseeing rubber plantations across Malaya. Harding and Evelyn married in 1959, the same year they moved to Singapore.

As a Chartered Surveyor, Harding spent his first four years in Singapore working as a Land Agent for the Lands Office of the FARELF (War Department/War Office). During this time, he was tasked with a joint project between the War Department, the Colonial Government, and the Singapore Improvement Trust (SIT) to relocate communities living in *attap* or "shack" housing on War Department ("Crown") land to new low-cost housing. Many of these communities were living in appalling conditions with no proper road access, sanitation, electricity or running water.

His report of April 1961 comprised the background, history and solutions to the above and highlighted three main drivers for the relocation of these communities: (i) where the land was required for development, (ii) where the communities were located close to Barracks, military installations or other buildings, and (iii) the resettlement of all families and communities to permanent low-cost housing.

New housing for the above would come with proper road access and sanitation, electricity and water (to be provided by the Government through the SIT), subject to compensation to those affected (e.g., loss of crops) to be paid by the War Department, who also met the cost of the new housing provided by the SIT. Indeed, although the SIT was responsible for the resettlement, the FARELF District Works Department was responsible for laying new roads (and drains) in new settlements and houses, which, in turn, would be erected by the Government.

The Colonial Government also recognised this problem, with the formation in 1955 of its own Working Party to look at the problem that the SIT faced in trying to gain possession of land for development that was occupied by so-called "squatters" — a misnomer that stuck over time, even though many of the occupants of this land had written agreements and paid rent to the War Department and private landowners.

In 1957, all parties agreed to follow the recommendations of the War Office/FARELF regarding the process for resettlement, as described above. The process faced the additional challenge of finding suitable War Department land for resettlement to take place — this problem would ultimately be resolved by the formation of the HDB in 1960 with its wide-ranging powers, and finally, in 1971, with the transfer of the British Military Assets to the Singapore Government.

The project was a success, with numerous communities relocated to new low-cost housing from the late 1950s onwards (see Figures 21.4. and 21.5.), but the sheer number of communities (and people in them) was continuously increasing. Indeed, Harding's report states that 67,000 families and 600,000 people (half the population of Singapore) lived in these communities on "War Department" land, which then occupied only 5% of the useable land on the island. Unfortunately, because the SIT had limited funds, the resettlement objective was only partly met by the time the SIT was replaced by the HDB in 1960. Ultimately, the HDB rehoused all these communities through its massive public housing programmes of the 1960s and 1970s.

FIGURE 21.4. (a) An illustration of a road and drainage installation for resettlement land. (b) An image of a new low-cost house at Chye Kay Road, Sembawang. Both illustrations were taken from the Report on Resettlement that A.I.G. Harding wrote for the War Department Land Office, FARELF in April 1961.

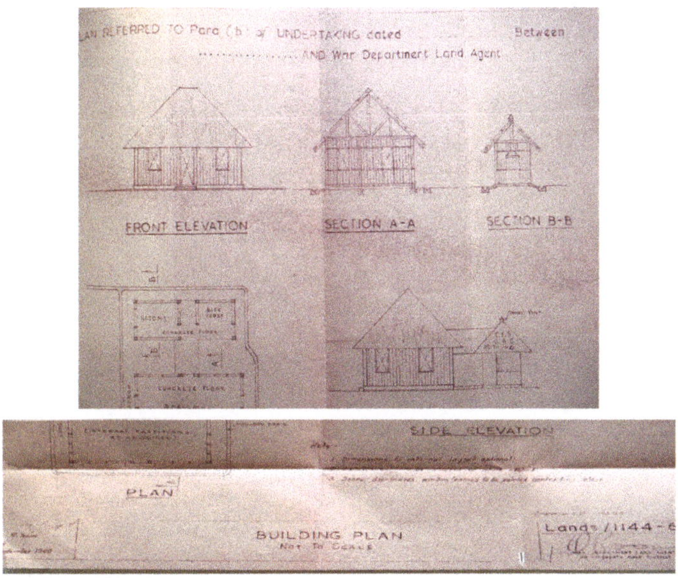

FIGURE 21.5 Samples of the Building Plan for one of the new low-cost houses. This is also taken from Harding's Report on Resettlement in April 1961.

In 1964, Harding joined Singapore Land Limited, a new, fast-growing commercial building company, where he held the position of Managing Director (1964–70). He was personally responsible for the concept and design of the Cuscaden House Hotel, which incorporated a rare underground nightclub called "The Eye". This club closed in 1972 for conversion to a cocktail bar due to the then overwhelming number of clubs and discotheques in Singapore. The Hotel also had a rooftop swimming pool with large viewing windows on the floor below — the first of its kind in Singapore. After several name changes, the Hotel was demolished in 2000 for the development of luxury condominiums.

Harding resigned from Singapore Land at the end of 1970, and it was announced in *The Straits Times* in February 1971 that he had set up a practice as an independent Chartered Surveyor ("Development Consultant, Estate Agent and Valuer") offering property management services in addition to the usual services of an estate agent, auctioneer and valuer.

He operated from his home at 37 West Coast Road. He returned to England in 1972, where he established his own surveying and estate agent practice in Tunbridge Wells. In his time in Singapore, he was regarded as an expert in land and property values, and to this end, he was called upon to give talks at Conferences about the Singapore property market.

Harding passed away in 1996 in England.

Others of note who became Chartered Surveyors at this time are **Ng Chee Yoong**, who qualified in 1967 and spent most of his career with the Public Works Department (PWD) of Singapore and then with CPG Corporation, which is, of course, part of the former PWD, and **Lim Kim Thian**, who qualified in 1968 and worked for Langdon Every & Seah for many years. -

Tan Choo Haw

As a final footnote to this book, a special mention goes to **Lieutenant Commander Tan Choo Haw** (see Figure 21.6.), who qualified in 1970 as a Chartered Surveyor but whose story started in the 1960s. He was a Royal

Malaysian Naval Officer from 1964 to 1972 and commanded various classes of patrol boats and Inshore Minesweepers and acted as Squadron Commander of two Patrol Craft Squadrons in the Royal Malaysian Navy. He served on the Royal Naval Survey Ship, *H.M.S. Dampier*, and Royal Australian Naval Survey Ships, *H.M.A.S. Paluma* and *H.M.A.S. Moresby*. He was also in command of the Royal Malaysian Naval Survey Ship, *K.D. Mutiara*.

FIGURE 21.6. Lieutenant Commander Tan Choo Haw at the time of his qualification as a Chartered Surveyor in 1970. During this time, he was serving in the Royal Malaysian Navy as a Hydrographic Surveyor. This photograph shows him testing a Heliograph used as a target for long-distance angle measurement with a theodolite in a Triangulation Control survey.

His training by the Royal Navy and the Royal Australian Navy in Hydrographic Surveying led to him attaining a 1st Class Certificate of Competence in Hydrographic Surveying in 1970.

The R.I.C.S. qualification was made known to him by the Royal Navy, and so he sat for and passed the R.I.C.S. Examinations as arranged through the R.I.C.S Singapore Branch in Singapore, starting with the Intermediate Examination (May 1967) and then the Final Examination in June 1970. Although this qualification was difficult to attain through part-time study whilst working, his time served in the Naval Hydrographic Services of Britain, Australia and Malaysia provided him with the practical experience and knowledge to make the Examinations easier for him.

By 1977, he was a Fellow of R.I.C.S. In addition to his R.I.C.S. qualifications, he became a member of the ISM (now Royal Institution of Surveyors Malaysia) in 1971 and the Singapore Institute of Surveyors (SIS) in 1972. He is now a Fellow of all three professional bodies.

Furthermore, Lieutenant Commander Tan was President of the Singapore Institute of Surveyors and Valuers (SISV) from 1985 to 1986, and in 2012, he was conferred a Distinguished Award by the SISV.

By 1973, he founded United Surveyors in Tunas Building on Anson Road, and he continues as Managing Director of that practice to this day, operating across South-East Asia, Taiwan, China and Timor Leste.

Among the many honours and recognitions he received was the Public Service Medal in 2011.

Lieutenant Commander Tan has had a long and successful career in professional practice and has received public recognition at the highest level.

Conclusion

A final word in this Book goes to Singapore-based members of the Chartered Institute of Auctioneers and Estate Agents (CIAEA) who become Chartered Surveyors in 1970 upon the merger of the CIAEA with the R.I.C.S. Some of them have already been mentioned in this book, but of those whose careers started in the 1960s and became Chartered Surveyors upon the above merger, the following individuals are noteworthy:

Kwok Chuon Wei

Kwek Chuon Wei's (see Figure 21.7) career started with the HDB as a Research Officer circa 1964 and then later as a Senior Research Officer. During this time, he contributed to many different public research works pertaining to the built environment. For example, he was co-author of the often-quoted publication entitled "Public housing and resettlement in Singapore: An overview". For a time, he was an area Estates Officer for Queensway. When he became a Chartered Surveyor, he moved to a role with the Central Provident Fund (CPF) as their Property Manager. Prior to his appointment, CPF used the services of Francis Lai Pang Fee of Lai Pang Fee & Co.

FIGURE 21.7. Kwok Chuon Wei.

Kwok was the CPF's Property Manager at the time of their 1976 move from Anson Road into the purpose-built CPF Building at 79 Robinson

Road. Kwok also supported the Monetary Authority of Singapore (MAS) on property matters from its inception in 1971. He was rewarded with the Public Administration Medal (Bronze) in 1977 for his extensive public service. Sadly, he died at a relatively young age in 1992. He was one of the early pioneers in finding ways to reduce office energy costs through improved building efficiencies.

Penu Mohan Singh — His career started with the HDB in the mid-1960s and continued into the 1970s, by which time he was an Estates Manager.

Ho Lum Khuan — Another HDB area Estates Officer who started with the HDB circa 1964. His area of responsbility was Bukit Merah.

Tang Soon Lee — After spending time with the HDB in Singapore in the mid to late 1960s as a Lands Officer, he returned to his native Malaysia to work for Talhar and Wong but has long since retired.

Wong Kum Sek — After he served the Government in the 1960s and early 1970s, he set up his own Valuation practice from the Golden Mile Complex, which he ran for many years.

Acknowledgements

I would like to thank the following individuals and organisations for their time spent on being interviewed and ongoing email correspondences, and their never-ending support as this book was being written:

- Mr. Seah Choo Meng and Mr. John Peacock, for all their encouragement and help and for providing me with a copy of *Quantifying Asia* (a book compiled by John), published in 2014 by Langdon & Seah Asia Company Ltd. Hong Kong. I would also like to thank them for their kind permission in reproducing photographs from that book.
- Marcus Lim, RICS APAC, Head of Partnerships, for all his help, patience and encouragement as I wrote this book.
- Cathy Linacre, the retired RICS Head of Reference Services, for her help and patience in searching the archives.
- Calista Lucy, Keeper of the Dulwich College archive, for releasing the photograph of Harold Carpmael on behalf of the College Governors.
- Sarah Semple, Jules Feldman, Rebecca Rott and Alain Farhi for their information and help on the life of Joseph Bernard Maschoieff (later known as Joe Marsh). Also, thanks to Charles Marsh for the information about his father.
- Andrea Ruddock, for the work on Henry Lake Coghlan and photographs of his grave.
- James Edwards, for the family photographs of Henry Lake Coghlan.
- Jamie Maule-ffinch, for the photograph of K.H. Maule-ffinch.

- Michael Pennefather, for information on John Pyne Pennefather, including his photograph.
- John Farrer and Caroline (Carol) Chow née Nicholas, for their kind help in relation to their grandfather, Wilfred Dunstan Nicholas.
- Paul Benedict Lincoln, for the information on John Albert Lincoln, including photographs.
- Helen Griffiths of Newport Reference Library, for her guidance and background on the early life of Leonard Langdon Williams.
- Paula Lermit-Webb, for her contribution to the Lermit story and for her contribution on behalf of Lermit family relatives.
- Anthony Cobbold of the Cobbold Family History Trust, for his advice and information on Harvey Chevallier.
- Stephen Leong, Chancery of the Roman Catholic Archdiocese of Singapore, for information on Church of Our Lady of Lourdes, Ophir Road.
- Libby Mills and Claudia Mills, for their help on Alexander Niven, including extracts from the diaries of Eileen Mills Niven and the Mills family.
- David Atkin-Berry and Andy Atkin-Berry for their help with providing family photographs of Herbert Courtenay Atkin-Berry.
- Gordon Snell, for his help and encouragement in the story of his father's time in Singapore.
- Charlotte Howard of Fox and Howard, Literary Agents for the late Margaret Shennan, for helping me to obtain permission from Margaret's son for the use of photographs from her book, *Out in the Midday Sun: The British in Malaya 1880–1960*.
- Coral Streader née Bidwell, for all her help and encouragement with her great grandfather Regent Bidwell's family history and associated photographs.
- Michael Gray, for information pertaining to Bidwell's time as a member of various Masonic Lodges in Singapore during the early 1900s.

- Mr. Ian D.R. Watson, for kindly allowing me to reproduce photographs in his possession of his late father and the Waters & Watson practice, as well as the photograph of Horace W. Langdon and Every (incorporating Waters & Watson).
- Edward Bottoms, Head of Archives at the Architectural Association, for his help on Bidwell.
- Mr. Yeo Soo Pio, for his continuing encouragement and valuable information from the 1960s and 1970s, along with the national context for that period.
- Mr. Teo Ching Ming, Amos, for his help in doing a review of key components of the final manuscript for historical and factual accuracy.
- Dr. Anamah (Ann) Tan for her help, contribution, encouragement and constant words of enthusiasm.
- Alice Goh, Office Director at NUS School of Real Estate, for the NUS information.
- The late Stuart Rix (1930–2020) and his son Stephen, for information on the 1957 and 1958 R.I.C.S. Dinners.
- Francis Lai, for his encouragement, as well as his help with information on members and R.I.C.S. activities in the 1960s and early 1970s and for his permission to use photographs in his ownership and possession.
- Mr. Tan Choo Haw, for sharing his career background and experiences.
- Paddy Walker-Taylor, for the information on Basil Paterson Walker-Taylor, including the book, *The Road to Singapore*, and the various photographs.
- Lim Chan Hoe, for his insights into his early career in the 1960s.
- Alec Audley, Alun Williams and Ron Cork from the 81 Squadron Association, for their help with RAF Tengah.
- The photograph of Ian Whyte was provided with kind permission from David Stephen.
- For information on R.W. Compton-Hall, the following individuals contributed: Alexander Michael Compton Hall, his parents Michael

Wilfred Compton-Hall and Pamela Sophia Elizabeth Compton-Hall (née Lewis), his daughter Sheila Louise Strong and his wife Felice Dianne Hall (née Wechsler).

- Charles Ho, for information on his early career and his time in Singapore, Brunei, etc.
- Jack LeRoy, retired partner of Davis LeRoy & Partners, for his information on the 1958 R.I.C.S. examinations held in Singapore.
- The Francis Bennion estate, for their kind agreement to allow me to quote from his diaries, which record the Singapore leg of the 1966 Presidential World Tour, on which the R.I.C.S. President Mr. Guy Biscoe was accompanied by his wife Sheila and the Secretary-General of the Institution, Mr. Francis Bennion.
- Pauline Goh and Willy Shee, for their time and valuable information on C.H. Williams and Peter Newman.
- Craig Oates for his help with information on and photographs of Basil E. Thompson, former City Engineer.
- Ron Taylor of the "Roll of Honour" (a website that is dedicated to the memories of the Far East Prisoners of War), for his help in relation to Gordon M. Lowden and the story of the *Rakuyō Maru* sinking on 12 September 1944. https://www.roll-of-honour.org.uk
- Susanna Jack, for her help with photographs of Aubrey Distin and his surveying colleagues at the Admiralty, Singapore, circa 1950–52.
- Guy Jubb, for information regarding his father's time in Singapore.
- D.R.W. (Robin) Tillyard and Judy Tillyard for the information on W.J.F. Tillyard.
- Michael Wee, for his recollections of Singapore in the 1960s and early 1970s and his career story.
- Libby Meredyth of the Surveying & Spatial Sciences Institute (SSSI) in Deakin ACT, for help on the information on Harry L. Ward.
- Mdm. Lily Ong, for stories about Frank, Vargeson, Wicks and Partners, which was later known as M.H. Goh, Wicks & Partners and eventually known as Colliers Goh & Tan.

- Brenda Vargeson for information and photographs pertaining to her late husband, Desmond Vargeson.
- Peter Rhodes and Barbie Brinkman (née Rhodes) for their information and photographs pertaining to their late father, R.N.S. Rhodes.
- Mr. Liu Thai Ker and his son Daniel, for the interview and insights on senior Chartered Surveyors working for the HDB in the 1970s and 1980s.
- Dr. Ian Chan and Hugh Chan, for information on and the photograph of Mr. Chan Wan Hang.
- Stephen Komlosy, for his help in relation to the career of his father Denis Komlosy and for allowing me to use family photographs for my book, and to Sandra "Sandy" Lincoln (daughter of Lincoln Page), for her help with additional information. Further thanks to Ho Weng Hin (Adjunct Assistant Professor, NUS Architecture) for his help in the loan and reproduction of Sandra Lincoln's photograph collection.
- Richard Burrells and his family, for their help with information on and providing photographs of Malcolm Burrells.
- Kerry Littlejohn, for information and photographs of her father, Rae Littlejohn.
- Dr. Ng, Bee Lim and Robin Pho for their kindness and help in the story of Ng Boon Ong.
- Geraldine Chilton, for information and photographs relating to her father, Gerald R. Chilton.
- June Morris, her daughters, Susan and Kim, and son-in-law, Ian Davies, for all their kind help with information and photographs of Derrick Morris.
- Esther Teo, Chief Operating Officer of the Royal Institution of Surveyors Malaysia, for her kind help in arranging for permission to use the photographs of J.H. McDonald and Sr. Yap Eng Bock.
- Sarah Williams, Editor of *Our Media* and Mark Lewis, for their help and information on the career of H. Selwyn Lewis in Singapore.

- Liz Godbolt and David Harding, for their kind help and assistance with the life and career of Adrian ("A.I.G.") Harding.
- David Robson, MA, RIBA, for the Edwards, Reid and Booth photograph and for additional information on the career of Sidney James Edwards, some of which has been referenced from *Geoffrey Bawa: The Complete Works*, published in 2002 by Thames & Hudson Limited.
- Anthony Thorp and the Directors of Thorstone Land & Property Limited of London, for the photograph and information on A.J.M. Davies.
- Mohd Talhar Abdul Rahman, co-founder and Chair of CBRE-WTW, for the photograph and information on Colin H. Williams.
- Mr. John Rosser for allowing me to use his photograph of Lt.-Col. N.F. Croft.
- Rabbi Nissim Zawady, for allowing me to visit the Chesed-El Synagogue and take photographs for my book.
- Singapore National Library Board archives and Singapore Press Holdings (SPH) for all their help and for the permissions to reproduce a number of photographs for this book.

Finally, Julian Davison's *Swan & Maclaren — A Story of Singapore Architecture* (published in 2020 by ORO Editions and The National Archives of Singapore) has been a major inspiration and information source. This masterpiece makes detailed references to several Chartered Surveyors such as Lermit, Bidwell, Taylor, Edwards and Atkin-Berry and includes archived architectural drawings of Lermit, Bidwell and Atkin-Berry. I will not seek to replicate any of Julian's excellent work. Instead, this book seeks to complement his book by focusing on different perspectives of the lives and careers of these talented individuals.

Endnotes

CHAPTER 1:

Information on the early history of Singapore has been referenced from the following sources:

- *The Singapore House 1819–1942* by Lee Kip Lin. Edited by Gretchen Liu, designed by Kathleen Lau. First published 1988 by Times Editions, reprinted in 1995 and published 2015 by Marshall Cavendish Editions and National Library Board. ISBN 9789814634014.
- *Contesting Space: Power Relations and the Urban Built Environment in Colonial Singapore* by Brenda S.A. Yeoh (published by Oxford University Press, 1996, and NUS Press, 2003, ISBN 9971692686, 9789971692681).
- *Singapore Real Property Guide*, Fourth Edition, Chapter 2, by Philip Motha and Belinda K.P. Yuen (published by Singapore University Press, 1999 (First Edition, 1979), ISBN 9971692244 (Case) and ISBN 997169218x (Paper)).
- *The Malay Annals* (*Sejarah Melayu*), printed for Longman, Hurst, Rees, Orme and Brown, 1821, and translated from Malay into English by John Leyden and with an introduction by Sir Stamford Raffles. The Annals are available to read digitally via BookSG, a Singapore Government Agency website (Digital Description: application/pdf, 35938 KB, 376 p.).
- Recent research into pre-1819 Singapore is referenced from the following sources: *Seven Hundred Years: A History of Singapore*

by Kwa Chong Guan, Derek Heng, Peter Borschberg, and Tan Tai Yong (published by Marshall Cavendish International (Asia), 2009, ISBN-13 9789814828109, "Review essay of 'Jacques de Coutre's And Matelieff's Singapore and Johor': Exploring sources on pre-modern history of Singapore" by Teddy Sim Y.H., National Institute of Education (Singapore) — HSSE Online **4**(2)17–35, and "Reading the traditional city of maritime Southeast Asia: Reconstructing the 19th century Port Town at Gelam-Rochor-Kallang, Singapore" by Imran bin Tajudeen, MA (Arch) Candidate, Department of Architecture, School of Design and Environment, published in the *Journal of Southeast Asian Architecture*, Vol. 8 (2005).

- A number of extracts have been taken from Arnold Wright's *Twentieth Century Impressions of British Malaya: Its History, People, Commerce, Industries, and Resources*, originally published in 1908 by Lloyd's Greater Britain Publishing Company, Limited.
- Information on Jackson and on the Singapore Stone comes from Singapore Infopedia, a Singapore Government Agency website: http://eresources.nlb.gov.sg/infopedia/articles/SIP_140_2005-01-20.html and http://eresources.nlb.gov.sg/infopedia/articles/SIP_43_2005-01-26.html
- Material has also been referenced in this Chapter from the book by C.M. Turnbull, *A History of Singapore 1819–1988*, Second Edition (published by Oxford University Press).
- The N. Pillai and A.L. Johnston information is referenced from Singapore Infopedia: http://eresources.nlb.gov.sg/infopedia/articles/SIP_771_2004-12-29.html and http://eresources.nlb.gov.sg/infopedia/articles/SIP_535_2004-12-27.html.
- Street name information including the Coleman Street/Lane information comes from the *What's in the Name? How the Streets and Villages in Singapore Got Their Names* by Ng Yew Peng (published by World Scientific Publishing, 2018, ISBN 9789813221390 (pbk)).

(i) G.D. Coleman information comes from various sources, including Singapore Infopedia: https://eresources.nlb.gov.sg/infopedia/articles/SIP_134_2004-12-10 and http://eresources.nlb.gov.sg/INFOPEDIA/articles/SIP_766_2004-12-14.html, plus http://eresources.nlb.gov.sg/infopedia/articles/SIP_134_2004-12-10.html

(ii) Information on G.D. Coleman, Lieutenant Jackson and J.T. Thomson comes from *A History of Singapore Architecture: The Making of a City* by Jane Beamish and Jane Ferguson (published by Graham Brash (Pte) Ltd, ISBN 9971947978) and from *Coleman's Singapore* by T.H.H. Hancock (published by the Malaysian Branch of the Royal Asiatic Society in association with Pelanduk Publications, 1986).

(iii) More information about Coleman's original house being demolished is in *The Straits Times*, 29 November 2015.

(iv) Information on Charles Alexander Dyce is from Jane Beamish and Jane Ferguson's *A History of Singapore* and the *Singapore Free Press*, 27 May 1961, page 6, held in the National Archives of Singapore on Microfilm Reel NL2519.

(v) G.D. Coleman, J.T. Thomson and other Land Surveyors from 1819 to 1871 and from 1871 to 1914: this was referenced from "Mapping Singapore: 1819–2014" an essay by Mok Ly Yng, FRGS, in the National Archives of Singapore. In turn, this essay references a number of historical documents, and also in Singapore Infopedia: http://eresources.nlb.gov.sg/infopedia/articles/SIP_818_2005-01-22.html and http://eresources.nlb.gov.sg/infopedia/articles/SIP_2_2004-12-17.html

(vi) The NLB e-resource attributes a number of references about Thomson to J. Hall-Jones and C. Hooi's *An Early Surveyor in Singapore: John Turnbull Thomson in Singapore, 1841–1853* (published by National Museum of Singapore, 1979), pp. 1–2 (Call no.: RSING 526.90924 THO).

(vii) The story on the Dalhousie Obelisk is from Singapore Infopedia: http://eresources.nlb.gov.sg/infopedia/articles/SIP_481_2005-01-20.html

(viii) The information on Thomson's last resting place and Thomson's early years is from: http://www.heritage.org.nz/the-list/details/2497 and from https://teara.govt.nz/en/biographies/1t97/thomson-john-turnbull

(ix) Information on McNair is referenced from Singapore Infopedia: http://eresources.nlb.gov.sg/infopedia/articles/SIP_1708_2010-10-10.html

(x) The section on the PWD comes from Arnold Wright's *Twentieth Century Impressions of British Malaya*.

(xi) The section on F.J. Pigott also comes from *Twentieth Century Impressions of British Malaya*. Pigott's statement on the 1839 Act is not entirely correct. Based on the author's research, it was property and land tax that was collected under the 1839 Act powers, not just the power to collect rent (this power was enhanced by Section 24 of the 1853 Act empowering the Municipal Committee to "fix and estimate and fix the rent of houses, shops, buildings and land"). The term or word "quit-rent" in Pigott's statement is confusing — the Crown sold leases on premiums with a continuing quit-rent paid to the Crown, and the quit-rent was treated as a land tax. However, when a private freeholder sells land on a lease at a premium with a continuing rent payment, the term "Ground rent" is used; it is not called a "quit-rent" and is not a tax — it is an annual payment to the private freeholder.

(xii) Supplemental information to the Pigott story comes from the *Singapore Free Press and Mercantile Advertiser*, 22 November 1884, page 13, which is in the Singapore National Library Board archives on Microfilm Reel NL1976, and *The Straits Times*, 4 October 1884, page 11, which is also in the Singapore National Library Board archives on Microfilm Reel NL1976.

(xiii) Information on John van Cuylenburg's death is from *The Straits Times*, 11 September 1940, page 2, which is held in the Singapore National Library Board archives.

(xiv) The Survey School is mentioned in various newspaper articles held on Microfilm in the Singapore National Library Board archives, including: *The Straits Times Weekly Issue*, 4 September 1889, page 9, which is available on Microfilm Reel NL5209; *The Straits Times Weekly Issue*, 10 September 1890, page 6, which is available on Microfilm Reel NL5210; the *Malaya Tribune*, 19 December 1890, page 2, on Microfilm Reel NL2113; *The Straits Times Weekly Issue*, 14 April 1891, page 9, on Microfilm Reel NL5211; the *Daily Advertiser*, 29 April 1891, page 2, on Microfilm Reel NL2430; the *Daily Advertiser*, 4 June 1891, page 2, which is available on Microfilm NL243; *The Straits Times Weekly Issue*, 23 June 1891, page 5, on Microfilm Reel NL5211; the *Daily Advertiser*, 15 September 1891, page 3, on Microfilm Reel NL2431; *The Straits Times Weekly Issue*, 29 September 1891, page 3, on Microfilm Reel NL5212; *The Straits Times Weekly Issue*, 19 April 1893, page 13, on Microfilm Reel NL5214; the *Daily Advertiser*, 5 May 1893, page 2, on Microfilm Reel NL2434; the *Daily Advertiser*, 2 July 1894, page 3, on Microfilm Reel NL2436; the *Singapore Free Press and Mercantile Advertiser* (Weekly), 19 December 1901, page 382, on Microfilm Reel NL1940; and finally the *Singapore Free Press and Mercantile Advertiser*, 20 June 1912, page 392, on Microfilm Reel NL1953.

(xv) Mapping and surveying information was referenced from Mok Ly Yng's "Mapping Singapore: 1819–2014": http://www.nas.gov.sg/blogs/offtherecord/wp-content/uploads/2016/06/Visualising-Space_Ch3.-Mapping-Singapore-1819-2014-by-Mr-Mok-Ly-Yng.pdf); and from Jane Beamish and Jane Ferguson's *A History of Singapore Architecture*.

(xvi) Treacher Technical School information comes from Arnold Wright's *Twentieth Century Impressions of British Malaya.*

CHAPTER 2:

(i) Information on early legislation and statute comes from *Law of Real Property & Conveyancing in Singapore* by N. Khublall (published by Longman, 1986, ISBN 99718999507) and from *Singapore Real Property Guide, Fourth Edition*, Chapter 2, by Philip Motha and Belinda K.P. Yuen (published by Singapore University Press, 1999 (First Edition, 1979), ISBN 9971692244 (Case) and ISBN 997169218x (Paper)).

(ii) Some of the Information regarding Alfred Lermit, including the Lermit-Annamalai partnership and the employment of Harvey Chevallier by Swan & Lermit, has been referenced from *Swan & Maclaren — A Story of Singapore Architecture,* Chapter 2, by Julian Davison (published by ORO Editions and The National Archives of Singapore, ISBN 9781935935476).

(iii) Some reference sources for information on Lermit, Tomlinson and Westerhout are (a) Swan & Maclaren's website (http://www.swanmaclaren.com/site/history/), (b) *What's in the Name? How the Streets and Villages in Singapore Got Their Names* by Ng Yew Peng (published by World Scientific Publishing, 2018, ISBN 9789813221390 (pbk)), and (c) *Twentieth Century Impressions of British Malaya: Its History, People, Commerce, Industries, and Resources* by Arnold Wright (originally published by Lloyd's Greater Britain Publishing Company, Limited, 1908).

(iv) Lermit's birthplace, marriage place, and son's birthplace were confirmed by Paula Lermit-Webb, who also provided additional family information on Lermit family relatives.

(v) Chevallier's birth and death places were confirmed by Anthony Cobbold of the Cobbold Family History Trust.

(vi) Lermit's residences in London just prior to coming to Singapore are mentioned in the London Electoral Registers 1847–1913, held in microfilms owned by the London Metropolitan Archives.

(vii) Lermit's arrival in Singapore is mentioned in *The Straits Times*, 11 June 1883, page 3, which is in the Singapore National Library archives, an article that is also available on Microfilm Reel NL5047, and his new business with Mr. Annamalai was mentioned in *The Straits Times*, 18 September 1883, page 2, which is available on Microfilm Reel NL05047.

(viii) Swan's life is referenced from the University of Glasgow's website: University of Glasgow :: Story :: Biography of Archibald Alexander Swan, https://www.universitystory.gla.ac.uk/biography/?id=WH16898&type=P

(i) Information on No. 38 Oxley Road comes from Singapore Government website sources.

(ix) The story of the Boat Quay disaster is mentioned in the *Singapore Free Press and Mercantile Advertiser* (Weekly), 5 July 1900, page 12, which is on Microfilm Reel NL1938.

(i) Some information regarding Alfred Lermit and his son A.A. Lermit has been referenced from the book by Julian Davison's *Swan & Maclaren.*

(ii) The Our Lady of Lourdes Church information comes from various sources: (i) information kindly provided by the Chancery of the Roman Catholic Archdiocese of Singapore, (ii) the Singapore Government Infopedia pages for the Church, (iii) a paper dated November 1998 by Jean Ernoult entitled "Les églises en fer de Borna et Fernan Vaz", and (iv) the *Singapore Weekly Herald*, 4 August 1888, page 10, which is held in the Singapore National Library Board archives on Microfilm Reel NL2880.

(iii) Lermit's death and the story of his son are in the *Malaya Tribune*, 18 December 1929, page 8, which is held in the Singapore National

Library Board archives on Microfilm Reel NL3945, and *The Straits Times*, 8 March 1936, page 7, which is held in the Singapore National Library Board archives on Microfilm Reel NL2364.

(iv) *Singapore Eurasians: Memories, Hopes And Dreams*, edited by Alexius A. Pereira, Myrna Braga-Blake and Ann Ebert-Oehlers (published by World Scientific Publishing, 2017).

(i) A.A. Lermit working at Swan & Maclaren and A.W. Lermit's likeness to Nestor are both from the book *One Hundred Years of Singapore, Volume II* by Walter Makepeace, Dr. Gilbert E. Brooke and Roland St. J. Braddell (published by John Murray, Albermarle Street, London, 1921). The information on Nestor comes from the *The Concise Dictionary of Greek and Roman Mythology* by Michael Stapleton (ISBN 0872260062).

(ii) The Telok Blangah case: this was mentioned in *The Straits Times*, 9 June 1906, page 5, which is held in the Singapore National Library Board archives on Microfilm Reel NL305.

(iii) The Adelphi Hotel information is referenced from Singapore Infopedia: http://eresources.nlb.gov.sg/infopedia/articles/SIP_1109_2006-04-07.html, and the Bras Basah Road information comes from the Government URA website: https://www.ura.gov.sg/Conservation-Portal/Explore/History?bldgid=BRBRD

(iv) Tanjong Pagar Land Company: this was referenced from the *Singapore Free Press and Mercantile Advertiser* (Weekly), 6 November 1889, page 559, which is held in the Singapore National Library Board archives on Microfilm Reel NL5607.

(v) Battery Road widening is mentioned in the *Singapore Free Press and Mercantile Advertiser* (Weekly), 13 August 1890, page 11, available on Microfilm Reel NL1928.

(vi) Singapore Land Company Limited: this is referenced from *The Straits Times Weekly Issue*, 19 July 1892, page 8, held in the Singapore National Library Board archives on Microfilm Reel NL5213.

(vii) The Boustead Institute and Lermit's involvement is mentioned in *The Straits Times Weekly Issue*, 29 June 1892, page 6, which is held in the Singapore NLB archives on Microfilm Reel NL 5213. His involvement in the construction of the Anglo-Chinese School is mentioned in *The Straits Times Weekly Issue*, 25 July 1893, page 11, which is on Microfilm Reel NL5214.

(viii) The Bukit Timah Road land auction was reported in the *Singapore Free Press and Mercantile Advertiser* (Weekly), 28 March 1901, page 201, which is held in the Singapore National Library Board archives on Microfilm Reel NL1939.

(ix) Charles Lermit's death is reported in the *Singapore Free Press and Mercantile Advertiser* (Weekly), 25 November 1909, page 1, which is held in the Singapore National Library Board archives on Microfilm Reel NL1950.

(x) The May 1904 merger was announced in *The Straits Times*, 2 May 1904, page 5, which is held in the Singapore National Library Board archives on Microfilm Reel NL296.

(xi) The Chapman incident was reported in the *Singapore Free Press and Mercantile Advertiser* (Weekly), 21 December 1905, page 387, which is held in the Singapore National Library Board newspaper archives on Microfilm Reel NL1945, and in the *Eastern Daily Mail and Straits Morning Advertiser*, 11 January 1906, page 3, which is on Microfilm Reel NL2975.

(xii) The 1906 Orchard Road case is mentioned in the *Eastern Daily Mail and Straits Morning Advertiser*, 22 May 1906, page 3, which is in the Singapore National Library Board archives on Microfilm Reel NL2976.

(xiii) The split in 1914 was announced in *The Straits Times*, 11 February 1914, page 3, which is in the Singapore National Library Board archives on Microfilm Reel NL380.

(xiv) Lermit's letter to *The Straits Times* on 6 May 1903, page 5, is in the Singapore National Library Board archives on Microfilm Reel

NL00292 in response to a report in *The Straits Times*, 17 April 1903, page 5, which is also on Microfilm Reel NL00292.

(xv) The Armenian Church Parsonage, Singapore: information taken from http://eresources.nlb.gov.sg/infopedia/articles/SIP_809_2004-12-23.html and from the *Free Press and Mercantile Advertiser* (Weekly), 26 October 1905, page 259, which is held in the National Archives of Singapore on Microfilm Reel NL1944.

(xvi) Swan & Lermit being approved for Government work: announced in the *Singapore Free Press and Mercantile Advertiser* (Weekly), 15 October 1889, page 454, which is in the Singapore National Library Board archives on Microfilm Reel NL5607.

(xvii) Lermit's obituary in *The Straits Times*, 19 December 1929, page 15, Microfilm Reel NL630, Singapore National Library Board.

(xviii) Lermit's wife and son's burial is referenced from the Lenches Church website.

(xix) Harvey Chevallier information from the World Statesmen website: http://www.worldstatesmen.org/Brunei.html, and from Anthony Cobbold of the Cobbold Family History Trust: https://family-tree.cobboldfht.com/people/view/9052, as well as the following sources: *The Straits Times*, 24 November 1913, page 9, which is in the Singapore National Library Board newspaper archives and available on Microfilm Reel NL00377, the *Singapore Free Press and Mercantile Advertiser* (Weekly), 24 May 1892, page 327, which is available on Microfilm Reel NL1929, *The Straits Times Weekly Issue*, 25 March 1891, page 5, which is available on Microfilm Reel NL5211, and the *Singapore Free Press and Mercantile Advertiser*, 2 October 1923, page 6, which is available on Microfilm Reel NL1688.

(xx) The Labuan report is referenced from *The Straits Times*, 28 July 1913, page 8, which is available on Microfilm Reel NL 00373 in the Singapore National Library Board archives.

(xxi) The 1913 Brunei river case was reported in *The Straits Times*, 27 November 1913, page 10, which is in the Singapore National Library Board archives on Microfilm Reel NL377.

(xxii) Photographs for H.L. Coghlan, Coghlan & Co.'s Singapore premises, the various Engineers/Surveyors of Singapore as of 1908 and S. Tomlinson are reproduced from *Twentieth Century Impressions of British Malaya: Its History, People, Commerce, Industries, and Resources* by Arnold Wright (originally published by Lloyd's Greater Britain Publishing Company, Limited, 1908). The photographs are reproduced on the understanding that the copyright is not applicable, given that the book was published in 1908. Anyone claiming copyright can contact the author through the Publishers.

(xxiii) The address for Powell & Co., Auctioneers: "The Directory & Chronicle for China, Japan, Corea, Indo-China, Straits Settlements, Malay States, Siam, Netherlands India, Borneo, the Philippines: With which are Incorporated 'The China Directory' and 'The Hong Kong List for the Far East' January 1, 1910, Hong Kong Daily Press Office".

(xxiv) Coghlan becoming a Fellow of the Surveyors' Institution is mentioned in *The Straits Times*, 21 November 1908, in the Singapore National Library Board archives on Microfilm Reel NL317.

(xxv) The birth of Coghlan's son is reported in the *Singapore Free Press and Mercantile Advertiser*, 16 August 1907, page 4, which is held in the Singapore National Library Board archives on Microfilm Reel NL5318.

(xxvi) Coghlan on the Municipal Council is mentioned in various newspaper articles of the time, including *The Straits Times*, 29 February 1908, page 7, which is in the Singapore National Library Board archives on Microfilm Reel NL00312 and *Eastern Daily Mail and Straits Morning Advertiser*, 24 July 1907, page 3, on Microfilm Reel NL2979.

(xxvii) The Craig Road action sale is mentioned in *The Straits Times*, 2 June 1911, page 6, which is held in the Singapore National Library Board archives on Microfilm Reel NL00355.

(xxviii) Coghlan v. Lintner & Co. is mentioned in the *Singapore Free Press and Mercantile Advertiser*, 22 December 1919, page 12, which is held in the Singapore National Library Board archives on Microfilm Reel NL1648.

(xxix) Coghlan as an assessor for a Government compensation case is in the *Singapore Free Press and Mercantile Advertiser* (Weekly), 3 October 1907, page 5, which is in the Singapore National Library Board newspaper archives on Microfilm Reel NL1947.

(xxx) The sale of Buitenzorg was mentioned in *The Straits Times*, 19 October 1911, page 6, which is held in the Singapore National Library Board newspaper archives on Microfilm Reel NL00357.

(xxxi) Coghlan's obituary is in the *Singapore Free Press and Mercantile Advertiser*, 20 January 1938, page 2, which is in the Singapore National Library Board newspaper archives on Microfilm Reel NL3592, and his wedding is captured in *The Straits Times*, 6 April 1903, page 5, which is available on Microfilm Reel NL00292.

(xxxii) Coghlan's family tree information came from Andrea Ruddick, with the birth of his son reported by the *Singapore Free Press and Mercantile Advertiser*, 17 May 1911, page 6, which is held in the Singapore National Library Board archives on Microfilm Reel NL3171, and his marriage is mentioned in *The Straits Times*, 3 April 1903, page 4, in the Singapore National Library Board archives, Microfilm Reel NL00292.

(xxxiii) Information on A. Arch comes from the *Singapore Free Press and Mercantile Advertiser* (Weekly), 8 October 1890, page 404, which is in the Singapore National Library Board archives on Microfilm Reel NL1928.

(xxxiv) Information on Koch comes from the Journal of E. V. (Edward Verrall) Lucas entitled "Who's who in the Far East, 1906–7", which is available through online sources, and the Arch information is from the 1891 Institution of Civil Engineers: Obituaries.

CHAPTER 3:

(i) Coral Streader (née Bidwell) has been a huge help in pulling together the information around the life and times of R.A.J. Bidwell. Coral also provided much of the information around his origins as well as his parents, his wife's family and his children.

(ii) Much of the information in this Chapter has also been crosschecked against Julian Davison's *Swan & Maclaren — A Story of Singapore Architecture* (published by ORO Editions and The National Archives of Singapore, ISBN 9781935935476). However, key information sources for this Chapter are from:

Black And White: The Singapore House, 1898–1941 by Julian Davison (published 2006 and again in 2014 by Talisman Publishing Pte Ltd, ISBN 9789810903282).

(iii) His membership of the Architectural Association was confirmed by Ed Bottoms, Head of Archives at the Architectural Association. Separately, Jason Canham of RIBA Library & Information Centre confirmed that Bidwell was not a member of RIBA.

(iv) His time with PWD Selangor comes from "The Bangunan Sultan Abdul Samad" by Gullick, J.M., *Journal of the Malaysian Branch of the Royal Asiatic Society*, **65**(1):27–38, 1992, and also from the website dedicated to the life and times of Arthur Benison Hubback of the PWD Selangor: https://www.abhubback.com. The information from Bidwell's Selangor PWD contract comes from the Malaysian National Archives.

(v) Bidwell is mentioned in *A History of Singapore Architecture: The Making of a City* by Jane Beamish and Jane Ferguson (published by Graham Brash (Pte) Ltd, ISBN 9971947978), and in Gullick, J.M.'s "The Bangunan Sultan Abdul Samad".

(vi) Information on the Sultan Abdul Samad Building is from Pusat Pengajian Seni, Universiti Sains Malaysia. Archived from the original *on 11 July 2015*. Retrieved 21 July 2016.

(vii) Bidwell and Selangor Cricket Club in 1893 is mentioned in the *Singapore Free Press and Mercantile Advertiser*, 14 September 1893, page 2, which is held in the Singapore National Library Board archives on Microfilm Reel NL3153.

(viii) Bidwell's role as Committee Honorary Secretary for St. Mary the Virgin Church (now known as St. Mary's Cathedral) Kuala Lumpur is mentioned in the *Singapore Free Press and Mercantile Advertiser* (Weekly), 20 February 1894, page 96, which is in the Singapore National Library Board archives on Microfilm Reel NL1935.

(ix) The story of his visit to Kuala Lumpur in 1897 is referenced from the *Singapore Free Press and Mercantile Advertiser* (Weekly), 13 April 1897, page 4, which is in the Singapore National Library Board archives on Microfilm Reel NL 1933.

(x) Lermit as architect of the Katz Brothers new Department store on Bonham Street was confirmed in the *Singapore Free Press and Mercantile Advertiser*, 14 December 1893, page 2, which is held in the Singapore National Library Board archives on Microfilm Reel NL 3154.

(xi) Bidwell's 1897 Jury role was mentioned in the *Mid-Day Herald and Daily*, 24 November 1897, page 3, which is held in the Singapore National Library Board archives on Microfilm Reel NL2394, and his involvement in the construction of the Katz Brothers Department Store is mentioned in the *Singapore Free Press and Mercantile Advertiser* (Weekly), 9 March 1897, page 5, available on Microfilm Reel NL1933.

(xii) The 1900 Methodist Girls' School opening is mentioned in the *Singapore Free Press and Mercantile Advertiser* (weekly), 1 March 1900, page 10, which is in the Singapore National Library Board archives on Microfilm Reel NL1938.

(xiii) His return to the UK in 1901 was recorded in *The Straits Times*, 8 July 1901, page 2, held in the Singapore National Library Board archives on Microfilm Reel NL00286.

(xiv) The Victoria Memorial Hall Committee meetings are reported in the *Singapore Free Press and Mercantile Advertiser* (Weekly), 6 July 1897, page 2, and also the *Singapore Free Press and Mercantile Advertiser* (Weekly), 29 June 1897, page 11. The design and construction is mentioned in detail in *The Straits Times*, 17 September 1903, page 5, which is held in the Singapore National Library Board archives on Microfilm Reel NL00293, and also in the *Singapore Free Press and Mercantile Advertiser* (Weekly), 15 September 1904, page 170, which is available on Microfilm Reel NL1943. Swan & Lermit's original involvement back in 1890 was mentioned in *The Straits Times Weekly Issue*, 14 January 1890, page 2, which is available on Microfilm Reel NL5209. Further mentions are in *The Straits Times*, 26 September 1902, page 4, as well as *The Straits Times*, 10 March 1903, page 5.

(xv) Information on the Raffles Place design and construction comes from the *Singapore Free Press and Mercantile Advertiser* (Weekly), 6 July 1897, page 8, the *Singapore Free Press and Mercantile Advertiser* (Weekly), 23 November 1899, page 9, *The Straits Times*, 9 August 1899, page 2, and *The Straits Times*, 8 April 1904, page 5, all of which are held in the Singapore National Library Board archives.

(xvi) The 1908 burglary at the Bidwell residence was reported in the *Singapore Free Press and Mercantile Advertiser* (Weekly), 7 May 1908, page 10, which is held in the Singapore National Library Board archives on Microfilm Reel NL1948, and the May 1908 District Court case is in *The Straits Times*, 7 May 1908, page 6, which is also in the library's archives on Microfilm Reel NL00314.

(xvii) The 1909 Australian cricket team visit to Singapore and Bidwell's involvement is in the Argus newspaper dated 23 December 1909, page 9, which is held the Australian Government archives (trove.

nla.gov.au). His Australian cricketing links is mentioned in the *Singapore Free Press and Mercantile Advertiser*, 5 February 1910, page 9. His involvement in the Singapore Cricket Club new pavilion is mentioned in *The Straits Times*, 11 January 1907, page 7, and 24 August 1907, page 8, as well as the *Singapore Free Press and Mercantile Advertiser* on 21 March 1907, page 186.

(xviii) Bidwell's house in Tanjong Katong is mentioned in *The Straits Times*, 27 June 1898, page 1, which is held in the Singapore National Library Board archives on Microfilm Reel NL00343.

(xix) Bidwell's involvement in the John Little Store renovations was mentioned in the *Singapore Free Press and Mercantile Advertiser* (Weekly), 22 September 1910, page 10, which is held in the Singapore National Library Board archives on Microfilm Reel NL1952.

(xx) Maclaren's illness is mentioned in the *Singapore Free Press and Mercantile Advertiser* (Weekly), 29 September 1910, which is held in the Singapore National Library Board archives on Microfilm Reel NL1952.

(xxi) Bidwell's involvement in the Singapore Sports Club stand and drainage works was mentioned in *The Straits Times*, 16 June 1910, page 7, which is held in the Singapore National Library Board archives on Microfilm Reel NL00326.

(xxii) Bidwell's involvement in the transformation of the old Town Hall to the Victoria Theatre is mentioned in *The Straits Times*, 1 February 1909, page 8, which is held in the Singapore National Library Board archives on Microfilm Reel NL00318.

(xxiii) The renovation of the Gallop Road houses was mentioned in *The Straits Times*, 7 December 2015.

(xxiv) The history of David McLeod Craik's time in Singapore is referenced from the Scottish Architects website: http://www.scottisharchitects.org.uk, as well as the Swan & Maclaren website: http://www.swanmaclaren.com/site/history/

(xxv) Information on the Telecoms Building and about "Glencaird" comes from the National Archives of Singapore Infopedia pages: http://eresources.nlb.gov.sg/infopedia/articles/SIP_179_2005-01-19.html and http://eresources.nlb.gov.sg/infopedia/articles/SIP_1091_2008-12-28.html, as well as from *The Straits Times*, 2 September 2007, which is in the Singapore National Library Board newspaper archives on Microfilm Reel NL28510.

(xxvi) Information on the Chesed-El Synagogue comes from the National Archives of Singapore Infopedia pages: https://eresources.nlb.gov.sg/infopedia/articles/SIP_841_2004-12-17.html

(xxvii) *The Singapore House 1819–1942* by Lee Kip Lin. Edited by Gretchen Liu, designed by Kathleen Lau (published by Times Editions, 1988, reprinted 1995 and Marshall Cavendish Editions and National Library Board, 2015, ISBN 9789814634014).

(xxviii) Bidwell's 1907 break from Swan & Mclaren: http://eresources.nlb.gov.sg/newspapers/Digitised/Article/easterndaily19071125-1.2.12.3

(xxix) The Teutonic Club information is referenced from the *Singapore Free Press and Mercantile Advertiser* (Weekly), 15 December 1898, page 11, which is in the Singapore National Library Board archives on Microfilm Reel NL1937 and *The Straits Times*, 21 May 1917, page 8.

(xxx) His ill health in 1901 is mentioned in *The Straits Times*, 8 July 1901, in the Singapore National Library Board archives on Microfilm Reel NL00286.

(xxxi) Reggie Bidwell is mentioned in *The Straits Times*, 23 January 1918, page 8, which is in the Singapore National Library Board archives on Microfilm Reel NL00439.

(xxxii) The Adelphi Hotel is mentioned in *The Straits Times*, 30 March 1903, page 5, which is in the Singapore National Library Board archives on Microfilm Reel NL00291.

(xxxiii) His marriage to Edith is mentioned in the *Singapore Free Press and Mercantile Advertiser* (Weekly), 17 August 1897, page 1, which is in the Singapore National Library Board archives on Microfilm Reel NL1934.

(xxxiv) Bidwell's involvement in the design of Eu Yan Sang Building 265 to 271 South Bridge Road is mentioned in *The Straits Times*, 22 July 2019.

(xxxv) His home and work address is from the *Singapore Free Press and Mercantile Advertiser* (Weekly), 7 May 1908, page 10, which is in the Singapore National Library Board archives on Microfilm Reel NL1948, and the *Singapore Free Press and Mercantile Advertiser* (Weekly), 3 May 1906, page 6, which is in the Singapore National Library Board archives on Microfilm Reel NL1945.

(xxxvi) The Oranje Hotel is referenced from the Hotel's website (BRIEF-HISTORY 2) (hotel-majapahit.com) and also from Julian Davison's book.

(xxxvii) His attempts to lease out his Dalvey Road property was mentioned in the *Singapore Free Press and Mercantile Advertiser*, 30 January 1913, which is in the Singapore National Library Board archives on Microfilm Reel NL1611.

(xxxviii) The Hotel L'Europe project is mentioned in the *Singapore Free Press and Mercantile Advertiser* (Weekly), 20 August 1908, page 117, which is on Microfilm Reel NL1948.

(xxxix) The St. Joseph Church information was in the *Singapore Free Press and Mercantile Advertiser* (Weekly), 24 August 1904, page 125, which is in the Singapore National Library Board archives on Microfilm Reel NL1943.

(xl) The YMCA Building project was mentioned in the *Singapore Free Press and Mercantile Advertiser*, 27 August 1906, page 4, which is in the Singapore National Library Board archives on Microfilm Reel NL5314.

(xli) Bidwell's involvement as architect of the Crescent Flats at Meyer Road was captured in the *Singapore Free Press and Mercantile*

Advertiser, 23 May 1912, page 330, in the Singapore National Library Board archives on Microfilm Reel NL1953 and also from Singapore Infopedia: https://eresources.nlb.gov.sg/infopedia/articles/SIP_616_2005-01-24.html

(xlii) The 1914 Supreme Court case was reported in the *Malaya Tribune*, 16 July 1914, page 10, which is in the Singapore National Library Board archives on Microfilm Reel NL2089, and his bankruptcy case was reported in *The Straits Times*, 9 January 1915 — page is in the National Archives of Singapore on Microfilm Reel NL00406.

(xliii) Bidwell's death is in the *Singapore Free Press and Mercantile Advertiser*, 9 April 1918, page 10, which is held in the Singapore National Library Board newspaper archives on Microfilm NL1639.

CHAPTERS 4 and 5:

Information for this Chapter was referenced from and cross-checked with the following sources:

(i) R.I.C.S. Year Books for names of individuals in Singapore from 1900 to 1924.

(ii) Information on Edgar Taylor is in the *Singapore Free Press and Mercantile Advertiser* (Weekly), 20 June 1901, page 5, which is in the Singapore National Library Board archives on Microfilm Reel NL1940.

(iii) Information on John Todd Rea comes from the Diary of Irish Architects website: https://www.dia.ie/architects/view/4491/REA-JOHNTODD and from *The Straits Times*, 8 January 1904, page 5, which is held in the Singapore National Library Board archives on Microfilm Reel NL00295.

(iv) Information on Atkin-Berry and Bone is in the *Malaya Tribune*, 1 May 1915, page 1, which is available in the Singapore National Library Board archives on Microfilm Reel NL2093.

(v) Information on H.G. Atkin-Berry's arrival in Singapore was in *The Straits Times*, 7 January 1935, page 10, which is in the Singapore National Library Board archives on Microfilm Reel NL1495.

(vi) House burglary: this is mentioned in the *Singapore Free Press and Mercantile Advertiser*, 6 May 1936, page 3, which is contained in the Singapore National Library Board archives on Microfilm Reel NL3580.

(vii) Information on the death of Herbert and Edith's son is taken from the East India Club Roll of Honour 1939–45.

(viii) Information on Harold Carpmael from Jon Baker's family history records.

(ix) 1910 Address for Carpmael: "The Directory & Chronicle for China, Japan, Corea, Indo-China, Straits Settlements, Malay States, Sian, Netherlands India, Borneo, the Philippines: With which are Incorporated 'The China Directory' and 'The Hong Kong List for the Far East' January 1, 1910, Hong Kong Daily Press Office".

(x) Carpmael's death was recorded in *The Straits Times*, 12 December 1927, page 10, which is held in the Singapore National Library Board archives on Microfilm Reel NL00561.

(xi) Carpmael Road origin is in the book: *What's in the Name? How the Streets and Villages in Singapore Got Their Names* by Ng Yew Peng (published by World Scientific Publishing, 2018, ISBN 9789813221390).

(xii) Carpmael's arrival in Singapore is mentioned in the *Singapore Free Press and Mercantile Advertiser* (Weekly), 4 January 1906, page 6, which is in the Singapore National Library Board archives on Microfilm Reel NL1945; his 1923 Report as Municipal Assessor is in *The Straits Times*, 11 April 1924, page 13, which is available on Microfilm Reel NL00517, and the 1918 landlord false returns is mentioned in the *Singapore Free Press and Mercantile Advertiser*, 14 March 1918, page 8, which is available on Microfilm Reel NL1962.

(xiii) Economy as at 1924–31: *A History of Singapore 1819–1988* (Second Edition) by C.M. Turnbull (published by Oxford University Press, ISBN 0195889436).

(xiv) The Sir Walter Shaw information is referenced from his Wikipedia page.

(xv) Information on the family history of J.A. Lincoln has kindly been provided by one of his descendants, Paul Benedict Lincoln.

(xvi) John Pyne Pennefather information has come from some of his descendants, including Michael Pennefather. Pennefather's appraisers' licence was mentioned in *The Straits Times,* 12 July 1924, which is held in the Singapore National Library Board archives on Microfilm Reel NL520.

(xvii) Pennefather is also mentioned in *Twentieth Century Impressions of British Malaya: Its History, People, Commerce, Industries, and Resources* by Arnold Wright (originally published by Lloyd's Greater Britain Publishing Company, Limited, 1908).

(xviii) Walton information: "The Directory & Chronicle for China, Japan, Corea, Indo-China, Straits Settlements, Malay States, Sian, Netherlands India, Borneo, the Philippines: With which are Incorporated 'The China Directory' and 'The Hong Kong List for the Far East' January 1, 1910, Hong Kong Daily Press Office".

(i) Wilfred Dunstan Nicholas information comes from the National Library Board "Print Heritage" pages and also the following sources: *The Straits Times*, 8 August 1950, page 5, which is in the Singapore National Library Board newspaper archives on Microfilm Reel NL02503, *The Straits Times*, 10 January 1957, page 10, which is available on Microfilm Reel NL01805, and the *Singapore Free Press and Mercantile Advertiser*, 19 January 1914, page 12, which is available on Microfilm Reel NL1618.

(ii) King's Dock information also from *The Straits Times*, 27 August 1913, page 10, which is held in the Singapore National Library Board archives on Microfilm Reel NL374.

(iii) The Neubronner information is taken from *Malaya Tribune*, 1 October 1919, page 6, which is held in the Singapore National Library Board archives on Microfilm Reel NL2121. Further information (including his photograph) are in Arnold Wright's *Twentieth Century Impressions of British Malaya: Its History, People, Commerce, Industries, and Resources*. The photograph is reproduced on the understanding that the copyright is not applicable because the book was published in 1908. Anyone claiming copyright can contact me through the Publishers.

(iv) Maule-ffinch (Mauleffinch) information comes from the following sources: Radio's Who's Who 1947, which is held in the Internet Archive; the family history pages of myheritage.com; the Kuala Lumpur Cenotaph; *The Straits Times*, 19 January 1904, page 3, which is held in the National Archives of Singapore on Microfilm Reel NL00295; *The Straits Times*, 9 July 1906, page 5, which is on Microfilm Reel NL00306; *The Straits Times*, 31 January 1908, page 7 and on Microfilm Reel NL00312; the *Singapore Free Press and Mercantile Advertiser* (Weekly), 14 July 1910, page 7 and on Microfilm Reel NL1951; the *Eastern Daily Mail and Straits Morning Advertiser*, 8 September 1905, page 4 and on Microfilm Reel NL2974; and *The Straits Times*, 26 July 1920, page 10, which is available on Microfilm Reel NL00469.

(xliv) John Taylor information is referenced from the R.I.C.S. Year Books, and also from the Lives of the First World War website: https://livesofthefirstworldwar.org/lifestory/4377928#facts and cross-checked against information in Chapters 24 and 25 of Julian Davison's *Swan & Maclaren — A Story of Singapore Architecture* (published by ORO Editions and The National Archives of Singapore, ISBN 9781935935476).

(xlv) J.A. Lincoln information is from one of his descendants, Paul Benedict Lincoln, and other sources are: The *Singapore Free Press and Mercantile Advertiser*, 17 June 1929, page 8, which is held in the National Archives of Singapore on Microfilm Reel NL1724; *The Straits Times*, 24 October 1911, page 9, which is available on Microfilm Reel NL0035; *The Straits Times*, 14 November 1911, page 8, which is held in the National Archives of Singapore on Microfilm Reel NL00358; "The Directory & Chronicle for China, Japan, Corea, Indo-China, Straits Settlements, Malay States, Sian, Netherlands India, Borneo, the Philippines: With which are Incorporated 'The China Directory' and 'The Hong Kong List for the Far East' 1908" and finally the *Singapore Free Press and Mercantile Advertiser*, 10 July 1914, page 6, which is available on Microfilm Reel NL1621.

(xlvi) George Willis Grey's Hong Kong career as an architect is mentioned on the heritage.hku.hk website, as well as the HKIA website (Journal Issue 45 03). Information on Grey during World War II has been provided by Tony Banham whose website www.hongkongwardiary.com contains a wealth of information on Hong Kong during World War II, and in particular, on the period of the Japanese Occupation.

(xlvii) Information on Grey and his role with Keys and Dowdeswell (as well as the formation of the Singapore Society of Architects with (amongst others) S.J. Edwards and W.J. Brown) is all referenced from the *Singapore Free Press and Mercantile Advertiser*, 6 May 1931, page 10, which is in the Singapore National Library Board archives on Microfilm Reel NL1749. Grey's 1926 job position (and his FSI status) is mentioned in *The Straits Times*, 3 September 1926, page 8, which is held in the Singapore National Library Board archives on Microfilm Reel NL 00546.

(xlviii) Bernard Dyke's first surveying job was announced in the *London Gazette*, 4 April 1911, page 2712, and his arrival in Singapore is mentioned in *The Straits Times*, 27 October 1923, page 8, which is held in the Singapore National Library Board archives on Microfilm Reel NL00511. The Naval Base project is mentioned in (inter alia) *Pacific Affairs* **5**(4)306–318, 1932.

(xlix) Information on Sidney James Edwards comes from the following sources: *The Straits Times*, 29 January 1929, page 8, held in the Singapore National Library Board archives on Microfilm Reel NL00574, the *Singapore Free Press and Mercantile Advertiser*, 8 May 1916, page 10, held in the Singapore National Library Board archives on Microfilm Reel NL1631, the *Malaya Tribune*, 12 September 1923, page 5, held in the Singapore National Library Board archives on Microfilm Reel NL3905, the *Malaya Tribune*, 11 March 1921, page 5, available on Microfilm Reel NL3891, and the *Malaya Tribune*, 15 November 1921, page 4, available on Microfilm Reel NL 3890.

(l) Information on Colombo Town Hall and the involvement of Edwards come from online Sri Lankan Government sources and is also mentioned in *The Straits Times*, 25 November 1922, page 10, held in the Singapore National Library Board archives on Microfilm Reel NL00500.

(li) The Booty & Edwards transformation to BEP Akitek comes from their website: http://www.bepakitek.com/about

(lii) Information on Walter James Brown's involvement in the City Hall project is taken from the *Dictionary of Irish Architects 1720–1940* website (with help from Eve McAulay who runs the website), the *Singapore Free Press and Mercantile Advertiser* (Weekly), 13 May 1925, page 1, held in the Singapore National Library Board archives on Microfilm Reel NL1986, and the *Malaya Tribune* 16 February 1924, page 8, Microfilm Reel NL3907.

(liii) J.R. Caseley information has been sourced from the annual Straits Settlements Blue Book for 1924 (copy in possession of Stanford University) and the *London Gazette*, 1 January 1949, page 11, as well as the *Malaya Tribune*, 25 April 1931, page 8, which is held in the Singapore National Library Board archives on Microfilm Number NL3955 and *The Straits Times*, 23 June 1930, page 12, which is held on Microfilm No. NL 00636.

(liv) The BMBC Caldecott story is in the *Malaya Tribune*, 2 March 1937, page 12, which is in the Singapore National Library Board archives on Microfilm Reel NL4055.

(lv) The story of the Caseley family and their escape from Singapore was captured by East Sussex Libraries and is held in the BBC archive: https://www.bbc.co.uk/history/ww2peopleswar/stories/28/a7971528.shtml. Further information is referenced from the *Singapore Free Press and Mercantile Advertiser* (Weekly), 19 October 1927, page 250, which is in the Singapore National Library Board archives on Microfilm Reel NL1991, the *Singapore Free Press and Mercantile Advertiser* (Weekly), 19 October 1927, page 250, on Microfilm Reel NL1991, the *Malaya Tribune*, 19 April 1928, page 10, on Microfilm Reel NL3932, *The Straits Times*, 24 October 1928, page 11, on Microfilm Reel NL00571, and finally in the Singapore Government Infopedia page: http://eresources.nlb.gov.sg/infopedia/articles/SIP_921_2004-12-10.html

(lvi) Information on H.C. Atkin-Berry comes from the sources listed below, as well as supplemental information that is referenced from Julian Davison's *Swan & Maclaren — A Story of Singapore Architecture*. Other sources are: The *Singapore Free Press and Mercantile Advertiser*, 3 May 1932, page 7, which is in the Singapore National Library Board archives on Microfilm Reel NL 1759, *The Straits Times*, 7 September 1937, page 12, on Microfilm Reel NL01527, *The Straits Times*, 2 June 1931, page 10, on Microfilm Reel NL00648, *The Straits Times*, 16 October 1939, page 10, on Microfilm Reel NL01786, Singapore Infopedia page: http://eresources.nlb.gov.sg/infopedia/articles/SIP_954_2005-01-10.html, and, finally, the *Malaya Tribune*, 20 October 1939, page 2, which is held in the Singapore National Library Board archives on Microfilm Reel NL2123.

(lvii) Atkin-Berry as a licensed appraiser is mentioned in the *Malaya Tribune*, 13 February 1932, page 8, which is held in the Singapore

National Library Board archives on Microfilm Reel NL3961 and his residence is mentioned in the *Singapore Free Press and Mercantile Advertiser*, 28 March 1935, page 2, on Microfilm Reel NL3566.

(lviii) The Supreme Court case is in the *Malaya Tribune*, 19 November 1930, page 10, which is in the Singapore National Library Board archives on Microfilm Reel NL3952; also the *Singapore Free Press and Mercantile Advertiser*, 21 November 1930, page 12, on Microfilm Reel NL1745, and, finally, *The Straits Times*, 28 November 1930, page 20, on Microfilm Reel NL00641.

(lix) Henry Gordon Atkin-Berry's death is reported in: www.roll-of-honour.org.uk/civilians/html/a_database_38.htm

(lx) St. Patrick's School is mentioned in *The Straits Times*, 5 April 1932, page 5, and is available on Microfilm Reel NL01444.

(lxi) Majestic Theatre is mentioned in *The Straits Times*, 11 July 1929, page 12, and is available on Microfilm Reel NL00625.

(lxii) Nunes and Medeiros Buildings are mentioned in *The Straits Times*, 5 October 1937, page 4, and are available on Microfilm Reel NL04150.

(lxiii) Union Building and Hong Kong Bank Building: in the *Malaya Tribune*, 15 November 1921, page 4, on Microfilm Reel NL3890, *The Straits Times*, 29 December 1924, page 9, on Microfilm Reel NL00525, and the *Singapore Free Press and Mercantile Advertiser* (Weekly), 30 September 1925, page 11, on Microfilm Reel NL1989.

(lxiv) Presbyterian Church, Prinsep Street: *Malaya Tribune*, 6 March 1930, page 7, and is available on Microfilm Reel NL3947.

(lxv) Chinese Methodist Church: *Malayan Saturday Post*, 2 February 1924, page 24, and is available on Microfilm Reel NL2253.

(lxvi) Seaman's Chapel: *Sunday Tribune* (Singapore), 10 July 1938, page 13, and is available on Microfilm Reel NL1441.

(lxvii) Lavender Street shophouses: *Morning Tribune*, 25 July 1936, page 4, and is available on Microfilm Reel NL2266.

(lxviii) *Gender Imperialism and Global Exchanges*, edited by Stephan F. Miescher, Michelle Mitchell and Naoko Shibusawa (published by Wiley Blackwell, 2015, ISBN 9781119052203), is the reference point for Cullen's scouting views.

(lxix) Cullen references in *Out in the Midday Sun — the British in Malaya 1880–1960* by Margaret Shennan (published by John Murray (Publishers) Ltd, 2000, and Monsoon Books Pte Ltd, 2015, ISBN 9789814625319 (Paperback) and 9789814625326 (e-book)).

(lxx) Chandos Eric Bone salary: information from the Straits Settlements Blue Book 1937 — copy in possession of Stanford University, and he is in the list of Licensed Surveyors in the *Victoria Government Gazette*, 14 January 1925 and 30 January 1962. He is also listed in the *London Gazette*, 4 April 1924, page 2801, and his story is captured in the Honour Roll 1914–18 published by the Department of Crown Lands and Survey & Office of Titles Publisher Department of Environment, Land, Water and Planning Published Victoria State Government, 2020 (Author: Rebecca Russell, Geographic Names Victoria, Land Use Victoria, Department of Environment, Land, Water and Planning).

(lxxi) Some of the Quarmby and Niven information comes from the Malayan Volunteers Group website and from *The Straits Times*, 17 September 1970, page 15, which is in the Singapore National Library Board archives on Microfilm Reel NL06609.

(lxxii) J.T. Turner information is taken from the *London Gazette*, 8 September 1903, page 5602. His wife's involvement in the Singapore Amateur Wireless Society and the Society's role is taken from (inter alia) the *Singapore Free Press and Mercantile Advertiser* (Weekly), 31 March 1926, page 201, which is held in the Singapore National Library Board newspaper archives on Microfilm Reel NL1987, and also the National Library Board website: http://www.nlb.gov.sg/biblioasia/2016/04/05/the-story-of-singapore-radio-1924-41/

(lxxiii) Frederick Leslie Forsaith information is from Frederick's descendants, David and Greg Forsaith as well as the following newspaper references: The *Singapore Free Press and Mercantile Advertiser*, 6 August 1930, page 83, which is available on Microfilm Reel NL1992 and the *Singapore Free Press and Mercantile Advertiser*, 12 March 1930, page 172, which is available on Microfilm Reel NL1992.

(lxxiv) Philip Roy Pank is mentioned in the *Malaya Tribune*, 26 August 1929, page 10, which is in the Singapore National Library Board archives on Microfilm Reel NL3943, the *Singapore Free Press and Mercantile Advertiser*, 26 June 1930, page 20, on Microfilm Reel NL1741, the *Singapore Free Press and Mercantile Advertiser*, 26 April 1930, page 20, on Microfilm Reel NL1739, and the *Singapore Free Press and Mercantile Advertiser*, 11 July 1929, page 16, on Microfilm Reel NL1724, and Philip Pank and Company's website.

(lxxv) Finally, some information in this Chapter about Eurasian families is referenced from *On Parade: Straits Settlements Eurasian Men who Volunteered to Defend the Empire 1862–1957* by Mary Anne Jansen, John Geno-Oehlers and Ann Ebert Oehlers (published by the Wee Kim Wee Centre, SMU, ISBN 987 9811187292).

CHAPTER 6:

The following sources were used for the information on L. Langdon Williams:

(i) *A Genealogy of Tropical Architecture, Colonial Networks, Nature and Technoscience* by Jiat-Hwee Chang (published by Routledge, 2016, ISBN-13 9780415840774).

(ii) "Habitation and the invention of a nation, Singapore 1936–1979" by Eunice Seng — a piece of work submitted in partial fulfilment of the requirements for the degree of Doctor of Philosophy under the Executive Committee of the Graduate School of Arts and Sciences.

(iii) The 1933 list of houses declared uninhabitable was in *The Straits Times*, 21 December 1933, page 4, which is contained in the Singapore National Library Board archives on Microfilm Reel NL1482.

(iv) The 1924 information on Williams is in Singapore Infopedia: http://eresources.nlb.gov.sg/infopedia/articles/SIP_2016-03-14_142655.html as well as in *The Straits Times*, 9 May 1924, Page 8, which is in the Singapore National Library Board archives on Microfilm Reel NL00518 and also *The Straits Times*, 3 March 1924, page 11, on Microfilm Reel NL00516.

(v) The report on the Court upholding the SIT appeal was reported in the *Singapore Free Press and Mercantile Advertiser*, 2 August 1935, page 6, and is held in the Singapore National Library Board archives on Microfilm Reel NL3568, and the 1936 Municipal Report information comes from the *Singapore Free Press and Mercantile Advertiser* (Weekly), 15 June 1927, page 37, and is held in the Singapore National Library Board archives on Microfilm Reel NL1991.

(vi) Langdon Williams becoming a Colonial Fellow of the Institution is mentioned in the *Malaya Tribune*, 3 December 1934, page 10, in the Singapore National Library Board archives on Microfilm Reel NL4039. His marriage in Singapore to Marion Galloway Black is mentioned in the *Singapore Free Press and Mercantile Advertiser* (Weekly), 17 August 1897, page 1, on Microfilm Reel NL1934.

(vii) *Singapore Struggle for Success* by John Drysdale. First published in 1984, reprinted in 1996 and a facsimile was printed by Marshall Cavendish (Asia) Pte Ltd in 2009.

(viii) The house on Pulau Ubin information is mentioned in various sources, including the National Parks Board website and a *Straits Times* article, 5 November 2016, as well as the *Sunday Tribune* (Singapore), 14 April 1935, page 8, which is in the Singapore National Library Board archives on Microfilm Reel NL1452, and the house off Singapore Golf Course Road was mentioned in *The Straits Times*, 6 March 1938, page 32, in the Singapore National Library Board archives on Microfilm Reel NL02368.

(ix) His holidays were covered in the *Morning Tribune*, 27 August 1938, page 2, on Microfilm Reel NL2312, and the *Singapore Free Press and Mercantile Advertiser*, 27 August 1938, page 3, on Microfilm Reel NL3596.

(x) The *Singapore Free Press* article on "Building Problems", July 27 1928, is a Digitised Article held in the Singapore National Library Board archives, and the Municipal Commissioners' 1928 report is reported in the *Singapore Free Press and Mercantile Advertiser* (Weekly), 2 October 1929, page 8, which is held in the Singapore National Library Board archives on Microfilm Reel 1990.

(xi) The Weisberg Report information comes from the final year dissertation entitled "Slums, squatters and urban redevelopment schemes in Bombay, Hong Kong, and Singapore, 1894–1960", written by Michael William Sugarman for Magdalene College in November 2017 for the degree of Doctor of Philosophy.

(xii) His Kuching trip was covered in the *Malaya Tribune*, 23 November 1940, page 3, which is available on Microfilm Reel NL2131.

(xiii) His 1931 SIT appointment was referenced from the *Malaya Tribune*, 26 January 1931, page 8, which is available on Microfilm Reel NL3953.

(xiv) His report to SIT was covered in *The Straits Times*, 29 April 1936, page 12, which is available on Microfilm Reel NL01510.

(xv) His funeral is mentioned in the *Singapore Free Press and Mercantile Advertiser*, 27 October 1941, page 5, which is available in the National Library Board archives on Microfilm Reel NL3614.

(xvi) The Empress Place plans are referenced from *The Straits Times*, 8 November 1930, Page 10, which is in the Singapore National Library Board archives on Microfilm Reel NL00641 and *The Straits*

Times, 25 August 1935, page 3, which is available on Microfilm Reel NL02361.

(xvii) Marion Williams' death is captured in the Roll of Honour for those who died in Japanese captivity kept by survivors and their descendants as well as descendants of those who died: https://www.encyclopedia.com/women/encyclopedias-almanacs-transcripts-and-maps/women-pows-sumatra-1942-1945

(xviii) The funding of SIT is mentioned in the *Singapore Free Press and Mercantile Advertiser*, 9 December 1936, page 11, which is available on Microfilm Reel NL3590, and the marriage of Jim Fraser to Alice Ross is mentioned in the *Singapore Free Press and Mercantile Advertiser* (Weekly), 9 January 1929, Page 5, on Microfilm Reel NL1989.

CHAPTER 7

The core information regarding the development of Singapore's Naval and RAF bases comes the Singapore Government Infopedia sources. In particular, the following Infopedia sources were referenced:
http://eresources.nlb.gov.sg/infopedia/articles/SIP_2013-07-19_113523.html
http://eresources.nlb.gov.sg/infopedia/articles/SIP_1771_2011-02-28.html
http://eresources.nlb.gov.sg/infopedia/articles/SIP_1771_2011-02-28.html

(i) Other general information is referenced from Chapter 4 of C.M. Turnbull's *A History of Singapore 1819–1988* (Second Edition) (published by Oxford University Press, ISBN 0195889436).

(ii) The Hunt information comes from the *Singapore Free Press and Mercantile Advertiser*, 29 July 1937, page 2, which is available on Microfilm Reel NL3591, and from *The Straits Times*, 20 August 1935, page 13, on Microfilm Reel NL01502.

(iii) The Watson christening is mentioned in the *Singapore Free Press and Mercantile Advertiser*, 25 September 1936, in the Singapore National Library Board archives.

(iv) Waters' travels and Hong Kong business is mentioned in *The Straits Times*, 4 April 1937, page 3, which is in the Singapore National Library Board archives on Microfilm Reel NL02366.

(v) Waters' 1937 Rotary Club talk: *The Straits Times*, 11 November 1937, page 12, which is available on Microfilm Reel NL04151 and the *Malaya Tribune*, 11 November 1937, page 13, which is on Microfilm Reel NL406.

(vi) Tovey's Royal Engineers record is taken from the ww2talk website and also from the National Library of Scotland's British Military Lists. Tovey's MBE award was reported in the *Malaya Tribune*, 29 October 1935, page 17, which is in the Singapore NLB archives on Microfilm Reel NL4048.

(vii) Honey's 1936 sailing incident is recorded in the *Singapore Free Press and Mercantile Advertiser*, 19 April 1937, page 3, which is available on Microfilm Reel NL3582, and *The Straits Times*, 6 April 1936, page 13, on Microfilm Reel NL01510.

(viii) The formation of R.I.C.S. Malaya Branch between 1935 and 1936 is referenced from the following newspaper articles in the Singapore National Library Board archives: *The Straits Times*, 12 February 1936, page 11, on Microfilm Reel NL01508, the *Singapore Free Press and Mercantile Advertiser*, 25 January 1935, page 8, on Microfilm Reel NL3513, the *Singapore Free Press and Mercantile Advertiser*, 7 February 1936, page 6, on Microfilm Reel NL3576, and *The Straits Times*, 11 June 1936, page 13, on Microfilm Reel NL01512.

(ix) 1936 Waters lecture to the Malaya Branch and the response to it: *The Straits Times*, 23 April 1936, page 16: article available at the Singapore National Library Board archives on microfilm reel NL1510 and Lee Kong Chian Reference Library. Also, the *Malaya Tribune*, 22 April 1936, page 9, article available at the Singapore National Library Board archives on microfilm reel NL3980, the *Malaya Tribune*, 24 April 1936, page 12, on microfilm reel NL3980.

(x) Waters and Watson information comes from *Quantifying Asia: The History of Langdon & Seah*, compiled by John Peacock (published by Langdon & Seah Asia Company Ltd, Hong Kong, ISBN 9789881367808).

(xi) 1937 "Second annual dinner": *The Straits Times*, 11 August 1937, page 17 — this article is available at the Singapore National Library Board archives on microfilm reel NL1526 and Lee Kong Chian Reference Library.

(xii) 1938 Annual Dinner: article in the *Singapore Free Press and Mercantile Advertiser*, 16 June 1938, page 5; this article is available at the Singapore National Library Board archives on microfilm reel NL3584.

(xiii) 1939 Annual Dinner: article in the *Singapore Free Press and Mercantile Advertiser*, 23 June 1939, page 7; this article is available at the Singapore National Library Board archives on microfilm reel NL3602.

(xiv) The information on Cheeseman comes from the following sources: http://westlandsprimary.blogspot.com/2018/04/harold-cheeseman.html and http://sumitrakanfk.com/edup3013/t3-development-of-education-in-malaysia/cheeseman-report/

(xv) Some information on Fred Snell has been referenced from *Out in the Midday Sun — the British in Malaya 1880–1960* by Margaret Shennan (published by John Murray (Publishers) Ltd, 2000, and Monsoon Books Pte Ltd, 2015, ISBN 9789814625319 (paperback) and 9789814625326 (e-book)), as well as from Gordon Snell and his publishers (The O'Brien Press, Dublin).

(xvi) Snell's football activity with the Loyals and about the Anson Road Stadium is mentioned in *The Straits Times*, 6 June 1939, page 18, which is available on Microfilm Reel NL04158 and *The Straits Times*, 25 June 1939, page 8, on Microfilm Reel NL02370 and Road Stadium, as well as the Singapore Government Infopedia (http://eresources.nlb.gov.sg/infopedia/articles/SIP_2016-09-29_160413.html).

(xvii) 1931 to 1937 R.I.C.S. Year Books were referenced in the RICS Library in Great George Street, London.

(xviii) Jones-Davies and Smith, Johor Government offices project (and Jones-Davies' temporary move to Johor) is mentioned in the *Singapore Free Press and Mercantile Advertiser*, 10 June 1939, page 8, available in the Singapore National Library Board archives. Article also available on Microfilm Reel NL3602, in the *Sunday Tribune* (Singapore), 3 September 1939, page 10, which is on Microfilm Reel NL1457, *The Straits Times*, 29 February 1940, page 2, on Microfilm Reel NL 01789, and finally the *Singapore Free Press and Mercantile Advertiser*, 1 March 1941, page 5.

(xix) Major Percy stories are from *The Straits Times*, 15 November 1937, page 13, in the Singapore National Library Board archives — the article is also available on Microfilm Reel NL4151 and from *British Civilians and the Japanese War in Malaya and Singapore, 1941–45* by Joseph Kennedy (published by The MacMillan Press Limited, 1987, ISBN 987033416037).

(xx) Old Supreme Court building information comes from (http://eresources.nlb.gov.sg/infopedia/articles/SIP_774_2005-01-10.html) on the Singapore Government Infopedia website, as well as *The Straits Times*, 19 January 1937, page 12, in the Singapore National Library archives on Microfilm Reel NL01519, and also *The Straits Times*, 4 April 1937, page 2, on Microfilm Reel NL 02366.

(xxi) Niven, Percy, Harris and Ward salaries: information from the Straits Settlements Blue Books 1937 and 1938 — copies in possession of Stanford University.

(xxii) Niven's escape story comes from the Mills family, from the Australian War Memorial website's article on Bill Reynolds, MBE, and also from the Malayan Volunteers Group website: https://www.malayanvolunteersgroup.org.uk/singapore-evacuation-ships.html

(xxiii) Polglase: http://www.rafcommands.com/archive/17628.php

(xxiv) 1939–40 Institution's Year Book (available in the RICS archives) has been referenced for the 1940 Branch Committee and other names, e.g., Quarmby. Quarmby's appointment to the Raffles Institution is mentioned in the *Singapore Free Press and Mercantile Advertiser* (Weekly), 19 January 1922, page 40, which is held in the Singapore National Library Board archives on Microfilm Reel NL1967 as well as the Straits Settlements Blue Book 1921 — copy in possession of Stanford University.

(xxv) Some of the birth, marriage and death dates for surveyors working in Singapore up to 1940 have been referenced from the International Genealogical Index (IGI) provided through International Genealogical Index (IGI).

CHAPTER 8:

(i) Information on J.B. Maschoieff (later Marsh) is from his son Charles Marsh as well as from interviews conducted with Joe by Sarah Semple, and with additional family history information provided by Jules Feldman, Alain Farhi and Rebecca Rott. I owe all of them a great deal of gratitude.

(ii) Information on Probationers in 1940 comes from the Institution's Year Book 1940 located in the RICS' archives in London.

(iii) The 1940 *Straits Times* article on their qualification was on 6 October 1940, page 5 — this article is in the National Archives of Singapore on Microfilm Reel NL2416.

(iv) The name change from Maschoieff to Marsh was mentioned in the *London Gazette*, 9 June 1944, page 2774.

(v) Some information about Eurasian families in this Chapter comes from the Muntok Peace Museum website: http://muntokpeacemuseum.org and also from *On Parade: Straits Settlements Eurasian Men Who*

Volunteered to Defend the Empire 1862–1957 by Mary Anne Jansen, John Geno-Oehlers and Ann Ebert Oehlers (published by the Wee Kim Wee Centre, SMU, 2018, ISBN 9879811187292).

(vi) The information relating to L. Nazareth comes from the *Singapore Standard*, 4 September 1953, page 2, which is in the Singapore National Library Board archives on Microfilm Reel NL2300.

(vii) John Peacock's *Quantifying Asia: The History of Langdon & Seah* provided some of information contained in this Chapter in relation to Seah Mong Hee and J.B. Maschoieff, backed up by discussions with John Peacock and Seah Choo Meng.

CHAPTER 9:

(i) The Job Advertisement was in the *Singapore Free Press and Mercantile Advertiser*, 2 August 1941, page 9, which is held in the Singapore National Library Board archives on Microfilm Reel NL3610.

(ii) The Captain Loveday trial information comes from numerous newspaper articles held in the Singapore National Library Board archives. Those referenced for this book are as follows:

- *Morning Tribune*, 5 September 1940, page 10, on Microfilm Reel NL2321.
- *The Straits Times*, 5 September 1940, page 11
- *The Straits Times*, 6 February 1941, page 10, on Microfilm Reel NL01802
- *Malaya Tribune*, 22 November 1940, page 3, on Microfilm Reel NL2131
- *The Straits Times*, 2 September 1940, page 4
- *Malaya Tribune*, 23 August 1940, page 5, on Microfilm Reel NL2130

- The *Singapore Free Press and Mercantile Advertiser*, 6 August 1940, page 5, on Microfilm Reel NL3608
- *The Straits Times*, 21 August 1940, page 16, on Microfilm Reel NL01796
- *The Straits Times*, 8 October 1940, page 10, on Microfilm Reel NL01798
- *Malaya Tribune*, 6 August 1936, page 5, on Microfilm Reel NL 3982
- *The Straits Times*, 11 June 1928, page 8, on Microfilm Reel NL00567
- The *Singapore Free Press and Mercantile Advertiser*, 28 June 1941, page 7, on Microfilm Reel NL3614
- *Malaya Tribune*, 8 August 1940, page 3, on Microfilm Reel NL2129
- *The Straits Times*, 4 May 1940, page 2, on Microfilm Reel NL 04159
- *The Straits Times*, 22 August 1940, page 11, on Microfilm Reel NL01796

(iii) Loveday's promotion to Lt. (Surveyors of Works) is mentioned in the December 1930 *London Gazette*.

(iv) Mrs. E. Loveday is mentioned on page 11, Miscellaneous Column 1 of the *Morning Tribune*, 8 September 1939, page 11, with the article being available in the Singapore National Library Board archives on Microfilm Reel NL2326, and in the *Malaya Tribune*, 23 September 1937, page 2, on Microfilm Reel NL4061.

(v) The secret microphone story comes from *The Straits Times*, 8 August 1940, page 16, which is held in the Singapore National Library Board archives on Microfilm Reel NL1796.

(vi) The Inquiry is featured in the *Malaya Tribune*, 4 July 1941, page 11, in the Singapore National Library Board archives on Microfilm Reel NL2135.

(vii) Croft's evidence was in *The Straits Times*, 27 August 1940, page 16, in the Singapore National Library Board archives on Microfilm Reel NL1796.

(viii) Soh Hun Swee's discharge is mentioned in the *Malaya Tribune*, 22 May 1940, page 2, which is held in the Singapore National Archive on Microfilm Reel NL2128.

(ix) The information on Dodd's conviction comes from *The Straits Times*, 21 January 1934, page 12, which is in the Singapore National Library Board archives on microfilm under NL 02358.

(x) Soh Hun Swee's evidence comes from the *Morning Tribune*, 9 August 1940, page 6, which is held in the Singapore National Library Board archives on Microfilm Reel NL2321.

(xi) Loveday's time in internment during World War II was confirmed by Ron Taylor whose excellent website (www.fepow.family) provides a wealth of information regarding Prisoners of War and internees, particularly in South-East Asia.

(xii) Information on Brigadier Pennycuick comes from the website: https://oddlysmallbridge.wordpress.com/2017/10/06/researching-the-young-royal-engineers-officers/ and from the *Singapore Free Press and Mercantile Advertiser*, 28 June 1941, page 7, as contained in the Singapore National Library Board archives on Microfilm Reel NL3614.

(xiii) The establishment of the Special Anti-Corruption Branch in Singapore is in the *Singapore Free Press and Mercantile Advertiser*, 2 April 1941, which is held in the Singapore National Library Board archives on Microfilm Reel 3613, and the Fowler information is in *The Straits Times*, 13 November 1950, page 1, held in the Singapore National Library Board archives on Microfilm Reel NL02505.

(xiv) The information on Henry Moore comes from: *Morning Tribune*, 2 December 1941, page 11, which is held in the Singapore National Library Board newspaper archives on Microfilm Reel NL2316 and the *Malaya Tribune*, 18 March 1941, page 3, on Microfilm Reel NL2133.

(xv) Information on Tom Kitching comes from *Life and Death in Changi: The Diary of Tom Kitching who Died in Japanese Hands in Singapore in 1944* by Brian Kitching (published by William Culross & Son Ltd., Perth and Scotland, 1998, ISBN 1873891350), and the Crickmay information comes from his letter to the RICS and his diaries, which are part of the World War II collections of the Imperial War Museum, London.

(xvi) Information on G.H. Sworder comes from the book on Brian Kitching's book (above) and also from Joseph Kennedy's *British Civilians and the Japanese War in Malaya and Singapore, 1941–45* (published by The MacMillan Press Limited, 1987, ISBN 987033416037), and a story in *The Straits Times*, 24 January 1946, page 3, which is contained in the Singapore National Library Board archives on Microfilm Reel NL04859.

(xvii) Jones-Davies' move from Singapore to the United States is contained in the Ellis Island and Other New York Passenger Lists, 1820–1957, which are on public record.

(xviii) Information on the deaths of both Waters and Cullen during World War II are contained in the Commonwealth War Graves Commission website: https://www.cwgc.org/find-war-dead, as well as in Margaret Shennan's *Out in the Midday Sun — the British in Malaya 1880–1960* (published by John Murray (Publishers) Ltd, 2000, and Monsoon Books Pte Ltd, 2015, ISBN 9789814625319 (paperback) and 9789814625326 (e-book)).

(xix) The Aeria family story in World War II is also told in the Muntok Peace Museum website: http://muntokpeacemuseum.org.

CHAPTER 10:

(i) The R.I.C.S. Malaya 1947 General Meeting information is referenced from *The Singapore Free Press*, 27 February 1947, page 5, which is contained in the Singapore National Library Board archives on Microfilm Reel NL3656

(ii) 1947 Royal Order: taken from *The Straits Times*, 17 August 1947, page 3, which is contained in the Singapore National Library Board archives on Microfilm Reel NL2419.

(iii) The J.W. Boyd-Walker position with Harper & Co. is from: "The Directory & Chronicle for China, Japan, Corea, Indo-China, Straits Settlements, Malay States, Sian, Netherlands India, Borneo, the Philippines, &c: With which are Incorporated 'The China Directory' and 'The Hong Kong List for the Far East' January 1, 1910, Hong Kong daily Press office".

(iv) 1948 student entry level for R.I.C.S. is taken from *The Straits Times*, 22 May 1948, page 7, held in the Singapore National Library Board archives on Microfilm Reel NL5228.

(v) 1948 Branch AGM: *The Straits Times*, 11 April 1948, page 5, which is in the Singapore National Library archives on Microfilm Reel NL2420.

(vi) 1949 examinations announced in the *Morning Tribune*, 9 June 1948, page 3, held in the Singapore National Library Board archives on Microfilm Reel 2347.

(vii) 1950 A.G.M of the Malaya Branch featured in *The Straits Times*, 22 April 1950, page 5, held in the Singapore National Library Board archives on Microfilm Reel NL2501.

(viii) Harry L. Ward's information comes from *The Australian Surveyor*, December 1982, Volume 31, No. 4 (also Volume 42, March 1997, Issue 1), with the kind permission of the Surveying & Spatial Sciences Institute (SSSI) in Deakin, ACT.

(ix) C.G. Burt's OBE award is from the Singapore Volunteer Corps website: http://sgvolunteerscorps.blogspot.com/2013_03_10_archive.html

(x) The FARELF compensation work is mentioned in the *Singapore Free Press*, 3 January 1948, page 5, which is available on Microfilm Reel NL3665.

(xi) Frank Angell's marriage was reported in *The Straits Times*, 12 July 1948, page 3, which is held in the Singapore National Library Board archives on Microfilm Reel NL5337.

(xii) Himely's family background is available in The Peerage.

(xiii) The visit of Horace W. Langdon to Singapore in 1948 was reported by the *Malaya Tribune*, 5 February 1948, page 8, which is available in the Singapore National Library Board archives on Microfilm Reel NL2144.

(xiv) Eric Watson's death was recorded in *The Straits Times*, 6 February 1956, page 6, which is in the Singapore National Library Board archives on Microfilm Reel NL1813. His memorial service is mentioned in *The Straits Times*, 27 February 1956, page 6, which is held in the Singapore National Library Board's archives on Microfilm Reel NL1814.

(xv) John Alan Kershaw information is the copyright of Susan Wilson: http://www.susansite.org/wilson/kershaw_wilson.php

(xvi) Dick Holmes information and the photograph of Langdon & Every, circa 1955, is reproduced with the kind permission of the editorial staff of the Building Journal (building.com.hk) whose special edition of August 2009 focused on the history of the Davis Langdon & Seah practice.

(xvii) The Singapore Badminton Hall Committee is mentioned in the *Singapore Standard*, 2 October 1951, page 11, on microfilm reel NL2293.

(xviii) Dick Holmes' plea for more quantity surveyors was featured in the *Singapore Free Press*, 15 May 1961, Page 3, contained within the Singapore National Library archives on Microfilm Reel NL2519.

(xix) The 1947 to 1958 details of qualified R.I.C.S. professionals in Singapore are referenced from the Institution's Year Books in the RICS archive in HQ, London.

(xx) Robert Neil's soccer exploits featured in the *Morning Tribune*, 26 September 1947, page 19, which is held in the Singapore National Library archives on Microfilm Reel NL2329. There are other references to his football exploits in the Singapore National Library Board archives.

(xxi) Ronald Ricketts' cricket exploits are in the *Sunday Tribune* (Singapore), 29 February 1948, page 8, which is in the Singapore National Library Board archives on Microfilm Reel NL1465. There are other references to his sporting exploits in the Singapore National Library archives.

(xxii) Basil Ernest Thompson's role as Chief Engineer is mentioned in *The Straits Times*, 28 June 1951, page 5, which is held in the Singapore National Library Board archives on Microfilm Reel NL02508. His departure in 1958 is mentioned in the *Singapore Free Press*, 5 June 1958 and is on Microfilm Reel NL2450, and his promotion to City Engineer was recorded in *The Straits Times*, 26 March 1956, which is on Microfilm Reel NL01814, and finally his departure is mentioned in *The Singapore Free Press*, 18 March 1958, page 1, which is available on Microfilm Reel NL2449.

CHAPTERS 11 and 12:

(i) Gordon Lowden is mentioned in *The Straits Times*, 7 March 1957, page 5, which is held in the Singapore National Library Board archives on Microfilm Reel NL1806, and also in the *Singapore*

Free Press, 23 September 1955, page 12, which is also held in the Singapore National Library Board archives on Microfilm Reel NL2444. The story of Gordon Lowden's experience on the *Rakuyō Maru* in 1944 (and his photograph) comes from the website (https://www.roll-of-honour.org.uk) managed by Ron Taylor. The "Roll of Honour" website is dedicated to the memories of the Far East Prisoners of War (FEPOW).

(ii) *Quantifying Asia: The History of Langdon & Seah*, compiled by John Peacock, provided some of the information in this Chapter on the Horace W. Langdon & Every story of this time.

(iii) The information about Barnett and Komlosy's involvement in the Town Planning Institute's Malayan Branch comes from *The Straits Times*, 18 January 1952, page 4, which is held in the National Archives of Singapore on Microfilm Reel NL02628.

(iv) Greig's CBE award is mentioned in the *London Gazette*, 8 June 1963, page 4813 and his War experience stories are kept in the Imperial War Museum in London: https://www.iwm.org.uk/collections/item/object/1030013407

(v) The 1951 R.I.C.S. Branch is mentioned in the *Singapore Free Press*, 16 April 1951, page 5, which is held in the Singapore National Library Board archives on Microfilm Reel NL3655 in their archives.

(vi) The story of Dr. Pearcy is taken from the following sources:

https://www.washingtonpost.com/archive/local/1980/07/10/dr-g-etzel-pearcy-dies/71b828a6-231d-49be-bde2-ca255d8b2072/?utm_term=.d880aad0ffcd and

https://www.nytimes.com/1974/04/14/archives/notes-sky-shuttle-to-londonwhen-here-and-there-just-38-states.html

(vii) H.L. Ward's retirement is in the *Singapore Standard*, 29 December 1952, which is held in the Singapore National Library Board archives on Microfilm Reel NL2284.

(viii) Scholarships from the City Assessors' Department were reported in *The Straits Times*, 25 February 1954, page 8 — this article is available in the Singapore National Library Board archives on Microfilm Reel NL 1766.

(ix) The Assistant Assessor's salary and job advertisement is in *The Straits Times*, 13 December 1951, page 2, is held in the Singapore National Library Board archives on Microfilm Reel NL2627.

(x) The Five Bright Girls article is in *The Straits Times*, 1 September 1950, page 10 — this is held in the Singapore National Library Board archives on Microfilm Reel NL2504.

(xi) Letter to *The Straits Times* on "Asians afraid of getting dirty", 27 February 1954, page 12, is in the Singapore National Library Board archives on Microfilm Reel NL1766.

(xii) Job Advertisement from the Singapore City Council: *The Straits Times*, 17 January 1955, page 6, which is available in the Singapore National Library Board archives and on Microfilm Reel NL1773.

(xiii) City Council, the new Mayor, and information about Francis Thomas: information from pages 160–80 in *Men in White: The Untold Story of Singapore's Ruling Political Party* by Sonny Yap, Richard Lim and Leong Weng Kam (published by Singapore Press Holdings, 2009, ISBN 9789814266246).

(xiv) Stevenson and other assessors resigning: *The Straits Times*, 12 November 1958, page 7, which is in the Singapore National Library Board archives on Microfilm Reel NL1803.

(xv) The information on James Khoo Cheok Chuan comes from interviews with people who knew him, as well as from the *Singapore Free Press*, 6 June 1960, page 7.

(xvi) Sanderson's arrival in Singapore was reported in *The Straits Times*, 12 October 1956, page 14, as contained in the Singapore National Library Board archives on Microfilm Reel NL1822.

(xvii) Bank of China building construction was reported in *The Straits Times* on 12 November 1950, page 11, as contained in the Singapore National Library Board archives on Microfilm Reel NL2423 and an earlier article about the project was in the *Morning Tribune* on 2 October 1940, page 16, which is available on Microfilm Reel NL2321. The information on economic growth in post-war Singapore comes from a paper entitled "Economic change in Singapore 1945–1977" by Cheng Siok Hwa of the Department of History, Nanyang University, Singapore. Finally, the construction of the Asia Insurance Building construction site is mentioned in the *Singapore Standard* on 25 June 1951, page 1, *The Straits Times*, 11 July 1951, page 7, and the *Singapore Standard*, 16 November 1951, page 9.

(xviii) Every's January 1958 interview in *The Straits Times*, 28 January 1958, page 12, is held in the Singapore National Library Board archives on Microfilm Reel NL3308.

(xix) Robert Lim Geok Eng's examination pass is in *The Straits Times*, 6 June 1957, page 5, and is in the Singapore National Library Board archives on Microfilm Reel NL1808.

(xx) 1957 newspaper report on the supply of lecturers to Singapore Polytechnic was in the *Singapore Free Press*, 28 March 1957, page 2, which is held in the Singapore National Library Board archives on Microfilm Reel NL5635.

(xxi) L.A. Balsom's RAF record is mentioned in the *London Gazette* supplement, 6 March 1956.

(xxii) The RAF Changi story in *The Straits Times*, 31 July 1968, page 12, which is contained in the Singapore National Library Board archives on Microfilm Reel NL5511.

(xxiii) Information on Walter Reeve has been provided kindly by his son, Brian Reeve. Further information on Reeve is in *The Straits Times*, 12 October 1966, page 12, which is available in the Singapore

National Library Board archives on Microfilm Reel NL12186 and his 1954 dinner was recorded in the *Tribune* on 11 May 1954, and it is also available in the Singapore National Library Board archives, and W.E. Reeve's letter to *The Straits Times*, 18 June 1966, page 15, which is available in the Singapore National Library Board archives on Microfilm Reel NL12182; finally, Reeve leaving the practice of Cheong Koon Seng: in *The Straits Times*, 15 July 1966, page 21, which is held in the Singapore National Library Board archive on Microfilm Reel NL12183.

(xxiv) Goodbourn's position in Langdon Every & Seah is mentioned in the *Building Journal* (building.com.hk) special edition of August 2009, which focused on the history of the Davis Langdon & Seah practice, and his Hong Kong legacy is in the HKIS Surveyors Times Issue dated November 2008 (ISSN 18182542).

(xxv) The City Council assessor resignations were recorded in *The Straits Times*, 12 November 1958, page 7, which is in the Singapore National Library Board archives on Microfilm Reel NL1803.

(xxvi) Thomson's departure from the City Council is recorded in *The Straits Times*, 24 May 1958, page 7, which is in the Singapore National Library Board archives on Microfilm Reel NL9037.

(xxvii) Information on Basil Walker-Taylor, his family and the 3 months prior to the Japanese occupation of Singapore come from his son Paddy Walker-Taylor and *The Road to Singapore: The Diary of Basil Walker-Taylor, Dec 1941 to Feb 1942*, which he edited and published in 2009 (ISBN 9780956193001).

(xxviii) Walker-Taylor's marriage was recorded in the *Malaya Tribune*, 28 April 1938, page 20, and *Morning Tribune*, 25 April 1938, page 21, which are held in the Singapore National Library Board archives on Microfilm Reel NL4067 and NL2314, respectively.

(xxix) Walker-Taylor's family roots have been kindly shared by Paddy Walker-Taylor.

(xxx) Walker-Taylor playing sport in Johore in 1928 is captured in the *Malaya Tribune*, 25 October 1928, page 10, which is contained in the Singapore National Library Board archives on Microfilm Reel NL3936.

(xxxi) Walker-Taylor's retirement is mentioned in *The Straits Times*, 23 March 1959, page 5, which is in the Singapore National Library Board archives on Microfilm Reel NL4020, and in the *Singapore Free Press*, 19 March 1959, page 5, on Microfilm Reel NL2454.

(xxxii) Change in Survey Department's Chief Surveyor 1959–1963: taken from Mapping and surveying information was referenced from "Mapping Singapore: 1819–2014" an essay by Mok Ly Yng, FRGS, in the National Archives of Singapore. This Essay, in turn, references a number of historical documents: http://www.nas.gov.sg/blogs/offtherecord/wp-content/uploads/2016/06/Visualising-Space_Ch3.-Mapping-Singapore-1819-2014-by-Mr-Mok-Ly-Yng.pdf

(xxxiii) MBSB Bank history is taken from their website: https://www.mbsb.com.my/corporate_about_profile.html

(xxxiv) Compton-Hall's RAF records and life information comes from his family, as mentioned in the Acknowledgements section of this book.

(xxxv) Malcolm Burrells' time with the State Bank of New South Wales is mentioned in *The Straits Times*, 23 August 1983, page 22, which is held on microfilm in the Singapore National Library Board archives on Microfilm Reel NL14147, and his time with Chestertons International is mentioned in *The Straits Times*, 11 February 1989, page 19, which is held on microfilm in the Singapore National Library Board archives on Microfilm Reel NL16471.

(xxxvi) CDL story: mentioned in *The Straits Times*, 31 December 1963, page 14, which is held on microfilm in the Singapore National Library Board archives on Microfilm Reel NL12152.

(xxxvii) Amber Mansions is mentioned in: http://eresources.nlb.gov.sg/infopedia/articles/SIP_587_2005-01-25.html

(xxxviii) R.M. (Max) Parker is mentioned in *The Straits Times*, 4 October 1960, page 13, which is held in the Singapore National Library archives on Microfilm Reel NL4148, and the *Singapore Free Press*, 15 August 1959, page 13, on Microfilm Reel NL2456.

(xxxix) The results of the "SURVEYORS EXAMINATIONS" was reported in the *Singapore Standard*, 16 July 1958, which is held in the Singapore National Library archives on Microfilm Reel NL2620.

CHAPTER 13:

(i) The transcript of Sir Robert Black's full speech to the 1957 Surveyors' Dinner is held in the Prime Minister's Office Singapore's (PMO) archives.

(ii) The 1957 Chair's report is in *The Straits Times* of 14 June 1957 as contained in the Singapore National Library Board archives — the article is also available there on Microfilm Reel NL1808.

(iii) Information on Sir Robert Black and the Federation of Malaya are referenced from C.M. Turnbull's *A History of Singapore 1819–1988* mentioned in previous Chapters, and also from the following Singapore National Library Board sources:

http://eresources.nlb.gov.sg/infopedia/articles/SIP_1675_2010-06-29.html

http://eresources.nlb.gov.sg/history/events/72e9ebe6-7a0f-4512-aa48-3da99d598525

CHAPTER 14:

(i) Lincoln Page's 1949 letter to J.M. Fraser is held in the National Archives of Singapore.

(ii) SIT and the Survey of Singapore: first mentioned in the *Singapore Free Press*, 29 December 1951, page 5, which is held in the

Singapore National Library Board archives on Microfilm Reel NL3660.

(iii) Komlosy interview in April 1952 was in *The Straits Times*, 5 April 1952, page 4, which is held in the Singapore National Library Board archives on Microfilm Reel NL2629.

(iv) Town Planning Institute Malaya Branch formation was reported in *The Straits Times*, 7 April 1952, page 5, a copy of which is in the Singapore National Library Board archives on Microfilm Reel NL2629.

(v) Kallang airfield site: report in *The Straits Times*, 21 October 1952, page 6, a copy of which is in the Singapore National Library Board archives on Microfilm Reel NL2632.

(vi) Update on the survey work: appeared in *The Straits Times*, 11 July 1952, page 7, which is held in the Singapore National Library Board archives on Microfilm Reel NL2631.

Further updates on the survey work during 1953 are in the *Singapore Free Press*, 7 and 28 January 1953, page 5, in the Singapore National Library Board archives on Microfilm Reel NL3670.

(vii) A transcript of Komlosy's Radio Malaya interview on Friday, 15 August 1958, is held in the National Archives of Singapore, Speeches @ Archives Online: http://www.nas.gov.sg/archivesonline/speeches/record-details/88bccafd-bcf7-11e6-b045-0050568939ad

(viii) The Industrial Study Group was mentioned in *The Straits Times*, 12 June 1953, page 10, which is held in the Singapore National Library Board archives on Microfilm Reel NL3304, and also in *The Straits Times*, 18 March 1953, page 7, on Microfilm Reel NL2510.

(ix) Playing fields quote from Komlosy — in *The Straits Times*, 5 July 1953, page 1, which is held in the Singapore National Library Board archives on Microfilm Reel NL2490, and also in *The Straits Times*, 27 January 1954, page 4, on Microfilm Reel NL1765.

(x) City Councillor J.M. Jumabhoy's comments are in the *The Straits Times*, 27 January 1953, page 5, Singapore National Library Board archives on Microfilm Reel NL2634.

(xi) Rural district committees: mentioned in *The Straits Times*, 22 August 1953, page 88, which is held in the Singapore National Library Board archives on Microfilm Reel NL3305.

(xii) The island-wide Traffic survey was reported in the *Singapore Free Press*, 24 April 1953, page 5, and held in the Singapore National Library Board archives on Microfilm Reel NL3676.

(xiii) Pepler's planned visit in July 1954 and his return to the UK was in the *Singapore Free Press*, 14 June 1954, page 1, held in the Singapore National Library Board archives on Microfilm Reel NL2438 and in the *Singapore Free Press*, 21 August 1954, page 1, held in the Singapore National Library Board archives on Microfilm Reel NL2439.

(xiv) The idea of a suburban railway system was mentioned in *The Straits Times*, 22 August 1954, page 11, held in the Singapore National Library Board archives on Microfilm Reel NL3692.

(xv) Pepler's arrival was reported in *The Straits Times*, 14 July 1954, page 4, held in the Singapore National Library Board archives on Microfilm Reel NL1769.

(xvi) Komlosy's Green Belt idea was mentioned in *The Straits Times*, 26 November 1954, page 7, held in Singapore National Library Board archives on Microfilm Reel NL1772.

(xvii) Komlosy's multi-storey car park idea was in the *Singapore Standard*, 14 December 1955, a copy of which is held in the Singapore National Library Board archives on Microfilm Reel NL9023.

(xviii) The first multi-storey carpark in Singapore comes from the Singapore Roots website: https://roots.sg/learn/collections/listing/1221637

(xix) Komlosy's return from leave was reported in *The Straits Times*, 6 August 1955, page 4, which is in the Singapore National Library Board archives on Microfilm Reel NL1820 and the findings of his UK home visit were recorded in *The Straits Budget*, 18 August 1955, page 11.

(xx) Pleasure Resort article is in *The Straits Times*, 12 March 1956, page 2, which is held in the Singapore National Library Board archives on Microfilm Reel NL1814.

(xxi) Master Plan inspection and approval were also reported in the *Singapore Standard* on 7 January 1956 and 9 August 1958, both held in the Singapore National Library Board archives on Microfilm Reels NL9024 and NL2621, respectively.

(xxii) Other general information was referenced from the Singapore National Library Board archives: http://eresources.nlb.gov.sg/history/events/ef5af33f-bc66-4080-81d4-013564ba3119

(xxiii) The 1954 New Delhi exhibition: mentioned in *The Straits Times*, 23 September 1953, page 4, held in the Singapore National Library Board archives on Microfilm Reel NL3306.

(xxiv) Smart's involvement in preparations for the New Delhi exhibition is from: "Habitation and the invention of a nation, Singapore 1936–1979" by Eunice Seng, a piece of work submitted in partial fulfilment of the requirements for the degree of Doctor of Philosophy under the Executive Committee of the Graduate School of Arts and Sciences.

(xxv) The 1954 New Delhi Exhibition was referenced from a paper entitled "Mid-century modernism: Jaqueline Tyrwhitt and the village center at the 1954 international exhibition of low-cost housing in New Delhi" by Ellen Shoshkes, Portland State University, in her article in *Traditional Dwellings and Settlements Review*, January 2012.

CHAPTER 15:

(i) The story of a career as a Chartered Surveyor was in a series of weekly articles by Chichak in the *Singapore Free Press* between December 1961 and January 1962. The reference to the office of the R.I.C.S. Singapore Branch is in the 5 January 1962 edition, page 6 — it is held in the Singapore National Library Board archives on Microfilm Reel NL2523.

(ii) Information on members, Board Chairs and Corresponding Secretaries from 1960 to 1975 have been taken from the R.I.C.S. Year Books, referenced in the RICS archives in Great George Street, London.

(iii) Goh Mia Hoon's position as is taken from the R.I.C.S. 1974/75 Yearbook.

(iv) The book *The Evolution of Singapore Real Estate: Journey to the Past and Future 1940–2015* (published by Knight Frank Pte Ltd, 2015, ISBN 9789810942731) has been referenced in this Chapter and in subsequent Chapters.

(v) Information on Dick Toh Chee Seng comes from interviews with members who knew him, and from his obituary in the HKIS magazine for members "Surveying" dated October 1992.

(vi) H.J. Frank's history is captured in *The Straits Times*, 14 February 1968, page 10, which is available on Microfilm Reel NL05383.

(vii) Some information on surveyors working for the HDB in this Chapter (and in Chapters 17 and 21) has been sourced from the HDB Annual Reports from 1961 to 1970.

CHAPTER 16:

The 1966 R.I.C.S. President's visit in 1966 to Singapore and meetings with members was captured in the diaries of then-R.I.C.S. Secretary-General Francis Bennion: the information is referenced in his diaries, which have been made available through the following, with the full permission of his executors, as confirmed by Will Vine of Everys (family solicitors).

CHAPTER 17:

(i) The information about the formation of REVA is referenced from *The Straits Times*, 8 April 1963, page 4, which is available in the Singapore National Library Board archives on Microfilm Reel NL12144.

(ii) The information on Lai Pang Fee and Simon Lim and the UK Hotels market is mentioned in an article in *New Nation*, 9 September 1977, page 4, which is in the Singapore National Library Board archives on Microfilm Reel NL9454.

(iii) One piece of information about the Malaya-Borneo Building Society comes from *The Straits Times*, 6 October 1957, page 8, which is available on Microfilm Reel NL0387.

(iv) The Phillips family tree is referenced from the "Burghers & Eurasians plus" family tree by David Mason on the Geneanet website, and additional information on Jack Phillips come from interviews with those who knew him, as well as from his obituary, *Australian Surveyor* **31**:5:360–61, 1983. DOI:10.1080/00050326.1983.10435021. His Government appointments in the 1970s are referenced from: am19770302s.pdf (nas.gov.sg) and https://eresources.nlb.gov.sg/newspapers/Digitised/Article/newnation19760922-1.2.15

CHAPTER 18:

Information comes from a series of interviews.

CHAPTER 19:

Information for this Chapter comes from the following sources:

(i) NUS 40th Anniversary Booklet published by the NUS Department of Building and Department of Real Estate, School of Design and Environment.

(ii) *Singapore's Real Estate: 50 Years of Transformation* by Seek Ngee Huat, Sing Tien Foo and Yu Shi Ming (editors) (by World Scientific Publishing, 2016, ISBN 9789814689250).

(iii) Colombo Plan history is referenced from: http://www.colombo-plan.org/index.php/about-cps/history/

(iv) The start of the Singapore Polytechnic courses was captured in *The Straits Times*, 19 August 1957, page 2, which is held in the Singapore National Library Board archives on Microfilm Reel NL1810, the *Singapore Free Press*, 28 March 1957, page 2, on Microfilm Reel NL5635, *The Straits Times*, 6 May 1958, page 10, on Microfilm Reel NL 3310, *The Straits Times*, 27 August 1961, page 5, on Microfilm Reel NL3883.

(v) The 1966 R.I.C.S. President's visit in 1966 to the Singapore Polytechnic was captured in the diaries of Francis Bennion, the R.I.C.S. Secretary-General (a role later renamed as the Chief Executive): the information is referenced in his diaries, which have been made available through the following, with the full permission of his executors, as confirmed by Will Vine of Everys: https://wayback.archive-it.org/org-467/20130122144833/http://francisbennion.com/list/webtitles.htm and https://wayback.archive-it.org/org-467/20130122145800/http://francisbennion.com/pdfs/fb/1966/1966-002-rics-world-tour.pdf

(vi) Polytechnic staff are mentioned in the Singapore Government's Infopedia pages: http://eresources.nlb.gov.sg/infopedia/articles/SIP_2019-04-15_102439.html

(vii) Professor Torrance information comes from an obituary notice from the Building magazine in the CIOB Archives.

(viii) Philip Motha and the Colombo Plan information comes from the *Indian Daily Mail*, 9 December 1951, which is available in the Singapore National Library Board archives on Reel No. NL 7319, and further background on him comes from the National Archives of Singapore's Oral History interview series with him.

CHAPTER 20:
Information for this Chapter comes from the following sources:

(i) IRAS information: (2003). *Enabling Digital Government Through E-Services: Second-Wave Re-Engineering in the Inland Revenue Authority of Singapore* by Neo, B.S. and Sia, S.K. (published by the Asian Business Case Centre, Nanyang Technological University, Singapore, 2003).

(ii) Property Tax, supplemental information: "Property tax in a city-state: The case of Singapore" by Mukul G. Asher and Amarendu Nandy in *Policy and Society* **22**(2):50–75, 2003. DOI:10.1016/S1449-4035(03)70019-8

(iii) Property Tax, additional information: *Property Tax in Singapore and Malaysia* by Leung Y.K. and M. Usilappan (published by Butterworths Asia, Singapore, 1997).

(iv) The Act of 1853 giving Municipal powers: source is *Singapore Free Press and Mercantile Advertiser*, 22 July 1853, pages 1–2, which is held in the Singapore National Library Board newspaper archives on Microfilm Reel NL 2202.

(v) Information on rent control comes from Chapter 12 of *Law of Real Property & Conveyancing in Singapore* by N. Khublall (published by Longman, ISBN 99718999507) and from Chapter 2, *Singapore Real Property Guide*, Fourth Edition by Philip Motha and Belinda K.P. Yuen (published by Singapore University Press, 1999 (First Edition, 1979), ISBN 9971692244 (case) and ISBN 997169218x (paper)).

(vi) Snell information is from the *Singapore Free Press*, 25 May 1932, page 10, which is held in the Singapore National Library Board archives on Microfilm Reel NL 1760 and the *Morning Tribune*, 21 July 1937, page 19, on Microfilm Reel NL 2262.

(vii) Singapore Rate-Payers Association concerns are mentioned in *The Straits Times*, 2 January 1953, page 4, which is held in the Singapore National Library Board archives on Microfilm Reel NL2634, and also in *The Straits Times*, 31 December 1952, page 7, on Microfilm Reel NL2633.

(viii) Aspinall's time as assessor is derived from the following sources: The *Singapore Free Press*, 24 November 1947, page 5, which is held in the Singapore National Library Board archives on Microfilm Reel NL 3664, the *Singapore Free Press*, 1 April 1948, page 5, on Microfilm Reel NL 3666, the *Singapore Standard*, 30 September 1952, on Microfilm Reel NL2286; *The Straits Times*, 7 July 1958, page 7, on Microfilm Reel 3311, *The Straits Times*, 14 February 1953, page 10, on Microfilm Reel 2634, and finally *The Straits Times*, 4 August 1958, page 6, on Microfilm Reel 3311. Aspinall's refusal to postpone his retirement was featured in *The Straits Times*, 23 February 1958, page 8, which is on Microfilm Reel NL2491.

(ix) Berkovits's career history is mentioned in *The Straits Budget*, 30 January 1946, page 6, and *The Straits Times*, 21 March 1946, page 3, both of which are which are in the Singapore National Library Board archives.

(x) Peter Forbes-Harriss's Valuation List 1962 is declared in *The Straits Times*, 4 August 1961, page 13, which is held in the Singapore National Library Board archives on Microfilm Reel NL12124, and the 1959 interim roles are mentioned in *The Straits Times*, 15 April 1959, page 13.

(i) The information about David Marshall is from the Singapore Government's Infopedia pages: http://eresources.nlb.gov.sg/infopedia/articles/SIP_283_2005-01-13.html

CHAPTER 21:

The individual stories come from personal interviews conducted by the author.

A.I.G. Harding information comes from interviews, family sources as well as *The Straits Times*, 22 February 1971, page 12, which is available in the Singapore National Library Board archives on Microfilm Reel NL06688 and also the Report on Resettlement, April 1961, written by A.I.G. Harding for the War Department Land Office, FARELF GHQ, Singapore, which is in the possession of the Harding family.

Bibliography

- Beamish, Jane and Jane Ferguson. (1989) *A History of Singapore Architecture: The Making of a City.* Graham Brash, Singapore.
- Bennion, Francis. (nd) Personal diaries, held in the Bennion family archives.
- Brooke, Gilbert E. and Roland St. J. Braddell. (1921) *One Hundred Years of Singapore, Volume II.* John Murray, Albermarle Street, London.
- Chang, Jiat-Hwee. (2016) *A Genealogy of Tropical Architecture, Colonial Networks, Nature and Technoscience.* Routledge.
- Davison, Julian. (2006) *Black and White: The Singapore House, 1898–1941.* Talisman Publishing, Singapore.
- Davison, Julian. (2020) *Swan and Maclaren: A Story of Singapore Architecture.* ORO Editions and National Archives of Singapore, Singapore.
- Drysdale, John. (1984). *Singapore Struggle for Success.*
- Gullick, J.M. (1992) "The Bangunan Sultan Abdul Samad". *Journal of the Malaysian Branch of the Royal Asiatic Society*, **65**(1):27–38.
- Hall-Jones, J. and C. Hooi. (1979) *An Early Surveyor in Singapore: John Turnbull Thomson in Singapore, 1841–1853.* National Museum of Singapore, Singapore.
- Hancock, T.H.H. (1986) *Coleman's Singapore.* Malaysian Branch of the Royal Asiatic Society in association with Pelanduk Publications.
- Imran bin Tajudeen. (2005) "Reading the traditional city of maritime Southeast Asia: Reconstructing the 19th century Port Town at

Gelam-Rochor-Kallang, Singapore". *Journal of Southeast Asian Architecture*, Vol. 8.

- Jansen, Mary Anne, John Geno-Oehlers and Ann Ebert Oehlers. (2018) *On Parade: Straits Settlements Eurasian Men Who Volunteered to Defend the Empire 1862–1957.* Wee Kim Wee Centre, Singapore Management University, Singapore.
- Kennedy, Joseph. (1987) *British Civilians and the Japanese War in Malaya and Singapore, 1941–45.* The MacMillan Press Limited.
- Khublall, N. (1986) *Law of Real Property & Conveyancing in Singapore.* Longman, Singapore.
- Kitching, Brian. (1998) *Life and Death in Changi: The Diary of Tom Kitching Who Died in Japanese hands in Singapore in 1944.* William Culross & Son Ltd., Perth and Scotland.
- Kwa, Chong Guan, Derek Heng, Peter Borschberg and Tan Tai Yong. (2009) *Seven Hundred Years: A History of Singapore.* Marshall Cavendish International (Asia), Singapore.
- Lee, Kip Lin. (1988) *The Singapore House 1819–1942.* Times Editions, Singapore.
- Miescher, Stephan F., Michelle Mitchell and Naoko Shibusawa, editors (2015) *Gender Imperialism and Global Exchanges.* Wiley Blackwell.
- Motha, Philip and Belinda K.P. Yuen. (1999) *Singapore Real Property Guide*, Fourth Edition. Singapore University Press, Singapore.
- National University of Singapore. (nd) *NUS 40th Anniversary Booklet.* Department of Building and Department of Real Estate, School of Design and Environment.
- Ng, Yew Peng. (2018) *What's in the Name? How the Streets and Villages in Singapore Got Their Names* (World Scientific Publishing, Singapore.
- North, Jennifer. *Building a Name — The History of the Royal Institution of Chartered Surveyors.* Arima Publishing, 2010.

- Peacock, John. (2014) *Quantifying Asia: The History of Langdon & Seah.* Langdon & Seah Asia Company Ltd, Hong Kong.
- Pereira, Alexius A., Myrna Braga-Blake and Ann Ebert-Oehlers, editors (2017) *Singapore Eurasians: Memories, Hopes And Dreams*, World Scientific Publishing, Singapore.
- Royal Institution of Chartered Surveyors. (nd) *The Royal Institution of Chartered Surveyors: Year Books 1883–1968 (and the Standard Method of Measurement of Building Works — For Use in the Federation of Malaya, Singapore and Borneo).* Royal Institution of Chartered Surveyors, London.
- Seek, Ngee Huat, Sing Tien Foo and Yu Shi Ming (editors). *Singapore's Real Estate: 50 Years of Transformation.* World Scientific Publishing, Singapore.
- Seng, Eunice. (nd) "Habitation and the invention of a nation, Singapore 1936–1979". Graduate School of Arts and Sciences.
- Shennan, Margaret (2000) *Out in the Midday Sun — the British in Malaya 1880–1960.* John Murray Publishers Ltd., United Kingdom.
- Sim, Y.H. Teddy. "'Jacques de Coutre's And Matelieff's Singapore and Johor': Exploring sources on pre-modern history of Singapore". National Institute of Education, Singapore.
- Sng, Lydia, Tan Tiong Cheng, Leung Yew Kwong and Knight Frank. (2015) *The Evolution of Singapore Real Estate: Journey to the Past and Future 1940–2015.* Knight Frank Pte Ltd, Singapore.
- Stapleton, Michael. (1986) *Dictionary of Greek and Roman Mythology.*
- Turnbull, C.M. (1989) *A History of Singapore 1819–1988* (Second Edition). Oxford University Press, United Kingdom.
- Walker-Taylor, Paddy, editor. (2009) *The Road to Singapore: The Diary of Basil Walker-Taylor, Dec 1941 to Feb 1942.*
- Wright, Arnold and H.A. Cartwright. (1908) *Twentieth Century Impressions of British Malaya: Its History, People, Commerce, Industries, and Resources.* Lloyd's Greater Britain Publishing Co. Ltd., London.

- Yap, Sonny, Richard Lim and Leong Weng Kam. (2009) *Men in White: The Untold Story of Singapore's Ruling Political Party*. Singapore Press Holdings, Singapore.
- Yeoh, Brenda. (2003) *Contesting Space: Power Relations and the Urban Built Environment in Colonial Singapore*. NUS Press, Singapore.

Image Credits

Every effort has been made to trace all sources and copyright holders of the images in the book before publication. If any have been inadvertently overlooked, the Publisher will ensure that full credit is given at the earliest opportunity.

The copyright of these images belong to the following:

Alex Hall, 273

Andrea Ruddick, 56 (bottom right)

Anthony Thorp and the Directors of Thorstone Land & Property Limited of London, 316 (right)

Bill Jones, 13, 14, 16, 38, 40 (top and bottom), 41, 53 (top left), 66 (top left), 66 (bottom), 69, 71, 74, 75, 76, 79, 85, 87, 89, 91, 103 (bottom), 112, 128, 129, 139, 153, 154, 171, 191 (right), 212, 251, 297 (bottom), 298, 306, 386, 408

Bill Jones and Sven Suhrbier, 134

Brenda Vargeson, 311 (middle and bottom)

Brian Reeve, 256, 257, 258, 348

Caroline (Carol) Chow née Nicholas and John Farrer, 101

Charles Marsh, 197

Coral Theresa Allen Streader (née Bidwell), 92, 93

Craig Oates, 236

David Atkin-Berry, 132

David Forsaith, 143

David Ong, 370 (bottom)

David Stephen, 255

Dawn Runnicles, James Runnicles and Antonia Parsons, 355

D.R.W. (Robin) Tillyard and Judy Tillyard, 316 (left)

Eddie Toh and Tony Toh, 320

Francis Lai Pang Fee, 353, 356, 358

Geraldine Chilton, 279, 332

Gordon Snell, 183 (left)

Guy Jubb, 328

G. Morham White, 237

Ian Chan, 322

Ian D.R. Watson, 180 (top and middle), 182, 221 (top and bottom)

Jacky Fox and Michael Speak, 325, 326

James Edwards, 52, 56 (bottom left)

Jamie Maule-ffinch, 113

John Farrer, 100

John Rosser, 271

Kerry Littlejohn, 323

Libby Mills and Claudia Mills, 191 (left)

Liz Godbolt, 409

Mark Lewis, 262, 263

Martin Riddett, 250

Michael and Lauren Pennefather, 96

Paddy Walker-Taylor, 265, 266, 268, 269, 270

Paul Benedict Lincoln, 97

Paula Lermit-Webb, 35, 46

Pauline Goh and Willy Shee, 371, 372

Peter Forbes-Harriss, 400

Peter Rhodes and Barbara, 310 (top and bottom), 370 (top)

Richard Burrells, 349

Ron Taylor, 233

Royal Institution of Surveyors Malaysia, 317, 369
Sallyann Loh, 359
Sandra Lincoln, 295, 296 (top and bottom), 297 (top)
Sharlyn Lim, 81
Stephen Komlosy, 282, 283, 285, 292
Stephen Rix, 276, 277, 278
Susanna Jack, 228 (top and bottom), 229
Tan Choo Haw, 413
The Morris Family, 245, 246, 261
Vera Teo and Elisabeth Teo-Pennell, 315
Victorian Institute school website, 401
81 Squadron Association, 168

These images were reproduced with the kind permission of:

Chung Chee Min and Ooi Boon Kheng, 401

Daryl Goh, 368 (right), 373, 374, 376

Dr. Anamah Tan, 379

Dr. Ng, 361, 363

Mohd Talhar Abdul Rahman, co-founder and Chair of CBRE-WTW, 318

National Archives of Singapore, 337, 365

Rabbi Nissim Zawady of Chesed-El Synagogue, 71

Richard Burrells, 329, 330 (top, middle, bottom), 343

Seah Choo Meng, 404 (left)

Singapore National Library Board archives and Singapore Press Holdings, 147, 167, 174, 196 (right), 226, 300, 312

Tan Cheng Yoon, 367 (top)

The Collection of David Robson, 125

The Governors, Dulwich College, 116

The Harding Family, 411 (top and bottom)

The late Ng Thin Wah's Family, 367 (bottom)

The Straits Times © SPH Media Limited, 196 (left), 388

These images were reproduced from the following publications:

Out in the Midday Sun — the British in Malaya 1880–1960 (with kind permission of the family of the late Margaret Shennan, obtained through Charlotte Howard of Fox and Howard, Literary Agents for the late Margaret Shennan): 141, 183 (right)

Quantifying Asia: The History of Langdon & Seah (with kind permission of Seah Choo Meng and John Peacock): 194, 249, 259, 312, 404 (right)

Twentieth Century Impressions of British Malaya: Its History, People, Commerce, Industries, and Resources: 19, 53 (top right), 66 (top right), 105

1975–76 Singapore Institute of Surveyors 7th Council photograph (with kind permission from Francis Lai Pang Fee, the owner of the photograph): 368 (left), 377

These images were reproduced from postcards in the possession of the Author:

Raffles Place, 84
Singapore Cricket Club, 72
The Boustead Institute, 39

Index

A

Aeria, Melville, 200
Alexander, A.C., 19
Anamah ("Ann") Tan, 379–381
Angell, Francis ("Frank"), 178, 180–183, **182**, 186, 216, 223
Arch, Arthur, 59
Aspinall, John, 216–217, 227, 393–398
Atkinson, C.F., 238
Atkin-Berry, Herbert, 132–135, **132**, 139–140, 175, 186
- as an Appraiser, 135–136
- Chinese (Institutional) Methodist Church, 139, **139**
- Presbyterian Church, 139, **139**
- Tanjong Pagar Railway Station, 134–135
- Telegraph House, 138, **139**
- Union Building, 138, **139**

B

Balsom, Leonard, 255
Barkshire, Charles, 109–110
Barnett, Charles, 234
Bennion, Francis, 339, **348**, 385
- diary records, 339–347

Berkovits, Karoly ("Stuart"), 398, 400, **400**
Bidwell, Regent, 61–65
- Adelphi Hotel, 74–75, **74**
- arrival in Singapore, 64
- as employee and Partner of Swan & Maclaren, 82–86
- Black and White houses, 77–81,
- Chesed-El Synagogue, 70–71, **71**
- children, 92–93
- Church of St. Joseph, 75–76, **75**
- later years and death, 86–90
- legacy, 94
- Methodist Girls School, 68
- Raffles Hotel, 66–68, **66**
- Singapore Cricket Club, New Pavilion, 72–73, **72**
- Stamford House, 76–77, **76**
- Teutonia Club, 65, **66**
- The Memorial Hall and Town Hall, 68–70, **69**
- The "Butterfly House", 86, **87**

Bills of Quantities, 165, 175–176, 181, 188
Biscoe, Guy, 339, **348**
- World Presidential Tour, 339–347, 385

Black, Robert, 242, 263, 275–276, 290
Bonfield, Terry, 306
Bone, Chandos, 121–122
Booty and Edwards, 123. *See* Edwards, Reid and Booth
Borschberg, Peter, 4

British military projects:
- Gillman Barracks, 162, **171**
- RAF Seletar, 162, 168–169
- RAF Tengah, 162, 166–168, **167**, **168**
- Sembawang Naval Base, 162

Brown, Walter, 127–128
Burrells, Malcom, 309, 329–331, **330**, 351, 372
Burt, Cornelius, 216, 225–226, **226**

C

Carpmael, Harold, 115–119, **116**
Carter, John, 299–301, **300**
Caseley, John, 128–132
Chan Wan Hang, 298, 322, **322**
Chartered Institute of Auctioneers and Estate Agents, 346, 379, 414
Chartered Surveyors:
- Pioneers, 57–59
- Eurasians, 28, 95–106
- locals, 193, 305, 390

Chee Yoong Ng, 412
Cheeseman, Harold, 178–179
Chevallier, Harvey, 49–51
Chilton, Gerald, 331–333, **332**
Chin Yun Howe, Lawrence, 200
Clayton, Glanville, 218
Coghlan, Henry Lake, 51–57, **52**. *See* H. L. Coghlan & Co.
Coleman, George, 5, 9, 10–14
- Armenian Church, 12
- Caldwell House, 13, **13**
- Coleman Bridge, 12
- career, 10–11
- No. 3 Coleman Street, 11
- Raffles Residence, 12
- Singapore Institution, 12
- St. Andrew's Church, 12
- Telok Ayer Market, 12
- The Arts House, 12, **13**

College of Estate Management, 383
Collier Goh & Tan, 352, 373
Colombo Plan scholarship, 383–384
Compton-Hall, Richard, 372–373, **373**
Cooke, Ernest, 237
Craik, David, 43, 111, 123
Crickmay, John, 209–210
Crisp, Denis, 307–308
Croft, Frederick, 205–206, 270–271, **271**
Cullen, Alexander, 140–141, **140**; 212

D

Davey, Harold, 234
Davis, Alan, 316, **316**. *See* W.J.F. Tillyard & Partners
Deans, Stuart, 307, 401
Death Railway, 210, 232
Distin, Aubrey, 227–229, **228**, **229**, 254
Downer, Robert, 230
Dyce, Charles Alexander, 32
Dyke, Bernard, 120

E

East India Company, 8–9, 17
Eden Hall, 264–265
Edwards, Reid and Booth, 125
Edwards, Sidney, 122–126, **125**. *See* Booty and Edwards
Every, Cecil Tom, 253. *See* Horace W. Langdon & Every

F

Farquhar, William, 6
Forbes-Harriss, Peter, 400
Forsaith, Frederick, 142–143, **143**

Fowler, E.V., 205, 207
Frank, Henry, 308, 309.
See Frank & Vargeson
Frank, Vargeson, Wicks & Partners, 309
Frank & Vargeson, 244–245, 246, 308–309. *See* Frank, Vargeson, Wicks & Partners
Franklin, James, 6
Fraser Jim, 155, 281, **300**

G

Godfrey, Stanley, 328–329, **330**, 351
Goh Mia Hoon, 309, 351–352, 372–373, **373**, 375, 401. *See* Collier Goh & Tan
Goodbourn, Ken, 259–260, **259**
Government scholarship, 240, 248, 383
Green Belt, 288, 290
Green, Henry, 230
green lungs, 288
Greig, Arnold, 230–231
Grey, George, 119–120

H

Harding, Adrian, 409, **409**, 412
Harding's Report, 409–411
Harding, Donald, 231
Harris, Alfred, 191
Heng Chong How, 398, 399
Higgs, Henry, 231
Hill, Eric, 173, 186
Hill, John, 298
Himely, Luis, 224
Ho Fook Weng, Charles, 407–408
Ho Lum Khuan, 415
Hollingsworth, G., 255
Holmes, Charles, 231–232, 249
Honey, Frederick, 169–170
Honeywall, Donald, 187
Hopkins, Norman, 186
Horace W. Langdon & Every, 164, 212, **221.** *See* Horace W. Langdon & Every (incorporating Waters & Watson)
Horace W. Langdon & Every (incorporating Waters & Watson), 164, 212. *See* Langdon & Every (Far East)
Howe Yoon Chong, 342
Hughes, John, 247
Hunt, Edgar, 166
H. L. Coghlan & Co., 52, **53**, 54–55

I

Ibbeston, Robert, 14
Inglis, Ronald, 169
Institute of Quantity Surveyors, 391
Institute of Surveyors, London, 9, 31
Institution of Surveyors (Federation of Malaya), 279–280. *See* Royal Institution of Surveyors Malaysia
Izambert, Alexandre, 37

J

Jackson, Philip, 6
Jackson Plan, 7
Jansen, William, 360–361
Jessamine, John, 218, 219–220, 224. *See* Jessamine & Marsh
Jessamine & Marsh, 199, 217
Johns, W.H., 271
Johnson, Herbert, 169
Johnston, Alexander Laurie, 32
Johnston's Pier, 32
Jones-Davies, Frederick, 186–187, 217
Jubb, Eric, 326, 327–328, **328**
Jumabhoy, J.M., 286

K

Kershaw, John, **221,** 232
Khoo Cheok Chuan, James, 247–248, **348**
Kitching, Tom, 201–211
Koch, John, 58
Komlosy, Denis, 281, **282**, **283**, **285**, 292–293
  1956 Master Plan, 290–292
  survey of Singapore and report, 283–284, 286–289
  Industrial Study Group, 284–285, 293
  Kallang airfield site, 285
Kwok Chuon Wei, 414–415, **414**

L

Laidlaw, D.H., 29
Lai Pang Fee, Francis, 319, **353**, 355–358, **358**, 390–391
Land Surveyors, 24, 29–30, 59
Lang, John, 173
Langdon Every & Seah, 312
Langdon, Horace, 220–221, **221**, 248, 253. *See* Horace W. Langdon & Every
Langdon & Every (Far East), 248–251, 312. *See* Langdon Every & Seah
  Asia Insurance Building, 251–252
  Bank of China, 250–251
Lau Fook Sing, 243, 313, **348**. *See* Lau Teo & Yong
Lau Teo & Yong, 313–314
Lee Kuan Yew, 336
Lermit, Alfred, 28, 34–47, **35**. *See* Swan & Lermit *and* Lermit & Westerhout
  Boustead Institute, 38, **39**
  children, 46
Lermit, Arthur, 28, 46
Lermit & Westerhout, 39
  Singapore Philatelic Museum, **38**, 40
  Parsonage, Armenian Church, 41, **41**. *See* Tomlinson & Lermit
LeRoy, Jack, 243
Lewis, Anthony, **349**, 355
Lewis, Selwyn, 242, 262, **262**, 398
Lim Ang Beng Hock, Eric, 387–388, 405
Lim Chan Hoe, 408–409
Lim Geok Eng, Robert, 241
Lim Peng Song, 371
Lim Poh Guan, 366
Lim Kim Thian, 412
Lincoln, John, 97–99, 97
  Fort Canning Lighthouse, 98
  Johnston's Pier Reclamation, 98
  Quarantine Station at St. John's Island, 98
  Singapore General Hospital, 98
  Tan Tock Seng Hospital, 98
Littlejohn, Rae, 323–324, **323**
Loh Jwee Siam, 298, 359–360, **359**
Loh Kum Seng, 247, 323, **348**
Loveday Corruption Trial, 202–208
Loveday, Robert Charles, 202–203, 208. *See* Loveday Corruption Trial
Low, James, 5
Lowden, Gorden, 232–233, **233**, 238, 242, 398
Lundon, Frank, 91, 142

M

Macclaren, James , 39
Malay Peninsula, 17, 24, 279
Malayan Institution of Surveyors, 197, 279. *See* R.C.I.S. Malaysia branch
Malayan Union, 280

Malayanisation process, 236, 241–242, 305, 397
Manasseh, Ezekiel, 91, 264
Marsh, Joe, 198–199, 217, 222. *See* Maschoieff, Joseph
Maschoieff, Joseph, 194, 197–199, **197**. *See* Marsh, Joe
Maule-ffinch, Knightley, 29, 113–115, **114**
McCallum, Henry, 19, 25, 27
McDonald, Johan, 317, **317**. *See* Suckling & McDonald
McNair, John, 18
Miksic, John, 3
Moniot, Jules, 18
Morris, Derrick, 242, 244–247, **245–246**. *See* Frank & Vargeson
Motha, M.J.P. ("Philip"), 389, 390
Murray, Alexander, 19

N

Neil, Robert, 229, **229**
Neubronner, Henry, 105–106
Newman, Peter, 371, **372**
Ng Boon Ong, 298, 361–363, **361, 363**, 390
Ng Thin Wah, 367–368, **367**
Nicholas, Wilfred, 27, 99–104, **100, 101**
- children, 101
- King's Dock project, 103–104, **103**
- Telok Ayer Reclamation, 104

Nivan, Alexander, 176, 186, 188–191, **191**
- later years and death, 190
- old Supreme Court building, 188
- World War II, 189–190

O

Oh Chong Guan, Michael, 331, 333
Ong Huck Jin, Justin, 370, **370**
Ong, Lily, 308–309, 352

P

Page, Lincoln, 281–283, **282**, **295**, 300
- Duchess Estate, 295, **296**
- Kay Siang Road houses, 296, **296**
- Princess House, 296, **297**

Pank, Philip, 141–142
Parker, Max, 298–299, **295**
Pearcy, Etzel, 238
Pennefather, George, 28
Pennefather, John, 28, 95–96, **96**
Pennycuick, James, 204, 205
Pepler, George, 283, 287–288
Percy, George, 178, 186, 187–188
Petrovitch, D.S., 135, 137
Phillips, Jack, 341
Pigott, F.J., 19–22, **19**
Pillai, Arumugam Annamalai, 36
Pillai, Naraina, 14, **14**
Planning Department, Singapore, 248
Polglase, Geoffrey, 127
Pollington, Edward, 173–174
Property Tax, 22–24, 399, 402
Public Works Department, Singapore, 17–18, **306**

Q

Quantity Surveyors, 161, 165, 175, 221
Quantity Surveyors Association, 114, 161
Quarmby, Charles, 173
Quinton, Daniel, 18

R

Raffles, Stamford, 3, 6, 7–8, 10
Rea, John, 109–111
Real Estate Valuers Association (Singapore), 340, 345–346, 351. *See* Singapore Institute of Valuers
Reed, Henry, 233
Reeve, Walter, 256–259, **256, 258**
Rent Control, 397, 398
Rhodes, Robert, 308, 309–310, **310**. *See* Frank & Vargeson
Ricketts, Ronald, 233
RICS Royal Charter, 30
Riley, Horace, 217
Robson, Graham, 218
Ross, Daniel, 6
Royal Institution of Surveyors Malaysia, 279
Runnicles, Alan, 355–356, **356**
Russell-Croucher, Donald, 255, 256
Russell-Croucher, Martin, 256
R.I.C.S. Examinations, 195, 220, 241, 352, 384–385
R.I.C.S. Malaysia branch, 279–280
R.I.C.S. Malaya branch: 1948 AGM, 219–220
  1957 Malaya Branch dinner, 275–277, **277**
  1958 Malaya Branch dinner, 277–279
  1959 Malaya Branch dinner, 278, **279**
  separation, 279–280. *See* R.I.C.S. Malaysia branch *and* R.I.C.S. Singapore branch
R.I.C.S. Singapore branch, 280, 305
  closure, 351
  1965 Branch dinner, 336, **337**

S

Sampson, W.J., 255
Sanderson, Peter, 249–250, **249**
Santry, Denis, 134, 139
Seah Choo Meng, 313, **353**, 403–405, **404**
Seah Mong Hee, 193–197, **194**, **196**, 199–200, 218, **221**, 234, 249, 312–313, **404**
Shah, Iskandar, 3
Sheath, Roland, 387
Side, Charles, 173, 218
Sims, Robert, 221, 224
Singapore City Council, 240, 286
Singapore Improvement Trust, 145–147
Singapore Institute of Valuers, 197, 257, 346, 351. *See* Singapore Institute of Surveyors and Valuers
Singapore Institute of Surveyors, 197, 340–341, 346, 351. *See* Singapore Institute of Surveyors and Valuers
Singapore Institute of Surveyors and Valuers, 346, 351
Singapore Naval Base Corruption Trial, 208–209
Singapore Polytechnic, 384, **385**
Singapore Professional Centre, 353, **353**
Singh, Penu Mohan, 415
Smart, William, 234, 276–277, 293, 294
  Dakota Crescent SIT flats, 297, **297–298**
  Duchess Estate, 295, **296**
  Princess Elizabeth Flats, 294, **295**
Smith, Alexander, 234
Smithson, Malcolm, 312, **312**

Snell, Charles Frederick ("Fred"), 127, 175, 177, 178, **183**, 185, 186; 1930 Report, 184
Speak, Edmund, 325–326, **325**
Standard Method of Measurement of Building Works, 334–336
Steer, Tom, 327
Stevenson, Thomas, 222, 234, 242
Stone, Leslie, 235
Suckling, Keith, 218, 235, 317, **349**, 351. *See* Suckling & McDonald
Suckling & McDonald, 317
Survey Department, Singapore, 17, 21, 266, **268–270**
  "Five Bright Girls", 239–240
  Annual Report, 265–266
Swan, Archibald Alexander, 36
Swan & Lermit, 36–37. *See* Swan & Mcclaren
  Our Lady of Lourdes Church, 37, **38**.
Swan & Maclaren, 36; 134–135, 136–139
  Bras Basah wing of Raffles Hotel court case, 42–43
Sworder, Gerald, 211

T

Tan Cheng Yoon, Andrew, 367, **367**
Tan Choo Haw, 412–414, **412**
Tan Kim Choon, 368–369, **368**
Tan Tian Boon, 298, 363–365, **365**, 390
Tang Soon Lee, 415
Taylor, John, 111–112
Teh Chiang Wan, 364–365
Teo Cheng Ann, Lionel, 313–315, **315**. *See* Lau Teo & Yong
Teo Hee Cher, 242–243
The Malay Annals, 3, 5–6
The Singapore Stone, 4–5
The Survey School, 24–29
Thomas, H.C., 372
Thompson, Basil, 235–236, **236**, 261
Thomson, John, 9, 14–17
  Cavenagh Bridge, 16, **16**
  Chinese Paupers' Hospital Building, 15
  Dalhousie Obelisk, 16, **16**
  Ellenborough buildings, 15
  European Seaman's Hospital, 16
  first Government Surveyor, 14–15
  Hajjah Fatimah Mosque, 16
  Horsburgh Lighthouse, 15
  Thomson Bridge, 16
Tillyard, Walter, 316–317, **317**. *See* W.J.F. Tillyard & Partners
Tiruchelvarayan, Chelliah, 247, 260, **261**, **353**
Toh Chee Seng ("Dick"), 200, 319–321, **320**
Toh Chin Chye, 336, 385
Tomlinson, Sam, 47–48
Tomlinson & Lermit, 40
Torrance, Victor, 388, **388**
Tovey, Leslie, 170–171
Treacher, William Hood, 29
Turner, Jabez, 143–144

U

Utama, Sang Nila, 5

V

valuers, 136, 161, 241
van Cuylenburg, John, **19**, 24–25
Vargeson, Desmond, 308, 311, **311**. *See* Frank & Vargeson
Vickery, David, 387

W

Walker-Taylor, Basil, 262–270, **265**
Walton, Edgar, 111
Ward, Frank Dorrington, 176, 190
Ward, Harry, 226, 227, 238, 239
Waters, David, 161, 163–165, 175, **180**, 186, 207–208, 212. *See* Waters & Watson
Waters & Watson, 163, 164, 178, **182**
Watson, Eric, 163–165, **180**, 216. *See* Waters & Watson
Wee Sek Key, Michael, 34, **349**, 389, 405–407
Weisberg, Hyman, 152
    Weisberg Building Policy Committee, 152
Westerhouse, Johannes, 48
White, George, 237, **237**
Whyte, Ian, 255, **255**
Wicks, William, 308–309, 373–373. *See* Wicks & Partners
Wicks & Partners, 308–309, 372. *See* Frank, Vargeson, Wicks & Partners
Widgery, Richard, 224
Williams, Colin, 317–319, 318.
Williams, Langdon, **147**, 147–150
    1936 Williams Report, 150–153
    career, 149–150
    death, 155
    Empress Place, 158–159
    living in Singapore, 154
    Tiong Bahru housing, 153, **153**
Willoughby, Wesley, 237
Winship, John, 232
Wolferstan Trower & Partners, 321
Wong Kum Sek, 415
Worley, Arthur, 178
Wright, Arnold, 5, 6, 19
W.J.F. Tillyard & Partners, 316

Y

Yap Eng Bok, 369, **369**
Yew Fan Tatt, 369–370, **370**
Yong, M.S., 314. *See* Lau Teo & Yong
Yoong Wah Pin, 401–402, **401**
Young, Arthur, 104
Young, Robert, **19**, 28

Ingram Content Group UK Ltd.
Milton Keynes UK
UKHW050849270423
420870UK00002B/47

9 789811 268441